STATISTICS

STATISTICS
Concepts and Controversies

Third Edition

David S. Moore
Purdue University

W. H. Freeman and Company
New York

Library of Congress Cataloging-in-Publication Data

Moore, David S.
    Statistics : concepts and controversies / David S. Moore—3d
  ed.
        p.   cm.
    Includes bibliographical references and index.
    ISBN 0-7167-2199-6
    1. Statistics.   I. Title.
  QA276.12.M66   1991   90-21704
  519.5—dc20   CIP

Printed in the United States of America

    6 7 8 9   VB   9 9 8 7 6 5 4

Does not He see my ways,
and number all my steps?
Job

But even the hairs of your head
are all numbered.
Jesus

Hell is inaccurate.
Charles Williams

Contents

To the teacher: Statistics as a liberal discipline

Statistics: Concepts and Controversies is a book on statistics as a liberal discipline, that is, as part of the general education of relatively nonmathematical students. I wrote and revised the book while teaching a class of freshmen and sophomores from Purdue University's School of Liberal Arts. My students come from many disciplines in the humanities and social sciences; they are fulfilling a dreaded mathematical sciences requirement. Some elect the course as preparation for a later course in statistical methods, but most will never again encounter statistics as a discipline. For such students, statistics is not a technical tool but part of the general intellectual culture that educated people share.

Statistics among the liberal arts

Statistics has a widespread, and on the whole well-earned, reputation as the least liberal of subjects. When statistics is praised, it is most often praised for its usefulness. Health professionals need statistics to read accounts of medical research; managers need statistics because efficient crunching of numbers will find its way to the bottom line; citizens need statistics to understand opinion polls and the Consumer Price Index. Because data are omnipresent, our propaganda line goes, everyone will find statistics useful, and perhaps even profitable. This is in fact quite true. But the utilitarian argument for studying statistics is at odds with the ideals of the liberal arts. The root idea of a liberal education is that it is general rather than vocational, that it aims to strengthen and broaden the mind rather than to prepare for a specific career.

A liberal education should expose students to fundamental intellectual skills, to general methods of inquiry that apply in a wide variety of settings. The traditional liberal arts present such methods: literary and historical studies; the political and social analysis of human societies; the probing of nature by experimental science; the power of abstraction and deduction in mathematics. The case that statistics belongs among the liberal arts rests on the fact that reasoning from uncertain empirical data is a similarly general and pervasive intellectual method. That statistical ideas are pervasive is surely true. Evidence that statistics is a general method of reasoning is provided by the research surveyed by a group of psychologists in a paper with the intriguing title "Teaching Reasoning."[1] The authors of this paper summarize numerous studies, both observational and experimental, that suggest that instruction in statistics helps us reason effectively about data and chance in everyday life.

Statistics: Concepts and Controversies is shaped, as far as the limitations of the author and the intended readers allow, by the view that statistics is an independent and fundamental intellectual method. The focus is on statistical thinking, on what others might call quantitative literacy or numeracy.

The nature of this book

There are books on statistical theory and books on statistical methods. This is neither. It is a book on statistical ideas and statistical reasoning, and on their relevance to public policy and to the human sciences from medicine to sociology. I have included many elementary graphical and numerical techniques to give flesh to the ideas and muscle to the reasoning. Students learn to think about data by working with data. But I hope that I have not allowed technique to dominate concepts. My intention is to teach verbally rather than algebraically, to invite discussion and even argument rather than mere computation, though some computation remains essential. The coverage is considerably broader than that of traditional statistics texts, as the table of contents reveals. In the spirit of general education, I have preferred breadth to detail.

Despite its informal nature, *Statistics: Concepts and Controversies* is a textbook. It is organized for systematic study and has abundant exercises. I hope that this organization will help students without deterring those admirable individuals who seek pleasure in uncompelled reading.

Teachers should be aware that the book is more difficult than its low mathematical level suggests. The emphasis on ideas and reasoning asks more of the reader than many recipe-laden methods texts.

Some mathematicians and statisticians who have taught from earlier editions noted the challenge of teaching nonmathematical material. I have tried to provide detailed help for teachers in the Instructor's Guide. Here are a few general suggestions. Try to establish a "humanities course" atmosphere, with much discussion in class. Many exercises require discussion; modify them to ask "Come to class prepared to discuss . . . ." Don't spend all your time at the board working problems, though some of that is necessary. Supplement this book with other material that is current (examples from the news) or of special interest to your students. Consider assigning readings from *Statistics: A Guide to the Unknown* for more elaborate examples of applied statistical reasoning.[2] Excerpts from the Annenberg/Corporation for Public Broadcasting video series *Against All Odds: Inside Statistics* offer an unparalleled opportunity to see statistics in use in the most varied settings.[3]

The third edition

I am pleased that the first two editions of *Statistics: Concepts and Controversies* have been widely read, and read by people from many disciplines and diverse backgrounds. I am grateful to many teachers and readers for helpful comments and suggestions. Jerrold Grossman of Oakland University, Gudmund Iversen of Swarthmore College, Dana Quade of the University of North Carolina, and Ralph Russo of the University of Iowa deserve special thanks for their detailed reviews. All will forgive me if at times I proved hard to sway.

The many topical examples and discussions have been systematically brought up to date for this third edition. Teachers who wanted more exercises will be happy, as there are over 150 additional exercises. Exercises now appear following each section, with additional review exercises at the end of each chapter. The material on data analysis (Chapters 5 and 6) has been expanded and reorganized. The thorough rewriting of this part of the book reflects the consensus among statisticians and statistics teachers on the importance of working with data.

Many users of the book teach students who are somewhat more advanced than my intended audience; the successive editions have added material to help meet their needs. But the original purpose remains: To present statistics to nonmathematical readers as an aid to clear thinking in personal and professional life.

NOTES

1. Richard E. Nisbett, Geoffrey T. Fong, Darrin R. Lehman, and Patricia W. Cheng, "Teaching Reasoning," *Science*, Volume 238 (1987), pp. 625–631.
2. Judith M. Tanur et al. (eds.), *Statistics: A Guide to the Unknown*, 3d ed. (Belmont, Calif.: Wadsworth, 1989).
3. For information on *Against All Odds: Inside Statistics*, write Intellimation, P.O. Box 1922, Santa Barbara, CA, 93116 or call 1-800-LEARNER.

Introduction:
What is statistics?

Most of us associate "statistics" with the words of the play-by-play announcer at the end of the sports broadcast, "And we thank our statistician, Alan Roth . . . ." We meet the statistician as the person who compiles the batting averages or yards gained. Statisticians do indeed work with numerical facts (which we call *data*), but usually for more serious purposes. The systematic study of data has now infiltrated most areas of academic or practical endeavor. Here are some examples of statistical questions.

- The Bureau of Labor Statistics reports that the unemployment rate last month was 7.5%. What exactly does that figure mean? How did the government obtain this information? (Neither you nor I were asked if we were employed last month.) How accurate is the unemployment rate given?

- The Gallup Poll reports that 42% of the American public currently approve the president's performance in office. Where did that information come from? How accurate is it?

- You may have heard that much of the evidence that links smoking to lung cancer and other health problems is "statistical." What kind of evidence is it?

- A medical researcher claims that taking aspirin regularly reduces the risk of a heart attack. How can an experiment be designed to prove or disprove this claim?

- Do gun control laws reduce violent crimes? Both proponents and opponents of stricter gun legislation offer numerical arguments in favor of their position; which of these arguments are sense and which are nonsense?

The aim of statistics is to provide insight by means of numbers. In pursuit of this aim, statistics divides the study of data into three parts:

    I. Producing data

    II. Organizing and analyzing data

    III. Drawing conclusions from data

This book is organized in three parts following this same pattern. In the first, we will look at statistical designs for producing good data. These concepts, though simple and almost entirely nonmathematical, are among the most important ideas in statistics. The second part of the book presents graphical and numerical tools and strategies for exploring data. This topic is now often called *data analysis*, the practical art of understanding what data say and communicating your understanding to others. The third part of the book is devoted to *statistical inference*. Inference uses the mathematics of probability to draw conclusions from data, and to accompany those conclusions by a formal statement of how confident we are that they are correct. Most statistics texts concentrate on inference, either on the many specific inference procedures available or on the mathematical theory that undergirds these procedures. We will avoid mathematics almost entirely and present very few inference procedures, concentrating instead on the reasoning of inference. I hasten to add that we will not leave the interesting business of drawing conclusions to the end of the book. Producing and analyzing data usually suggest conclusions, though not always correct conclusions. We will have much to say about these informal inferences in the first two parts of the book.

Your goals in reading this book should be threefold. First, reach an understanding of statistical ideas in themselves. Statistical concepts and modes of reasoning are major intellectual accomplishments that are worthy of your attention. Second, acquire the ability to deal critically with numerical arguments. Many persons are unduly credulous when numerical arguments are used; they are impressed by the solid appearance of a few numbers and do not attempt to penetrate the substance of the argument. Others are unduly cynical; they think numbers are liars by nature and never trust them. Numerical arguments are like any others. Some are good, some are bad, and some are irrelevant. A bit of quantitative sophistication will enable you to hold your own against the number-slinger. Third, gain an understanding of the impact of statistical ideas on public policy and in other areas of academic study.

In reading this book you will learn new facts and acquire new skills. You will, for example, learn what the Consumer Price Index measures and how to describe straight-line relationships between two variables by scatterplots, correlation, and regression. But the emphasis is on statistical reasoning. The facts and skills are needed to give flesh and muscle to the reasoning. Statistics is not a collection of facts and recipes; it is an intellectual method for gaining knowledge from data.

Data *vary*. Individual people, animals, or things are variable; repeated measurements on the same individual are variable. Conclusions based on data, whether the inference is formal or informal, are therefore *uncertain*. Statistics faces the variability and uncertainty of the world directly. Statistical reasoning can produce data whose usefulness is not destroyed by variation and uncertainty. It can analyze data to separate systematic patterns from the ever-present variation. It can form conclusions that, while not certain—nothing in the real world is certain— have only a little uncertainty. More important, statistical reasoning allows us to say just how uncertain our conclusions are.

Statistical ideas and techniques emerged only slowly from the struggle to work with uncertain data. Almost two centuries ago astronomers and surveyors faced the problem of combining many observations which, despite the greatest care, did not exactly match. Their efforts to deal with variation in their data produced some of the first statistical techniques. As the social sciences emerged in the nineteenth century, old statistical ideas were transformed and new ones invented to describe the variation in individuals and societies. The study of heredity and of variable populations in biology brought more advance. The first half of the twentieth century gave birth to statistical ideas for producing data and to formal inference based on mathematics. By mid-century it was clear that a new discipline had been born. As all fields of study place more emphasis on data and increasingly recognize that variability in data in unavoidable, statistics has become a central intellectual method. Every educated person should be acquainted with statistical reasoning. Reading this book will enable you to make that acquaintance.

Part I

Collecting data

Before numbers can be used for good or evil, we must produce them. Of course we could make up data, a common enough practice. Or we could base our conclusions on a few individual cases that happen to come to our attention. This is called *anecdotal evidence* because we often tell a story or anecdote that illustrates our conclusion, without stopping to ask whether the story is typical. The incidents that stick in our memory are often unusual. We remember the airplane crash that kills several hundred people and so fail to notice that data on all flights show that flying is in fact much safer than driving.

> **Example 1.** Homeless people are visible in any large city, sleeping on heat vents and clutching the shopping bags that hold their few possessions. How many homeless are there? Our chance encounters with the homeless are a poor basis for estimating their numbers and condition. In the absence of carefully designed data collection, estimates of the number of homeless in the U.S. ranged as high as 3 million; for the city of Chicago, estimates of 25,000 were common. These numbers, though frequently quoted by the news media, were simply guesses based on anecdotes and sympathy. The first attempts to make a careful count gave much lower estimates: about 350,000 homeless nationwide and between 2000 and 3000 in Chicago on any given night. The Chicago study was especially convincing because it involved a thorough search of several hundred city blocks chosen at random as well as visiting all the shelters in the city.[1]

Leaving invention and anecdote aside, many statistical studies are based on *available data*, that is, data not gathered specifically for the study at hand, but lying about in files or records kept for other reasons. Available data are often valuable. The data collected by the Bureau of the Census, for example, give a detailed picture of population and housing in the U.S. Census data are essential for social science research as well as for businesses looking for trends among their customers. But available data do not always fit our needs. Because the census is based on questionnaires sent to every household in the country, it counts only people in dwelling units. The census, useful as it is for many purposes, gives no information about the homeless.

Example 2. The American Cancer Society, in a booklet called "The Hopeful Side of Cancer," claimed that about one in three cancer patients is now cured, while in 1930 only one in five patients was cured. That's encouraging. But where does this encouraging estimate come from? From the state of Connecticut. Why Connecticut? Because it is the only state that kept records of cancer patients in 1930. It is a matter of available data. But Connecticut is not typical of the entire nation. It is densely populated but has no large cities, and has fewer blacks than the national average. Cancer death rates are higher in large cities than in rural locations, and higher among blacks than among whites. Data from Connecticut alone do not necessarily mirror national trends.

Using available data requires judgment and caution. The 350,000 cases in the Connecticut Tumor Registry may not be typical of national trends. But suppose we want to study the relationship between cancer and asbestos in water-supply pipes. The registry of every cancer case in Connecticut since 1930 can be matched with information about localities whose pipes contain asbestos. The registry has been used for over 200 studies of the occurrence of cancer and the effectiveness of various treatments. Not all sources of available data are so useful. Hospital records of cancer cannot tell how common the disease is, since, unlike a state registry, they have no clear geographical boundaries. "Facts" about crime based on police records are notoriously unreliable, since many crimes are never reported to the police. Good data, like other good things, are usually the fruit of systematic effort.

Historians must rely on available data. The rest of us can make an effort to obtain data that bear directly on the questions we wish to ask. Such data are obtained by either *observation* or *experimentation*. Observation is passive: The observer wishes to record data without interfering with the process being observed. Experimentation is active: The experimenter attempts to completely control the experimental situation. The difference is illustrated by the work of Tycho Brahe and Galileo at the beginning of the Scientific Revolution. Brahe devoted his life to recording precisely the positions of stars and planets and left records from which Kepler deduced his laws of planetary motion. This was observation. Galileo studied motion under the influence of gravity by rolling balls of various weights down inclined planes of various lengths and angles. This was experiment.

Neither observation nor experimentation is as simple as you might think, especially when we turn from stars and inclined planes to political opinions and the effectiveness of drugs. The first part of this book concerns the statistical ideas used to arrange observations or experiments. Statistics gives *designs* (patterns or outlines) for collecting data that can be applied in any area of study. Chapter 1 studies the design of *samples* (selecting units for observation), and Chapter 2 presents the design of *experiments*. Each chapter explores key statistical ideas, important examples of their use, and other topics, including the ethical problems of collecting data about people. Chapter 3 completes the topic of collecting data by addressing the process of *measurement* by which numbers are finally obtained.

NOTES

1. The Chicago Homeless Study is described in P. H. Rossi, J. D. Wright, G. A. Fisher, and G. Wills, "The Urban Homeless: Estimating Composition and Size," *Science*, Volume 235 (1987), pp. 1336–1339.

Chapter 1

Sampling

Boswell quotes Samuel Johnson as saying, "You don't have to eat the whole ox to know that the meat is tough." That is the essential idea of sampling: to gain information about the whole by examining only a part. Here is the basic terminology used by statisticians to discuss sampling:

Population — the entire group of objects about which information is wanted.

Unit — any individual member of the population.

Sample — a part or subset of the population used to gain information about the whole.

Sampling frame — the list or units from which the sample is chosen.

Variable — a characteristic of a unit, to be measured for those units in the sample.

Notice that population is defined in terms of our desire for information. If we desire information about all U.S. college students, that is our population even if students at only one college are available for sampling. It is important to define clearly the population of interest. If you seek to discover what fraction of the American people favor a ban on private ownership of handguns, you must specify the population exactly. Are all U.S. residents included in the population, or only citizens? What minimum age will you insist on? In a similar sense, when you read a preelection poll, you should ask what the population was: all adults, registered voters only, Democrats or Republicans only?

The distinction between population and sample is basic to statistics. Some examples will illustrate this distinction and introduce some major uses of sampling. These brief descriptions also indicate the variables to be measured for each unit in the sample. They do not state the sampling frame. Ideally, the sampling frame should be a list of all units in the population. But, as we shall see, obtaining such a list is one of the practical difficulties in sampling.

Example 1. *Public opinion polls*, such as those conducted by the Gallup and Harris organizations, are designed to determine public opinion on a variety of issues. The specific variables measured are responses to questions about public issues. Though most noticed at election time, these polls are conducted on a regular basis throughout the year. For the Gallup Poll,

Population: U.S. residents 18 years of age and over.

Sample: About 1500 persons interviewed weekly.

Example 2. *Market research* is designed to discover consumer preferences and usage of products. Among the better-known examples of market research is the television-rating service of Nielsen Media Research. The Nielsen ratings describe how many households watch each program and the age, sex, and race of the viewers. The ratings determine how much advertisers will pay to sponsor a program and ultimately whether or not the program remains on the air. Note that a unit here is a household, not an individual person.

Population: All 90 million U.S. households that have a television set.

Sample: About 4000 households that agree to use a "people meter" to record the TV viewing of all people in the household.

Example 3. *The decennial census** is required by the constitution. An attempt is made to collect basic information (number of occu-

---

Decennial means every 10 years. A census has been taken every 10 years since 1790. For more information on sampling by the Bureau of the Census, see Morris H. Hansen and Barbara A. Bailar, "How to Count Better: Using Statistics to Improve the Census," in J. M. Tanur et al. (eds.), *Statistics: A Guide to the Unknown* (Belmont, Calif.: Wadsworth, 1989).

pants, age, race, sex, family relationship, etc.) from each household in the country. Much other information is collected, but only from a sample of households that receive a longer questionnaire.

Population: All 93 million U.S. households.

Sample: The entire population (as far as possible) for basic information; only 17% of all households for other information.

Example 4. *Acceptance sampling* is the selection and careful inspection of a sample from a large lot of a product shipped by a supplier. On the basis of this, a decision is made whether to accept or reject the entire lot. The exact acceptance-sampling procedure to be followed is usually stated in the contract between the purchaser and the supplier.

Population: A lot of items shipped by the supplier.

Sample: A portion of the lot that the purchaser chooses for inspection.

Example 5. *Sampling of accounting data* is a widely accepted accounting procedure. It is quite expensive and time-consuming to verify each of a large number of invoices, accounts receivable, spare parts in inventory, and so forth. Accountants therefore use a sample of invoices or accounts receivable in auditing a firm's records, and the firm itself counts its inventory of spare parts by taking a sample of it.* A good example of this business use of sampling is the procedure for settling accounts among airlines for interline tickets. The passenger who takes a trip involving two or more carriers pays the first carrier, which then owes the other carriers a portion of the ticket cost. It is too expensive for the airlines to calculate exactly how much they owe each other, so only a sample of tickets is examined and accounts are settled on that basis.

Population: All interline air tickets purchased in a given month.

Sample: About 12% of these tickets used to settle accounts among airlines.

---

*For more information, see John Neter, "How accountants save money by sampling" in *Statistics: A Guide to the Unknown.*

There are many more uses of sampling, some bordering on the bizarre. For example, a radio station that plays a song owes the song's composer a royalty. The organization of composers (called ASCAP) collects these royalties for all its members by charging stations a license fee for the right to play members' songs. But how should this income be distributed among the members of ASCAP? By sampling: ASCAP tapes about 60,000 hours from the 53 million hours of local radio programs across the country each year. The tapes are shipped to New York, where ASCAP employs monitors (professional trivia experts who recognize nearly every song ever written) to record how often each song was played. This sample count is used to split royalty income among composers, depending on how often their music was played. Sampling is a pervasive, though usually hidden, aspect of modern life.

1. The need for sampling design

A *census* is a sample consisting of the entire population. If information is desired about a population, why not take a census? The first reason should be clear from the examples we have given: If the population is large, it is too expensive and time-consuming to take a census. Even the federal government, which can afford a census, uses samples to collect data on prices, employment, and many other variables. Attempting to take a census would result in this month's unemployment rate becoming available next year rather than next month.

There are also less obvious reasons for preferring a sample to a census. In some cases, such as acceptance sampling of fuses or ammunition, the units in the sample are destroyed. In other cases a relatively small sample yields more accurate data than a census. This is true in developing nations that lack adequate trained personnel for a census. Even when personnel are available, a careful sample of an inventory of spare parts will almost certainly give more accurate results than asking the clerks to count all 500,000 parts in the warehouse. Bored people do not count accurately.

The experience of the Census Bureau itself reminds us that a more careful definition of a census is "an *attempt* to sample the entire population." The bureau estimates that the 1980 census missed 1.4% of the American population. These missing persons include an estimated 5.9% of the black population, largely in inner cities. A number of cities sued the bureau, alleging that undercounting of their minority population had cost them federal aid dollars. The Census Bureau won the court cases, but the question of whether and how to correct the undercount remains unanswered. So a census is not foolproof, even with the legal and financial resources of the government behind it. Nevertheless, only a census can give detailed information about every small area of the population. For example, block-by-block population figures are required to create election districts with equal population. It is the main function of the decennial census to provide this local information.

So sample we must. Selecting a sample from the units available often seems simple enough, but this simplicity is misleading. If a supplier of oranges sold your company several crates per week, you would be wise to examine a sample of oranges in each crate to determine the quality of the oranges supplied. You find it convenient to inspect a few oranges from the top of each crate. But these oranges may not be representative of the entire crate if, for example, those on the bottom are damaged more often in shipment. Your method of sampling might even tempt the supplier to be sure that the rotten oranges were packed on the bottom with some good ones on top for you to inspect. Selection of whichever units of the population are easily accessible is called *convenience sampling*. Samples obtained in this way often are not representative of the population and lead to misleading conclusions about the population.

> **Example 6.** Convenience sampling occurs in situations less obvious than squeezing the oranges on the top of the crate. Interviews at shopping malls, for example, are often used by companies seeking information about consumer tastes. Mall interviews are a cheap way to talk to many consumers quickly. But shopping mall customers are more affluent than the population as a whole, and more of them are teenagers or elderly. The result will be a sample that systematically overrepresents some parts of the population (the prosperous, teenagers, and the elderly) and underrepresents others. The opinions of such a convenience sample may be very different from those of the population as a whole.

When a sampling method produces results that consistently and repeatedly differ from the truth about the population in the same direction, we say that the sampling method is *biased*. Convenience samples are often biased. A common form of convenience sampling, one almost sure to lead to strong bias, is *voluntary response*. A voluntary response sample chooses itself by responding to questions asked by mail or during television broadcasts. People who feel strongly about an issue, particularly people with strong negative feelings, are more likely to take the trouble to respond. The opinions of this group are often very different from those of the population as a whole. Here is an example of voluntary response.

> **Example 7.** In 1987, Shere Hite published a best-selling book called *Women and Love*. The author distributed 100,000 questionnaires through various women's groups, asking questions about love, sex, and relations between women and men. She based her book on the 4.5% of the questionnaires that were returned. The women who responded were fed up with men and eager to fight them. For example, 91% of those who were divorced said that they had initiated the divorce. The anger of women toward men became the theme of the book, but angry women are much more likely to return the questionnaire and so are overrepresented in Shere Hite's data. Her voluntary response sample gives almost no information about women in general. One reviewer rightly called the book "social science fiction."

Section 1 exercises

1.1. When conversation turns to the pros and cons of wearing automobile seat belts, Herman always brings up the case of a friend who survived an accident because he was not wearing seat belts. The friend was thrown from the car and landed on a grassy bank, suffering only minor injuries, while the car burst into flames and was destroyed. Explain briefly why this anecdote does not provide good evidence that it is safer not to wear seat belts. What kind of data would you prefer as evidence?

1.2. Organic gardeners think that garden plants can sometimes be protected from insects by placing them near other plants that the insects prefer. A letter to the editor of *Organic Gardening* magazine (August 1980) said,

> *Today I noticed about eight stinkbugs on the sunflower stalks. Immediately I checked my okra, for I was sure that they'd be under attack. There wasn't one stinkbug on them. I'd never read that stinkbugs are attracted to sunflowers, but I'll surely interplant them with my okra from now on.*

Explain briefly why this anecdote does not provide good evidence that sunflowers attract stinkbugs away from okra. In your explanation, suggest some reasons why all of the bugs might be on the sunflowers.

1.3. A sociologist wants to know the opinions of employed adult women about government funding for day care. She obtains a list of the 520 members of a local business and professional women's club, and mails a questionnaire to 100 of these women selected at random. Only 68 questionnaires are returned. What is the population in this study? What is the sampling frame? What is the sample?

1.4. Different types of writing can sometimes be distinguished by the lengths of the words used. A student interested in this fact wants to study the lengths of words used by Tom Wolfe in his novels. She opens a Wolfe novel at random and records the lengths of each of the first 250 words on the page.

What is the population in this study? What is the sample? What is the variable measured?

In each of Exercises 1.5 to 1.8, briefly identify the *population* (what is the basic unit and which units fall in the population?), the *variables* measured (what is the information desired?), and the *sample*. If the situation is not described in enough detail to identify the population completely, complete the description of the population in a reasonable way. Be sure that from your description it is possible to tell exactly when a unit is in the population and when it is not.

Moreover, each sampling situation described in Exercises 1.5 to 1.8 contains a serious source of probable bias. In each case, state the *reason* you suspect that bias will occur and also the *direction* of the likely bias. (That is, in what way will the sample conclusions probably differ from the truth about the population?)

1.5. A member of Congress is interested in whether constituents favor a proposed gun control bill. His staff reports that letters on the bill have been received from 361 constituents and that 283 of these oppose the bill.

1.6. A national newspaper wanted Iowa's reaction to President Bush's agricultural policy in early 1990. A reporter interviewed the first 50 persons willing to give their views, all in a single voting precinct. The headline on the resulting article read "Bush Policies Disenchant Iowa," and the reporter wrote that Bush would lose an election in Iowa if one were held then.

1.7. A flour company wants to know what fraction of Minneapolis households bake some or all of their own bread. A sample of 500 residential addresses is taken, and interviewers are sent to these addresses. The interviewers are employed during regular working hours on weekdays and interview only during those hours.

1.8. The Miami Police Department wants to know how black residents of Miami feel about police service. A questionnaire with several questions about the police is prepared. A sample of 300 mailing addresses in predominantly black neighborhoods is chosen, and a police officer is sent to each address to administer the questionnaire to an adult living there.

1.9. Advice columnist Ann Landers once asked her female readers whether they would be content with affectionate treatment by men with no sex ever. Over 90,000 women wrote in, with 72% answering "Yes." Many of the letters described unfeeling treatment at the hands of men. Explain why this sample is certainly biased. What is the likely direction of the bias? That is, is that 72% probably higher or lower than the truth about the population of all adult women?

1.10. Want to sample public opinion instantly and cheaply? Your telephone company will set up special telephone numbers that record how many calls are made to each number. Announce your question on TV, give one number for "yes" and another for "no," and wait. No word is needed from respondents; they just dial a number. The telephone company will add a small charge to the phone bills of those who call. The first major use of call-in polling was by the ABC television network in October of 1980. At the end of the first Reagan-Carter presidential election debate, ABC asked its viewers which candidate won. The call-in poll proclaimed that Reagan had won the debate by a 2 to 1 margin. But a random survey by CBS News showed only a 44 to 36% margin for Reagan, with the rest undecided. Why are call-in polls likely to be biased? Can you suggest why this bias might have favored the Republican Reagan over the Democrat Carter?

2.   Simple random sampling

The bias in convenience samples is due to human choice. The statistician's remedy for this bias is to eliminate human choice by allowing impersonal chance to choose the sample. The result is a *simple random sample*. The essential idea is to give each unit in the sampling frame the same chance to be chosen for the sample as any other unit. For reasons to be explained later, the precise definition is slightly more complicated.

> A *simple random sample* of size *n* is a sample of *n* units chosen in such a way that every collection of *n* units from the sampling frame has the same chance of being chosen.

We will abbreviate simple random sample as *SRS*. Notice that the definition concerns a property of the method for choosing the sample:

An SRS is obtained by a method that gives every possible sample of size n the same chance of being the sample actually chosen. An SRS has a clear advantage over a convenience sample: It is fair, or *unbiased*. No part of the sampling frame has any advantage over any other in obtaining representation in the sample. The definition of an SRS is designed to correct the overrepresentation of one part of the population often produced by convenience sampling.

Very well then, if an SRS is so useful a commodity, how do we actually obtain one? One way is to use *physical mixing*: Identify each unit in the sampling frame on an identical tag, mix the tags thoroughly in a box, then draw one blindly. If the mixing is complete, every tag in the box has the same chance of being chosen. The unit identified on the tag drawn is the first unit in our SRS. Now draw another tag without replacing the first. Again, if the mixing is thorough, every remaining tag has the same chance of being drawn. So every pair of tags has the same chance of being the pair we have now drawn; we have an SRS of size 2. To obtain an SRS of size n, we continue drawing until we have n tags corresponding to n units in the sampling frame. Those n units are an SRS of size n.

Physical mixing and drawing convey clearly the idea of an SRS. You should now grasp what it means to give each unit and each possible set of n units the same chance of being chosen. Physical mixing is even practiced on some occasions. But it is surprisingly difficult to achieve a really thorough mixing, as those who spend their evenings shuffling cards know. Physical mixing is also awkward and time-consuming. There is a better way.

Picture a wheel (such as a roulette wheel) rotating on a smooth bearing so that it does not favor any particular orientation when coming to rest. Divide the circumference of the wheel into 10 equal sectors and label them 0, 1, 2, 3, 4, 5, 6, 7, 8, and 9. Fix a stationary pointer at the

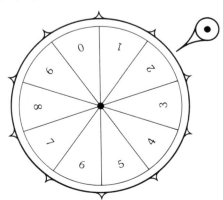

wheel's rim and spin the wheel. Slowly and smoothly it comes to rest. Sector number 2, say, is opposite the pointer. Spin the wheel again. It comes to rest with, say, sector number 9 opposite the pointer. If we continue this process, we will produce a string of the digits 0, 1, . . . , and 9 in some order. On any one spin, the wheel has the same chance of producing each of these 10 digits. And because the wheel has no memory, the outcome of any one spin has no effect on the outcome of any other. We are producing a table of random digits.

> A *table of random digits* is a list of the 10 digits 0, 1, 2, 3, 4, 5, 6, 7, 8, and 9 having the following properties:
>
> 1. The digit in any position in the list has the same chance of being any one of 0, 1, 2, 3, 4, 5, 6, 7, 8, and 9.
> 2. The digits in different positions are independent in the sense that the value of one has no influence on the value of any other.

Table A on page 430 is a table of random digits. The division into groups of five digits and into numbered rows makes the table easier to read and use but has no meaning. The table is just a long list of digits having properties 1 and 2. The table of random digits was produced by a very careful physical mixing process, much more elaborate than the wheel I used to illustrate random digits.

We can think of random digits as the result of someone else's careful physical mixing; our goal is to use their work in choosing an SRS rather than doing our own physical mixing. To use the table, we need the following facts about random digits, which are consequences of basic properties 1 and 2:

> 3. Any *pair* of digits in the table has the same chance of being any of the 100 possible pairs, 00, 01, 02, . . ., 98, and 99.
> 4. Any *triple* of digits in the table has the same chance of being any of the 1000 possible triples, 000, 001, 002, . . ., 998, and 999.
> 5. And so on for groups of four or more digits from the table.

How to use Table A to choose an SRS is best illustrated by a sequence of examples.

Example 8. A dairy products manufacturer must select an SRS of size 5 from 100 lots of yogurt to check for bacterial contamination. We proceed as follows:

(a) Label the 100 lots 00, 01, 02, . . ., 99, in any order.

(b) Enter Table A in any place and read systematically through it. We choose to enter at line 111 and read across:

$$81486 \quad 69487 \quad 60513 \quad 09297 \quad \cdots$$

(c) Read groups of two digits. Each group chooses a label attached to a lot of yogurt. Our SRS consists of the lots having the labels

$$81, 48, 66, 94, 87.$$

This example illustrates the basic technique: Give the units in the sampling frame numerical labels and use Table A to choose an SRS of these labels. *It is essential that each label consist of the same number of digits.* In the example, each label consists of two digits. By property 3 of random digits, each pair of digits in Table A is then equally likely to choose any label. Two-digit labels are adequate for a sampling frame containing between 11 and 100 units. (If no more than 10 units must be labeled, one-digit labels can be used. If 101 to 1000 units must be labeled, three-digit labels are needed.) *Always use as few digits as possible in labels.* That is why we labeled the first lot of yogurt 00 instead of 01; there are 100 labels 00, 01, . . ., and 99, so 100 units can be labeled with two digits if we start with 00. It is good practice to begin at zero rather than one even when not all labels are needed.

Suppose that the line of Table A used in the example had read

$$81486 \quad 68186 \quad 60513 \quad 09297 \quad \cdots$$

The first three lots chosen are those with labels 81, 48, and 66, as before. The next pair of digits in the table is 81. Because lot 81 is already in the sample, we *ignore repeated groups of digits* and go on to choose lots 86 and 60 to complete an SRS of size 5. You are now ready for a more difficult example.

Example 9. An SRS of size 5 must be chosen from a group of 300 students who have volunteered to take part in a psychology experiment. We proceed as follows:

(a) List the students in some order, such as alphabetically. This list is the sampling frame.

(b) Label the 300 students 000, 001, . . ., 299.

(c) Enter Table A at, say, line 116 and read across in three-digit groups. Each three-digit group 000 to 299 chooses a student; each three-digit group 300 to 999 is not a label and is ignored. The result is

144	592 ignore
605 ignore	631 ignore
424 ignore	803 ignore
716 ignore	510 ignore
362 ignore	253
	and so on.

This example illustrates the fact that we must *ignore unused labels*. But we examined 10 three-digit groups and succeeded in choosing only two students. It is more efficient to give several labels to each unit in the sampling frame, being sure to give each unit the same number of labels. Let us redo the last example.

Example 10. An SRS of size 5 must be chosen from a group of 300 students who have volunteered to take part in a psychology experiment. We proceed as follows:

(a) List the students in some order.

(b) Label the first student 000, 300, and 600.
Label the second student 001, 301, and 601.
.
.
.
Label the last student 299, 599, and 899.

(c) Enter Table A at line 116 and read across in three-digit groups. Choose a student for the sample if any of his labels occurs. The result is

144	student 144 is in the sample
592	student 292 is in the sample
605	student 005 is in the sample
631	student 031 is in the sample
424	student 124 is in the sample.

Notice that the three labels given each student in Example 10 are

1 The original label, between 000 and 299;

2 The original label plus 300 (the number of units in the sampling frame); and

3 The original label plus 600.

So label 592 is the same as label $592 - 300 = 292$, and label 605 is the same as label $605 - 600 = 005$. The labels 900 to 999 are not used and are ignored if we come upon them in Table A.

Section 2 exercises

1.11. Use the table of random digits (Table A) to select an SRS of 3 of the following 25 volunteers for a drug test. Be sure to say where you entered the table and how you used it.

Agarwal	Garcia	Petrucelli
Andrews	Healy	Reda
Baer	Hixson	Roberts
Berger	Lee	Shen
Brockman	Lynch	Smith
Chen	Milhalko	Sundheim
Frank	Moser	Wilson
Fuest	Musselman	
Fuhrmann	Pavnica	

1.12. Your class in Ancient Ugaritic Religion is poorly taught and the students have decided to complain to the dean. They decide to choose 4 of their number at random to carry the complaint. The class list appears below. Choose an SRS of 4 using the table of random digits beginning at line 145.

Anderson	Gutierrez	Patnaik
Aspin	Green	Pirelli
Bennett	Harter	Rao
Bock	Henderson	Rider
Breiman	Hughes	Robertson
Cochran	Johnson	Rodriguez
Dixon	Kim	Siegel
Edwards	Landis	Tompkins
Fuller	Laskowsky	Vandegraff
Grant	Olds	Williams

1.13. A food processor has 50 large lots of canned mushrooms ready for shipment, each labeled with one of the lot numbers below.

A1109	A2056	A2219	A2381	B0001
A1123	A2083	A2336	A2382	B0012
A1186	A2084	A2337	A2383	B0046
A1197	A2100	A2338	A2384	B1195
A1198	A2108	A2339	A2385	B1196
A2016	A2113	A2340	A2390	B1197
A2017	A2119	A2351	A2396	B1198
A2020	A2124	A2352	A2410	B1199
A2029	A2125	A2367	A2411	B1200
A2032	A2130	A2372	A2500	B1201

An SRS of 5 lots must be chosen for inspection. Use Table A to do this, beginning at line 139.

1.14. An SRS of 25 of 440 voting precincts in a metropolitan region must be chosen for special voting-fraud surveillance on election day. Explain clearly how you would label the precincts. Then use Table A to choose the SRS, and list the precincts you selected. Enter Table A at line 117.

1.15. Which of the following statements are true of the table of random digits, and which are false?
  (a) There are exactly four 0s in each row of 40 digits.
  (b) Each pair of digits has chance 1/100 of being 00.
  (c) The digits 000 can never appear as a group, because this pattern is not random.

1.16. The following page contains a population of 80 circles. (They might represent fish in a pond or tumors removed in surgery.) Do a sampling experiment as follows:
  (a) Label the circles 00, 01, . . ., 79 in any order, and use Table A to draw an SRS of size 4.
  (b) Measure the diameter of each circle in your sample. (All of the circles have diameters that are multiples of 1/8 inch; in decimal form the possible diameters are $1/8 = 0.125$, $1/4 = 0.25$, $3/8 = 0.375$, $1/2 = 0.5$, $5/8 = 0.625$, $3/4 = 0.75$, and $7/8 = 0.875$. Record your results in decimal form.) Then use a calculator to compute the *mean* diameter of the four circles in your sample. The mean of the diameters d_1, d_2, d_3, and d_4 is the ordinary average

$$\frac{d_1 + d_2 + d_3 + d_4}{4}$$

(c) Now repeat steps (a) and (b) three more times (four times in all), using a different part of Table A each time. Was any circle chosen more than once in your four SRSs? How different were the mean diameters for the four samples?

(d) Now draw an SRS of size 16 from this population using a part of Table A not yet used in this exercise. Measure the diameters of the 16 circles in your sample and find the mean (average) diameter.

Your results for SRSs of size 4 and 16 can be combined with those of the rest of the class to produce two pictures like the one in Figure 1-2 in the next section. These pictures show less sampling variability for means of samples with size 16 than for samples with size 4.

1.17. Another way to sample the population of circles in Exercise 1.16 is to close your eyes and drop your pencil point at random onto the page. Mark the circle that you hit. Do this until you have hit four circles.

(a) Is this (at least approximately) an SRS of size 4? Why or why not?

(b) Think of these circles as cross sections of trees at chest height above the ground. Foresters who want to estimate the total volume of wood in a woodlot use a sampling method that has the same effect as dropping a pencil point into the circles. Explain why this is done.

1.18. Figure 1-1 (page 20) is a map of a census tract in Cedar Rapids, Iowa. Census tracts are small, homogeneous areas averaging 4000 in population. On the map, each block is marked with a Census Bureau identification number. Use Table A beginning at line 125 to choose an SRS of five blocks from this tract.

3.   Population information from a sample

The advice columnist Ann Landers once asked her readers, "If you had it to do over again, would you have children?" She received nearly 10,000 responses, almost 70% saying "No!" Many of the letters included heart-rending tales of the torments inflicted by children on their parents. Now this is an egregious example of voluntary response. How egregious was suggested by a professional nationwide random sample commissioned by *Newsday*. That sample polled 1373 parents and found that 91% *would* have children again. It is, you see, quite possible for a voluntary response sample to give 70% "No" when the truth about the population is close to 91% "Yes."

511 · NEW ST. · 513 COOK ST. · 512 AVE. · 104 · 103 COOK · 102 ST. · 101
105 LAWRENCE · 106 · 107 · 108 · 109
510 CHENERY ST. · 509 ST. · 113 · COLLEGE · 112 · 111 CANEDY · 110 G.M.&O.
507 · 508 ST.
506 FAYETTE · 505 AVE. · 114 · 115 · 116 · 117 SCARRITT · 118
503 WILLIAMS · 504 ST.
502 ALLEN · 501 ST. · 205 · 204 ALLEN · 203 ST. · 202 ST. · 201 ST.
409 VINE · HENRIETTA · 410 ST. · 206 ST. · 207 SPRING · 208 · 209 VINE · 210 ST.
408 · 407 · (STATE HWY. 4) · 215 · 214 · 213 · 212 · 211
AVE. · AVE. · 404 · 405 CEDAR · 406 · 301 · 302 PINE · 305 ST. · 306 2ND · 307 PINE · 303 · 304 ST. · 308 3RD
403 LOWELL · 402 · WHITTIER · 401 PASFIELD · 317 COLLEGE · 316 SPRUCE · 313 · 1ST · 312 · 309 ST. · 315 · 314 · 311 · 310 ST.

"Hey, Pops, what was that letter you sent off to Ann Landers yesterday?"

But is *Newsday's* sample result really convincing evidence that about 91% of all parents would have children again? As a newspaper reporter put it, "Far be it from us to question the validity of any statistic that we read in the papers, but in 1974 there were 54,917,000 families in America. This means we are talking somewhere in the neighborhood of a 1-in-50,000 sampling."[1] Leaving aside the reporter's poor arithmetic (1373 out of 54,917,000 families is 1 in 40,000 not 1 in 50,000), he has raised a perceptive question. We know why convenience sampling is unreliable, but why is an SRS reliable, especially when so small a fraction of the population is sampled?

Well, an SRS has no bias. In this case, the *Newsday* poll gave all parents the same chance of responding, rather than favoring those who were mad enough at their children to write Ann Landers. But lack of favoritism alone is not enough when we are asked to draw conclusions about 55 million families from information about only 1400 of them. We need to think more carefully about the process of gaining information about a population from a sample, starting with some new vocabulary.

The conclusions we wish to draw from a sample usually concern some numerical characteristic of the population, such as the fraction of American parents who would have children again, the average lifetime of General Electric 60-watt standard light bulbs, or the fraction of Princeton alumni who approve of coeducation. As always, we must distinguish between population and sample.

A *parameter* is a numerical characteristic of the *population*. It is a fixed number, but we usually do not know its value.

A *statistic* is a numerical characteristic of the *sample*. The value of a statistic is known when we have taken a sample, but it changes from sample to sample.

Put simply, parameter is to population as statistic is to sample. Both parameters and statistics are numbers. The distinction lies entirely in whether the number describes the population (then it is a parameter) or a sample (then it is a statistic). The fraction of all American parents who would have children again is a parameter describing the population of parents. Call it p. Alas, we do not know the numerical value of p. We usually use a sample statistic to estimate the unknown value of a population parameter. *Newsday*, in an attempt to estimate p, took a sample of 1373 parents. The fraction of the sample who would have children again is a statistic. We call this statistic $\hat{p}$, pronounced "p hat." If 1249 of this sample of size 1373 would do it again, then

$$\hat{p} = \frac{1249}{1373} = 0.91$$

It is reasonable to use the sample proportion $\hat{p} = 0.91$ as an estimate of the unknown population proportion p, and that is exactly what *Newsday* did. But if *Newsday* took a second sample of size 1373, it is almost certain that there would *not* be exactly 1249 positive responses. So the value of $\hat{p}$ will vary from sample to sample. This is called *sampling variability*.

Sampling distributions

Aha! So what is to prevent one random sample from finding that 91% of parents would have children again and a second random sample from finding that 70% would not? After all, we just admitted that the statistic $\hat{p}$ wanders about from sample to sample. We are saved by a second property of random sampling, a property even more important than lack of bias: A sample statistic from an SRS has a predictable pattern of values in repeated sampling. This pattern is called the *sampling distribution* of the statistic. Knowledge of the sampling distribution allows us to make statements about how far the sample proportion $\hat{p}$ is likely to wander from the population proportion p owing to sampling variability.

To illustrate a sampling distribution, let us do an experiment. I have a box containing a large number of round beads, identical except for color. These beads are a population. The proportion of dark beads in the box is

$$p = 0.20$$

and this number is a parameter describing this population of beads. You can think of the beads as representing all the parents in the country, and the dark beads as representing those who would not have children again. I also have a paddle with 25 bead-sized indentations in it, so when I thrust it into the beads in the box, it selects a sample of 25 beads. If the beads in the box are well mixed, this is an SRS of size 25. Ask yourself a few questions about this SRS of size 25 from a population containing 20% dark beads.

- How many dark beads do you expect to appear in the sample?
- If you take several SRSs, do you expect to find a sample with 25 dark beads? One with no dark beads? One with as many as 15 dark beads?

You might reasonably expect about 20% of the beads in the sample to be dark, that is, about 5 dark beads among the 25 beads in the sample. But we will not always get exactly 5 dark beads. If we get, say, 4 dark beads, then the statistic

$$\hat{p} = \frac{4}{25} = 0.16$$

is still a good estimate of the parameter $p = 0.20$. But if we draw a sample with 15 dark beads, then

$$\hat{p} = \frac{15}{25} = 0.60$$

is a very bad estimate of p. How often will we get such poor estimates from an SRS?

I carried out this bead-sampling experiment 200 times and recorded the number of dark beads in each sample. (I was careful to return the sample to the population and stir the population after each repetition.) The results are shown in a table and pictorially in Figure 1-2. None of the

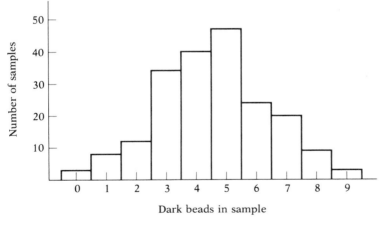

Number of dark beads in sample	0	1	2	3	4	5	6	7	8	9
Sample proportion $\hat{p}$	0	0.04	0.08	0.12	0.16	0.20	0.24	0.28	0.32	0.36
Number of samples with this outcome	3	8	12	34	40	47	24	20	9	3
Fraction of samples with this outcome	0.015	0.040	0.060	0.170	0.200	0.235	0.120	0.100	0.045	0.015

Figure 1-2. Results of 200 simple random samples of size 25 when $p = 0.20$.

200 samples contained more than 9 dark beads. The sample proportion $\hat{p}$ did indeed vary from sample to sample: It ranged from 0 (no dark beads) to 0.36 (9 dark beads) when all 200 samples were examined. But estimates as bad as $\hat{p} = 0$ or $\hat{p} = 0.36$ (remember that the true p is 0.20 for this population) did not occur often. Of the 200 samples, 56% had either 4, 5, or 6 dark beads ($\hat{p}$ of 0.16, 0.20, or 0.24) and 83% had 3, 4, 5, 6, or 7 dark beads ($\hat{p}$ between 0.12 and 0.28).

Errors in sampling

In our experiment, we knew p. If p were not known, the same facts would hold. We cannot guarantee that the sample statistic $\hat{p}$ is close to the unknown p (because of sampling variability), but we can be confident that it is close (because most of the time an SRS gives a $\hat{p}$ close to p). So the results of an SRS not only show no favoritism but tend to be repeatable from sample to sample. We need a final bit of vocabulary to describe the fact that lack of repeatability (the sample result wanders all over the barnyard) is as serious a flaw in a sampling method as is favoritism.

Because a sample is selected for the purpose of gaining information about a population, by "error in a sample" we mean an incorrect estimate of a population parameter by a sample statistic. Two basic types of error are associated with any method of collecting sample data.

> *Bias* is consistent, repeated divergence of the sample statistic from the population parameter in the same direction.

> *Lack of precision* means that in repeated sampling the values of the sample statistic are spread out or scattered; the result of sampling is not repeatable.

A common misunderstanding is to confuse bias in a sampling method with a strong trend in the population itself, especially if that trend is a reflection of prejudice or bias in the ordinary sense of that word. If, for example, 93% of a population of corporate personnel directors are opposed to the federal government's affirmative action hiring program, this is not bias in the statistical sense. It is simply a fact about this population.

We can think of the true value of the population parameter as the bull's-eye on a target, and we can think of the sample statistic as a bullet fired at the bull's-eye. *Bias* means that our sight is misaligned and we shoot consistently off the bull's-eye in one direction. Our sample values do not center about the population value. *Lack of precision* means that repeated shots are widely scattered on the target; that is, repeated samples do not give similar results but differ widely among themselves. This target illustration of the results of repeated sampling is shown in Figure 1-3.

Notice that high precision (repeated shots are close together) can accompany high bias (the shots are consistently away from the bull's-eye in one direction). Notice also that low bias (the shots center on the bull's-eye) can accompany low precision (repeated shots are widely

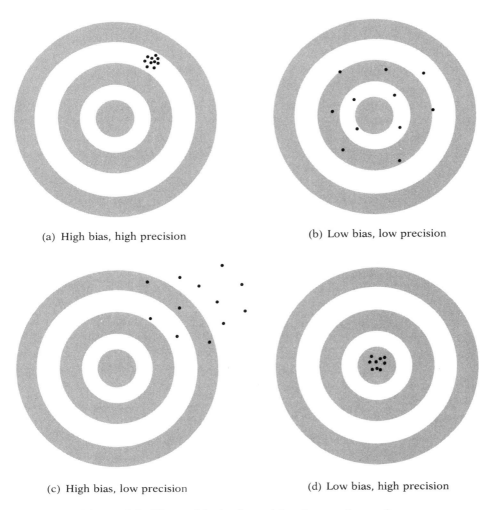

(a) High bias, high precision   (b) Low bias, low precision

(c) High bias, low precision   (d) Low bias, high precision

Figure 1-3. Bias and lack of precision in sample results.

scattered). A good sampling scheme, like a good shooter, must have both low bias and high precision.

The sampling distribution of a statistic describes both its bias and its precision. For example, the precision of $\hat{p}$ as an estimator of p in Figure 1-2 can be expressed by a statement such as "56% of SRSs of size 25 have a value of $\hat{p}$ within ±0.04 of the true value of p." The shape of the distribution of values shown in Figure 1-2 is typical of an SRS. These distributions can be studied mathematically to save us the work of experimentation. As you might guess, such studies (and experiments as well) show that *increasing the size of the sample increases the precision of sample statistics.* If in our experiment we had used samples of size 100, about 74% of these samples would have sample proportions $\hat{p}$ within ±0.04 of p. If, like *Newsday*, we take samples of size 1373, fewer than 2 in 10,000 will fail to have $\hat{p}$ within ±0.04 of the truth about the population. We can be nearly certain that *Newsday's* sample result is close to the true proportion p of all parents who would have children again.

Only one other fact about precision is needed to apply the coup de grace to that newspaper reporter's skepticism about 1-in-40,000 samples: *The precision of a sample statistic does not depend on the size of the population as long as the population is much larger than the sample.* In other words, the pattern of results from repeatedly thrusting my 25-bead paddle into a large box of beads does not depend on whether the box contains 1000 beads (as it did in my experiment) or 1,000,000 beads. The precision does depend on how many beads the paddle selects (the sample size) and, to a lesser extent, on the fraction of p of dark beads in the population.

This is good news for *Newsday*. Their sample of size 1373 has high precision because the sample size is large. That only 1 in 40,000 in the population were sampled is irrelevant. It is almost certain—Ann Landers to the contrary—that close to 91% of American parents would have children again. But the fact that the precision of a sample statistic depends on the size of the sample and not on the size of the population is bad news for anyone planning an opinion poll in a university or small city. For example, it takes just as large an SRS to estimate the proportion of Ohio State University students who favor legalizing marijuana as to estimate with the same precision the proportion of all U.S. residents 18 and over who favor legalization. That there are about 53,000 Ohio State students and over 185 million U.S. residents 18 and over does *not* mean that equally precise results can be obtained by taking a smaller SRS at Ohio State.

The facts acquired here are the foundation for an understanding of the uses of sampling. In review, those facts are as follows:

1. Despite the *sampling variability* of statistics from an **SRS**, the values of those statistics have a *known distribution* (that is, a known pattern) in repeated sampling.

2. When the sampling frame lists the entire population, simple random sampling produces *unbiased* estimates: the values of a statistic computed from an **SRS** neither consistently over-estimate nor consistently underestimate the value of the population parameter.

3. The *precision* of a statistic from an **SRS** depends on the size of the sample and can be made as high as desired by taking a large enough sample.

I have a closing confession to make: *Newsday* did not use an SRS to refute Ann Landers. Opinion polls use more complicated sampling methods; just how complicated will be seen in Section 6 of this chapter. But these sampling methods are based on random selection and so share the three basic properties listed. Our conclusions about the *Newsday* poll stand.

Section 3 exercises

Each boldface number in Exercises 1.19 to 1.22 is the value of either a *parameter* or a *statistic*. In each case, state which it is.

1.19. The Bureau of Labor Statistics announces that last month it interviewed all members of the labor force in a sample of 55,800 households, of whom **6.5%** were unemployed.

1.20. A carload lot of ball bearings has an average diameter of **2.503** centimeters. This is within the specifications for acceptance of the lot by the purchaser. But the acceptance sampling procedure happens to inspect 100 bearings from the lot with an average diameter of **2.515** centimeters. This is outside the specified limits, so the lot is mistakenly rejected.

1.21. A telephone sales outfit in Los Angeles uses a device that dials residential phone numbers in that city at random. Of the first 100 numbers dialed, **47** are unlisted numbers. This is not surprising, because **52%** of all Los Angeles residential phones are unlisted.

1.22. Voter registration records show that **68%** of all voters in Marion County, Indiana are registered as Republicans. To test a random-digit dialing device, you use the device to call 150 randomly chosen residential telephones. Of the registered voters contacted, **73%** are registered Republicans.

1.23. Figure 1-2 is a graph of the values of a sample statistic in 200 samples when the population parameter has the same value. Bias and lack of precision can be seen pictorially in such a graph of the sampling distribution as well as in the target illustration of Figure 1-3. Label each of the sampling distributions in Figure 1-4 as high or low bias and as high or low precision.

1.24. Just before a presidential election, a national opinion polling firm increases the size of its weekly sample from the usual 1500 people to 4000 people. Does the larger random sample lessen the bias of the poll result? Does it improve the precision of the result?

1.25. A management student is planning a project on student attitudes toward part-time work while attending college. She develops a questionnaire, and plans to ask 25 randomly selected students to fill it out. Her faculty advisor approves the questionnaire, but suggests that the sample size be increased to at least 100 students. Why is the larger sample helpful?

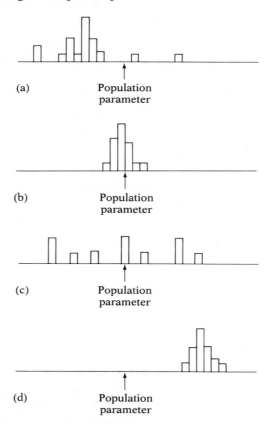

(a)   Population
             parameter

(b)   Population
             parameter

(c)   Population
             parameter

(d)   Population
             parameter

Figure 1-4.

1.26. An agency of the federal government plans to take an SRS of residents in each state to estimate the proportion of owners of real estate in each state's population. States range from California (over 29 million) to Wyoming (about 500,000) in number of inhabitants.

    (a) Will the precision of the sample proportion change from state to state if an SRS of size 2000 is taken in each state? Explain your answer.

    (b) Will the precision of the sample proportion change from state to state if an SRS of 1/10 of 1% (0.001) of the state's population is taken in each state? Explain your answer.

1.27. The *New York Times* conducted a national opinion poll on women's issues in June, 1989. The sample consisted of 1025 women and 472 men contacted by randomly selecting telephone numbers. One question was

    Many women have better jobs and more opportunities than they did 20 years ago. Do you think women have had to give up too much in the process, or not?

Forty-eight percent of the women and thirty-three percent of the men in the sample said that women have had to give up too much. The *New York Times* publishes complete descriptions of its polling methods. Here is part of the description for this poll. (From the *New York Times* of August 21, 1989.)

    In theory, in 19 cases out of 20 the results based on the entire sample will differ by no more than three percentage points in either direction from what would have been obtained by seeking out all adult Americans.

    The potential sampling error for smaller subgroups is larger. For example, for men it is plus or minus five percentage points.

Explain why the margin of error is larger for conclusions about men alone than for conclusions about all adults.

1.28. Not every household fills out the same form in the decennial census of the United States. Basic questions about the number of people in the household and their age, sex, race, and so on appear on all the forms. Other questions appear on only a sample of 17% of the forms. The Census Bureau publishes summary statistics for various geographic areas. For the questions that appear on all forms, these statistics are published for areas with as few as 100 households. But for questions appearing on the 17% sample, the bureau does not publish statistics for areas with fewer than about 2000 households. Can you think of possible reasons for this policy?

1.29. Let us illustrate sampling variability in a small sample from a small population. Ten of the twenty-five club members listed below are female. Their names are marked with asterisks in the list. You must choose five club members at random to receive free trips to the national convention.

Agassiz	Darwin	Herrnstein	Myrdal	Vogt*
Binet*	Epstein	Jimenez*	Perez*	Went
Blumenbach	Ferri	Lombrosco	Spencer*	Wilson
Chase*	Goddard*	Moll*	Thomson	Wong
Cuvier*	Hall	McKim*	Toulmin	Yerkes

(a) Draw five club members at random 20 times, using a different part of Table A each time. Record the number of females in each of your samples. Make a graph like Figure 1-2 to display your results. What is the average number of females in your 20 samples?

(b) Do you think the club members should suspect discrimination if none of the five tickets go to women?

1.30. Random digits can be used to *simulate* the results of random sampling. Suppose that you are drawing simple random samples of size 25 from a large box of beads and that 20% of the beads in the box are dark. To simulate this experiment, let 25 consecutive entries in Table A stand for the 25 beads in your sample. The digits 0 and 1 stand for dark beads, while other digits stand for light beads. This is an accurate imitation of the sampling experiment because 0 and 1 make up 20% of the 10 equally likely digits. Simulate the results of 50 samples by counting the number of 0s and 1s in the first 25 entries in each of the 50 rows of Table A. Make a graph like Figure 1-2 to display the results of your 50 samples. Is the truth about the population (20% dark, or 5 in a sample of 25) near the center of your graph? What are the smallest and largest counts of dark beads you obtained in your 50 samples? What percent of your samples had either 4, 5, or 6 dark beads? Compare the results of your 50 samples with the 200 samples described in Figure 1-2.

4.   Confidence statements

The results of an SRS are *random* in the special sense in which statisticians use that word: The outcome of a single sample is unpredictable, but there is a definite pattern of results in the long run. The results of tossing a coin are also random, as are the sexes of human or animal offspring. Random phenomena are described by the laws of probability, which allow us to calculate how often each outcome will occur in the long run. Statisticians use probability to calculate the sampling distributions of statistics. That's much faster than the actual repeated sampling of beads that we did in Section 3. The most important reason for the deliberate use of chance in collecting data is that we can then apply the laws of probability to draw conclusions from the data.

We will have much to say about probability later (Chapter 7), and we will even give the recipes for some methods of statistical inference based on probability (Chapter 8). But one form of statement based on probability is so commonly used to describe the accuracy of opinion polls and other samples that every newspaper reader and TV viewer should be familiar with it. Often the statement is casual. For example, a news report on the shifts in public attitudes in response to a presidential speech said,

> *The poll, conducted by Gordon Black Associates of Rochester, N.Y., is based on 714 interviews with adults randomly selected from across the USA. Results before the speech have a margin of error of 5.5 percent. Post-speech results have a 5 percent margin of error.* (From *USA TODAY*, November 1, 1983.)

That 5 percent margin of error describes the accuracy of the poll, but it leaves much unsaid. As connoisseurs of data, we want to know the whole story.

A sample statistic is *unbiased* if, on the average over many samples, the statistic is a correct estimate of the population parameter of interest. Simple random sampling produces unbiased statistics, and so do the more complex sampling designs used by polling organizations. That is, the values of the statistic, though they vary in repeated sampling, have a sampling distribution that is centered on the parameter value. The shape of the sampling distribution depends on the exact sample design and on the size of the sample, but can be calculated using the laws of probability. This sampling distribution permits a statement of the precision of the statistic.

> **Example 11.** In November 1989, the Gallup Poll asked 1234 adults, "Do you expect that the overall quality of your own life will be better by the year 2000?" Here the parameter of interest is the proportion p of all U.S. adults who expect life to be better in the

future. In the sample, 950 answered "Yes." The statistic used to estimate p is therefore

$$\hat{p} = \frac{950}{1234} = 0.77$$

Gallup Poll statisticians, from their knowledge of the sampling distribution of $\hat{p}$, describe the precision of the statistic as follows: *In 95% of all possible samples, the statistic $\hat{p}$ will take a value within ±0.04 of the parameter p.* The news article about the poll says this in less technical language: "The poll had a margin of sampling error of plus or minus four percentage points." (Reported in the *New York Times*, January 1, 1990.)

We made similar statements about our bead-sampling experiment on the basis of the sampling distribution in Figure 1-2 on page 23. If, for example, $p = 0.75$ is the true proportion of U.S. adults expecting their lives to get better, then 95 out of 100 Gallup Poll samples will produce sample proportions in the range 0.75 ± 0.04, or 0.71 to 0.79. The Gallup Poll will miss the truth by more than 0.04 in only 5% (1 in 20) of their samples.

This statement of precision has two parts: *a level of confidence* (95%) and a *margin of error* (± 0.04). The level of confidence says how often in the long run the margin of error will be met. We could demand higher confidence (99%), or settle for lower confidence (90%). But we cannot achieve 100% confidence for any margin of error that does not allow $\hat{p}$ to range all the way from 0 to 1. There is always some chance—a very small chance, to be sure—that 1234 persons chosen at random will all be optimistic about the future ($\hat{p} = 1$) or all deeply pessimistic ($\hat{p} = 0$). The news report left out the level of confidence that should accompany its margin of error. A naive reader might imagine that the sample proportion $\hat{p}$ is *certain* to be within ±0.04 (4%) of the true p. The connoisseur knows that polling organizations almost always make 95% confidence statements, and fills in the missing confidence level.

The margin of error that is met 95% of the time is a direct measure of precision. The Gallup $\hat{p}$ that is within ±0.04 of the truth with 95% confidence is more precise than the *USA TODAY* $\hat{p}$ that is within only ±0.05 with 95% confidence. *The smaller the margin of error at 95% confidence, the greater the precision.* The greater size of Gallup's sample (1234 versus 714 for USA TODAY) accounts for its smaller margin of error. Table 1-1 give the margins of error of $\hat{p}$ in 95% confidence statements for the Gallup Poll's sampling procedure. Since Gallup's sampling methods are typical of those used by modern opinion polls, you can use the table to judge the precision of most poll results appearing in the press, provided that the size of the sample is reported.

One last step: Saying that $\hat{p}$ falls within ±0.04 of p 95% of the time is the same as saying that p (which is unknown) is within ±0.04 of $\hat{p}$ (which is known) 95% of the time. Not knowing the true p, we cannot say whether the Gallup Poll of November 1989 was one of the 95% that hit or one of the 5% that miss. But, since 95% of all such polls do hit within ±0.04 of the true p, we say that we are 95% *confident* that between 0.73 and 0.81 of all adults believe that their lives will be better in the year 2000. (Check that: The sample gave $\hat{p} = 0.77$, and the margin of error was ±0.04. So we are confident that p falls between $0.77 - 0.04 = 0.73$ and $0.77 + 0.04 = 0.86$.)

A *confidence statement* turns a long-run fact about the sampling distribution of a statistic into a statement of our confidence in the result of a single sample. The usual form of a 95% confidence statement about a parameter estimated by an unbiased statistic is

With 95% confidence, the parameter lies in the range

statistic ± margin of error

This is shorthand for

The sampling distribution of the statistic is such that in 95% of all possible samples, the statistic falls within ± *margin of error* of the true parameter value.

> **Example 12.** If the Gallup Poll interviews 1514 people and finds that 53% of them oppose a longer school year, what confidence statement can we make?
>
> Table 1-1 shows that the margin of error for a sample of size 1500 never exceeds 3 percentage points (±3%). Since the margin of error changes slowly with sample size, we can use this value for a sample of 1514 people. We are 95% confident that between 50% (that's 53% − 3%) and 56% (53% + 3%) of adult Americans oppose lengthening the school year.

Table 1-1 shows that the margin of error, although it does not depend on the size of the population, does depend a bit on the parameter p. You can always use the largest margin of error given in the column for your sample size for a conservative confidence statement. The largest margin of error in each column is accurate for questions on which the popula-

Table 1-1. Precision of the sampling procedure used by the Gallup Poll as of 1972*

Population percentage	Sample size						
	100	200	400	600	750	1000	1500
Near 10	7	5	4	3	3	2	2
Near 20	9	7	5	4	4	3	2
Near 30	10	8	6	4	4	4	3
Near 40	11	8	6	5	4	4	3
Near 50	11	8	6	5	4	4	3
Near 60	11	8	6	5	4	4	3
Near 70	10	8	6	4	4	4	3
Near 80	9	7	5	4	4	3	2
Near 90	7	5	4	3	3	2	2

SOURCE: George Gallup, *The Sophisticated Poll Watcher's Guide* (Princeton Opinion Press, 1972), p. 228.

*The table shows the range, plus or minus, within which the sample percentage $\hat{p}$ falls in 95% of all samples. This margin of error depends on the size of the sample and on the population percentage p. For example, when p is near 60%, 95% of all samples of size 1000 will have $\hat{p}$ between 56% and 64%, because the margin of error is $\pm 4\%$.

tion is close to being evenly divided. For example, a sample of 1000 persons gives a margin of error of $\pm 4\%$ when p is between 30% and 70%. Little is lost in using this margin of error for all questions. Remember that Table 1-1 is for a complex national sample. The margins of error for simple random samples are a bit *smaller* than those in Table 1-1. The details are in Chapter 8, should you actually have to produce confidence statements. But you are already a sophisticated consumer.

Section 4 exercises

1.31. A news article on a Gallup Poll noted that "28 percent of the 1548 adults questioned felt that those who were able to work should be taken off welfare." The article also said, "The margin of error for a sample size of 1548 is plus or minus three percentage points." (From the *New York Times*, June 29, 1977.) Explain to someone who knows no statistics what "margin of error" means here.

1.32. A poll of 1000 voters uses the Gallup Poll's sampling design. Although 52% of the sample say they will vote for Ms. Caucus, the polling organization announces that the election is too close to call. Why is this the proper conclusion?

1.33. A television commentator, talking about the 1988 presidential election, said

*The final opinion poll before the election showed that 53%
would vote for Bush. The poll had a margin of error of only
±2%, so it was certain that Bush would win.*

Why was it *not* certain that Bush would win?

1.34. You ask a random sample of 1493 adults, chosen by the Gallup
Poll's sampling method, "Are you afraid to go out at night within a
mile of your home because of crime?" Of your sample, 672 say
"Yes." Make a confidence statement about the percent of all adult
Americans who fear to go out at night because of crime.

1.35. A random sample of 1028 adults, chosen by the Gallup Poll's
sampling method, are asked "Do you favor an amendment to the
Constitution that would permit organized prayer in public
schools?" Of the sample, 678 say "Yes." Make a confidence statement about the percent of all adults who favor a school prayer
amendment.

1.36. A national poll of 1433 adults finds that 46% feel that the country's
future will be better than the present. No margin of error is given
in the news account you read. If the poll used a method much like
Gallup's (and most do), what confidence statement can you make
based on the information given?

1.37. A news report says, "The latest Harris survey on crime in America
indicates that 26% of Americans feel less safe on the streets than
they did a year ago." Can you make a confidence statement about
this result? If so, make one. If not, explain why you cannot.

1.38. The final Gallup Poll prior to a presidential election interviews
3509 people. This is a larger sample size than appears in Table 1-1.
Is the margin of error of the result more or less than 3 percentage
points? Why?

1.39. Though opinion polls usually make 95% confidence statements,
other confidence levels are in common use. The monthly unemployment rate, for example, is based on a Census Bureau survey of
about 55,800 households. The margin of error in the announced
unemployment rate is about two-tenths of one percentage point
with 90% confidence. Would you expect the margin of error for
95% confidence to be smaller or larger? Why?

5. Sampling can go wrong

The conclusions of the last two sections seem too good to be true. Most
of us are aware of times when samples led to erroneous results, as when
opinion polls predicted that Thomas Dewey would defeat Harry Truman

in the 1948 presidential election. Alas, the glowing conclusions of the last two sections say only that errors from one source can be made small by properly conducted random sampling. These are *random sampling errors*, the deviations between sample statistic and population parameter caused by chance in selecting a random sample. When we are sampling beads or ball bearings or laboratory rats, random sampling error is our main problem, and it is an easy problem to deal with. That was the good news of Section 3. But, when we turn from sampling beads to sampling public opinion, consumer preferences, or personal habits, other sources of error become more serious than random sampling error in all but the smallest samples.

> *Sampling errors* **are errors caused by the act of taking a sample. They cause sample results to be different from the results of a census.**
>
> *Nonsampling errors* **are errors not related to the act of selecting a sample from the population and that might be present even in a census.**

Sampling errors

Random sampling error is one kind of sampling error. *Nonrandom sampling errors* arise from improper sampling and have nothing to do with the chance selection in an SRS. One source of nonrandom sampling error is a biased sampling method, such as convenience sampling. Another common source of nonrandom sampling error is a sampling frame that differs systematically from the population. Even an SRS from such a frame will give biased conclusions about the population. Here are some examples of sources of sampling error:

> **Example 13.** *Telephone directories.* In the 1970s, the two principal television-rating services selected their sample households at random from telephone directories. This sampling frame omits households having no telephones and those with unlisted numbers. Roger Rice, a San Francisco television station operator, charged in 1974 that television ratings were biased: They underestimated the audience for programs of special interest to minorities because the sampling system underrepresented the poor.[2] Rice noted that in the San Francisco–Oakland area, about 62% of households had listed telephones, 25% had unlisted phones, and 13% had no telephone service. Only the 62% with listed telephones could participate in rating TV programs. Census data reported that about 57% of the

population was white, but the sample for one of the rating services was 78% white. Hispanics made up 14% of the population but only 1.2% of the rating-service sample.

Example 14. *Who answers the phone?* Most sample surveys are conducted by telephone. Random-digit-dialing equipment places calls at random to residential phones, both listed and unlisted. The 7% of American households that have no phone are omitted from the sampling frame; this is a source of sampling error. Some telephone surveys form their sample by interviewing the person who answers the phone on a single call to each household selected; if no one answers, the random-digit-dialer moves on to another number. This is another source of sampling error, because these easy-to-reach people differ from the population as a whole. For one thing, they are more likely to be women. In fact, the New York Times/ CBS News Poll reports that only 37% of the people reached by their telephone surveys on the first call to a household are men. It also appears that Democrats are easier to reach than Republicans. One survey during the 1984 presidential campaign gave Reagan a 3 percentage point lead over Mondale after one call, but found his lead to be 13 percentage points after many callbacks contacted almost the entire original sample.[3]

Nonsampling errors

Nonsampling errors, those which plague even a census, are often rooted in the perversity and complexity of human behavior. I will mention four types of nonsampling error: missing data, response errors, processing errors, and the effect of the data-collection procedure.

Missing data are due to the inability to contact a subject or to the subject's refusal to respond. If subjects who cannot be contacted or who refuse to respond differ from the rest of the population, bias will result. For example, a survey conducted entirely during ordinary working hours will fail to contact households in which all adults work. Refusal to respond to survey questions is becoming more common. Though estimates vary, it appears that about 15% of persons sampled refuse to respond to nongovernmental surveys. In large cities the refusal rate is higher, probably 20 to 25%. This growing emphasis on privacy is in many ways praiseworthy, but it threatens even random samples with the same bias caused by voluntary response.

Response errors concern the subjects' responses. The subjects may lie about their age or income. Or they may remember incorrectly when asked how many packs of cigarettes they smoked last week. Or subjects who cannot understand a question may guess at an answer rather than show ignorance. Questions that ask subjects about their behavior during a fixed time period are particularly prone to response errors due to faulty memory. Many people who last visited a doctor 15 months ago will say "Yes" when asked whether they have visited a doctor during the past year.

Processing errors are mistakes in such mechanical tasks as doing arithmetic, entering responses into a computer, or coding the data (that is, assigning numerical values to responses for data-processing purposes). These can be minimized by rechecking every mechanical task.

The effect of the method used to collect data can be large. Sometimes this effect clearly leads to errors in conclusions drawn about the population, as when black subjects speak less openly about their views on race relations to a white interviewer than they would to a black interviewer. The *timing* of a survey often affects its results: The federal government, to the great frustration of almost everyone, collects data on farm labor in its monthly household survey only in December. As a result, the typical "seasonal farm worker" is a white, male college student who can be located in December to answer questions about his summer job. Hundreds of thousands of Mexican farmworkers, many in the United States illegally, seem to disappear.[4]

One important decision about the method of collecting data once a sample is chosen concerns the *exact wording of questions*. It is surprisingly difficult to word questions so that they are completely clear. A survey that asked about "ownership of stock" found that most Texas ranchers owned stock, though probably not the kind traded on the New York Stock Exchange. What is more, small changes in the wording of questions can significantly change the responses received. In fact, it is easy to introduce a definite bias into a sample survey by slanting the questions. A favorite trick is to ask if the subject favors some policy as a means to a desirable end: "Do you favor banning private ownership of handguns in order to reduce the rate of violent crime?" and "Do you favor imposing the death penalty in order to reduce the rate of violent crime?" are loaded questions that draw positive responses from persons worried about crime.

Finally, a survey taker must decide whether to use *mail, telephone, or personal interviews* to contact subjects. Each method has its own strengths and weaknesses. Mail surveys are the least expensive and reach households without phones. But mail surveys often have low response rates. Because people with strong feelings on an issue are most likely to respond, a voluntary response bias can result. Mail surveys are reliable only when nonrespondents are followed up, if necessary with personal

"Do I own any stock, Ma'am? Why, I've got 10,000 head out there."

visits. The U.S. decennial census mails out its questionnaires, then makes great efforts to hunt down the households who don't reply.

Telephone surveys are fast and economical, and the use of computer random-digit-dialing equipment makes it possible to automatically dial a random sample of residential telephones, including those with unlisted numbers. For these reasons, telephone interviewing is now used by almost all opinion polls and by political polling operations. But 7% of American households have no telephone — a source of sampling error in telephone surveys. As Example 14 notes, selecting a sample by calling randomly chosen telephone numbers only once is another source of sampling error. Careful telephone surveys plan to make many calls to reach members of the sample who are rarely at home. Failure to reach some respondents is then a nonsampling error because even a census can fail to contact some people. Once the call goes through, the response rate for telephone surveys is about 10% less than for face-to-face contact.

Personal interviews allow a skillful interviewer to establish rapport with the subject. This results in a higher response rate and clearer communication of the questions and probably introduces less bias than other methods of collecting data. But personal contact, including the repeat visits needed to contact people who are rarely at home, is very expensive. Another disadvantage is that it is harder to supervise the interviewers to keep the questioning the same for all subjects than it is in a telephone interview.* Personal contact is the choice of government agencies collecting such information as unemployment rates, and of most university survey workers.

Here are some examples of nonsampling errors.

> **Example 15.** *Race of interviewer.* In 1968, one year after a major racial disturbance in Detroit, a sample of black residents were asked:
>
> *Do you personally feel that you can trust most white people, some white people, or none at all?*
>
> Of those interviewed by whites, 35% answered "Most," while only 7% of those interviewed by blacks gave this answer. (Many questions were asked in this study. Only on some topics, particularly black-white trust or hostility, did the race of the interviewer have a strong effect on the answers given. Most sample surveys try to match the race of the interviewer to that of the subject.)[5]

> **Example 16.** *Wording of questions.* In 1980, the New York Times/ CBS News Poll asked a random sample of Americans about abortion. When asked "Do you think there should be an amendment to the Constitution prohibiting abortions, or shouldn't there be such an amendment?" 29% were in favor and 62% were opposed. (The rest were uncertain.) The same people were later asked a different question: "Do you believe there should be an amendment to the Constitution protecting the life of the unborn child, or shouldn't there be such an amendment?" Now 50% were in favor and only 39% were opposed.[6]

> **Example 17.** *Telling the truth?* In 1989, New York City elected its first black mayor and the state of Virginia elected its first black governor. In both cases, samples of voters interviewed as they left

---

*There is a surprising variation in the responses obtained by different interviewers. For an account of this interviewer effect on responses to the decennial census, see Morris H. Hansen and Barbara A. Bailar, "How to Count Better: Using Statistics to Improve the Census," in *Statistics: A Guide to the Unknown*.

their polling places predicted larger margins of victory than the official vote counts. The polling organizations were certain that some voters lied when interviewed because they felt uncomfortable admitting that they had voted against the black candidate.

Example 18. *Processing error.* The initial release of 1980 census data on income showed Stumpy Point, N.C., (pop. 205) to be a haven of the rich, with an average household income of $84,413. Stumpy Point had no doctors and no lawyers, and only half of the adults finished high school. Sure enough, the Census Bureau soon corrected the average income to $22,773. What happened? The incomes reported on census forms are coded in tens of dollars for entry into the computer. That is, an $8000 income is entered as 0800. Whoever coded Stumpy Point slipped up and entered $8000 as 8000 in many cases. The computer, following instructions, read this as $80,000 and declared Stumpy Point wealthy.[7]

The moral of this presentation of possible errors in conclusions drawn from samples is not that sampling is unreliable and untrustworthy. The moral is that great care is required in sampling human subjects and that the statistical idea of using a randomly chosen sample is not the cure for all possible ills. (It *is* an essential part of the cure, and for sampling ball bearings or accounting data it is most of the cure.) When you read or write an account of a sample survey, be sure that answers to the following questions are included:

- What was the *population*? That is, whose opinions were being sought?
- How was the sample *selected*? Look for mention of random sampling or probability sampling.
- What was the *size* of the sample? It is even better to give a measure of precision, such as the margin of error within which the results of 95% of all samples drawn as this one was would fall.
- How were the subjects *contacted*?
- *When* was the survey conducted? Was it just after some event that might have influenced opinion?
- What were the *exact questions* asked?

The code of ethics of the American Association for Public Opinion Research (AAPOR) requires disclosure of this information. The major opinion polls always answer these questions in their press releases when announcing the results of a poll. But editors and newscasters have the bad habit of cutting out these facts and reporting only the sample results. Worse, new methods for conducting telephone surveys have brought a proliferation of state or regional polls run by newspapers or broadcasting stations. CBS News counted 147 of these in 1980 and found that fewer than half used reliable methods. If a politician, an advertiser, or your local TV station announces the results of a poll without complete information, be skeptical.

Section 5 exercises

1.40. Which of the following are sources of *sampling error* and which are sources of *nonsampling error*? Explain your answers.
   (a) The subject lies when questioned.
   (b) A typing error is made in recording the data.
   (c) Data are gathered by asking people to mail in a coupon printed in a newspaper.

1.41. Each of the following is a source of error in a sample survey. Label each as *sampling error* or *nonsampling error* and explain your answers.
   (a) The telephone directory is used as a sampling frame.
   (b) The subject cannot be contacted in three calls.
   (c) Interviewers choose people on the street to interview.

1.42. How bad is crime? It depends on whose data you look at. The annual FBI publication *Crime in the United States* says that there were 91,110 forcible rapes in 1987. The FBI data list all crimes that are reported to law enforcement agencies and then reported by local law enforcement agencies to the FBI. The other main source of crime data is the National Crime Survey, a sample of 10,000 households per month conducted for the Bureau of Justice Statistics. The survey asks people if they have been victims of a crime. According to the survey, there were about 141,000 ± 6000 rapes in 1987. These two sets of data appear almost side by side in the *Statistical Abstract of the United States.*
   (a) Why does the Census Bureau report many more rapes than the FBI?
   (b) No margin of sampling error can be attached to the FBI data. Why not?
   (c) Each source of data may be subject to nonsampling errors. What are some important sources of nonsampling error for the FBI report? For the National Crime Survey?

1.43. On the next page is part of a newspaper report of a Gallup poll, reprinted from the *New York Times* of January 1, 1990. At the end

90's, in Poll: A Good Life Amid Old Ills

MICHAEL R. KAGAY

As Americans look to the year 2000, most of them anticipate a better life for themselves, but at the same time they foresee a worsening of many of the nation's social and economic problems, according to a new Gallup Poll.

Seventy-seven percent of the 1,234 adults polled said they expected the overall quality of their own life to be better by 2000. Similarly, 77 percent anticipated that their family life would be better in 10 years' time. Seventy-four percent said their financial situation would be better. Eighty-two percent of employed adults also predicted their job situation would improve in 10 years.

Somewhat smaller majorities of Americans also anticipated that by 2000 people would be spending more time on leisure and recreation (68 percent), and more time with their families (58 percent). A minority said people would be spending more time on jobs (38 percent) or household chores (13 percent).

The poll, conducted by telephone Nov. 16–19 had a margin of sampling error of plus or minus four percentage points.

The participants' optimism about their own lives was accompanied by a more pessimistic outlook on many current social and economic problems. Large majorities expected by 2000 to see increases in the rate of inflation (74 percent), the crime rate (71 percent), poverty (67 percent), homelessness (62 percent), and environmental pollution (62 percent).

of Section 5 are some questions that should be answered by a careful account of a sample survey. Which of these questions does this newspaper report answer, and which not? Give the answers whenever the article contains them.

1.44. Market research is sometimes based on samples chosen from telephone directories and contacted by telephone. The sampling frame therefore omits households having unlisted numbers and those without phones.

(a) What groups of people do you think will be underrepresented by such a sampling procedure?

(b) How can households with unlisted numbers be included in the sample?

(c) Can you think of any way to include in the sample households without telephones?

1.45. We have seen that the method of collecting the data can influence the accuracy of sample results. The following methods have been used to collect data on television viewing in a sample household:

(a) *The diary method.* The household is asked to keep a diary of all programs watched and who watched them for a week, then mail in the diary at the end of the week.

(b) *The roster-recall method.* An interviewer shows the subject a list of programs for the preceding week and asks which programs were seen.

(c) *The telephone-coincidental method.* The household is telephoned at a specific time and asked if the television is on, which program is being watched, and who is watching it.

(d) *The automatic-recorder method.* A device attached to the set records what hours the set is on and which channel it is turned to. At the end of the week, this record is removed from the recorder.

(e) *The people meter.* Each member of the household is assigned a numbered button on a hand-held remote control. Everyone is asked to push their button whenever they start or stop watching TV. The remote control signals a device attached to the set that keeps track of what channel the set is tuned to and who is and is not watching at all times.

For each method, discuss its advantages and disadvantages, especially any possible sources of error associated with each method. Method (e) is now the most commonly used; it replaced a combination of methods (a) and (d). Do you agree with that choice? (Do not discuss choosing the sample, just collecting data once the sample is chosen.)

1.46. You wish to determine whether students at your school think that faculty members are sufficiently available to students outside the classroom. You will select an SRS of 200 students.

   (a) What is the exact population? (Will you include part-time students? Graduate students?)

   (b) How will you obtain a sampling frame?

   (c) How will you contact subjects? (Is door-to-door interviewing allowed in campus residence halls?)

   (d) What specific question or questions will you ask?

1.47. The monthly unemployment rate is estimated from the Current Population Survey (CPS), a random sample of about 56,000 households each month conducted by the Bureau of the Census. It would be cheaper to choose one random sample and reinterview the adults in the chosen households each month for a year or more. This is called using a *panel*. Because we are interested most of all in changes over time for employment and unemployment, a panel also has the advantage of following these changes for a group of people over time. Can you think of any disadvantages of the panel method? (In fact, the CPS uses a modified panel method. The sample is changed every month, but not completely. Once chosen, a household stays in the sample for four months, drops out for eight months, then returns for four more months before finally dropping from the sample.)

1.48. We have seen that the exact wording of questions can influence sample survey results. Two basic types of questions are *closed questions* and *open questions*. A closed question asks the subject to choose one or more of a fixed set of responses. An open question allows the subject to answer in his or her own words; the responses are written down verbatim by the interviewer and sorted later. For example, here are an open and closed question on the same issue, both asked by the Gallup Poll within a few years of each other [recorded in *Public Opinion Quarterly*, Volume 38 (1974–1975), pp. 492–493]:

> OPEN: In recent years there has been a sharp increase in the nation's crime rate. What steps do you think should be taken to reduce crime?

> CLOSED: Which two or three of the approaches listed on this card do you think would be the best ways to reduce crime?

> > Cleaning up social and economic conditions in our slums and ghettos that tend to breed drug addicts and criminals.

> > Putting more policemen on the job to prevent crimes and arrest more criminals.

> > Getting parents to exert stricter discipline over their children.

> > Improving conditions in our jails and prisons so that more people convicted of crimes will be rehabilitated and not go back to a life of crime.

> > Really cracking down on criminals by giving them longer prison terms to be served under the toughest possible conditions.

> > Reforming our courts so that persons charged with crimes can get fairer and speedier justice.

What are the advantages and disadvantages of open and closed questions? Use the example just given in your discussion.

1.49. Comment on each of the following as a potential sample survey question. If any are unclear or slanted, restate the question in better words.

    (a) Does your family use food stamps?

    (b) Which of these best represents your opinion on gun control?

        (1) The government should confiscate our guns.

        (2) We have the right to keep and bear arms.

(c) In view of escalating environmental degradation and predictions of serious resource depletion, would you favor economic incentives for recycling of resource-intensive consumer goods?

1.50. Public reaction to a proposal often varies with its source. Try the
following: Tell several of your friends (don't burden yourself with
selecting an SRS) that you are collecting opinions for a course. Ask
some,

> *Thomas Jefferson said, "I hold that a little rebellion, now and
> then, is a good thing, and as necessary in the political world
> as storms are in the physical." Do you generally agree or
> generally disagree with this statement?*

Ask others the same question, but replace "Thomas Jefferson said"
with "Lenin said." (Be sure to ask each privately. To avoid bias,
randomize the question you ask each person by tossing a coin.)
Record the opinions you obtain and be prepared to discuss your
results.

6. More on sampling design

By now you have absorbed the message that a reliable sample survey
depends both on statistical ideas (random sampling) and on practical
skills (wording questions, skillful interviewing, etc.). When our goal is to
sample a large human population, using an SRS is good statistics but too
expensive to be good practice. First, a sampling frame is rarely available.
Second, if we choose an SRS of 1500 U.S. residents for a public opinion
poll, it would be a bit expensive to send interviewers off to Beetle,
Kentucky, and Searchlight, Nevada to find the lucky persons chosen.
Even telephone interviewing is easier and cheaper when the numbers to
be called are clustered in a few exchanges. The solution to these practical difficulties is to use a sampling design more complicated than an SRS
and "to sample not people but the map."[8] A government agency or
university survey office that plans to conduct personal interviews usually
proceeds somewhat as follows:

First, select a sample of counties and then a sample of townships or
wards within each county chosen. There is no difficulty in obtaining lists
of counties and wards to serve as sampling frames for these two steps.
Now, using a map or aerial photograph as the sampling frame, select a
sample of small areas (such as city blocks) within each ward chosen
earlier. Finally, select a sample of households in each small area chosen.
This can be done by sending interviewers to the area to compile a list of

households if none is available. Interview one adult from each household selected.

Such a *multistage sampling design* overcomes the practical drawbacks of an SRS. We do not need a list of all U.S. households, only a list of households in the small areas arrived at by sampling counties, then wards, then areas within wards. Moreover, all the households in the sample are *clustered* within these few small areas; this makes it much cheaper to collect the data.

The sample selected at each stage of a multistage design may be an SRS, but other types of random samples are also used. Households in a block or telephone numbers in an exchange are often selected by a *systematic random sample*. A starting point is chosen at random, then every third or tenth unit in geographical or numerical order is selected. See Exercise 1.59 for details. Systematic samples are fast, require no frame, and make geographical spread certain if the units are in geographical order. But there are pitfalls. A colleague of mine, working as an interviewer, was once instructed to visit every third address in a Chicago neighborhood. The block contained three-story walk-up apartments, so all the addresses visited were on the third floor. These households are poorer than those on the lower floors.

The earlier stages in a multistage sample often select a *stratified random sample* of counties or telephone exchanges. To obtain a stratified random sample, proceed as follows:

Step 1. Divide the sampling frame into groups of units, called *strata*. The strata are chosen because we have a special interest in these groups within the population or because the units in each stratum resemble each other.

Step 2. Take a separate SRS in each stratum and combine these to make up the stratified random sample.

The strata in a stratified random sample must be chosen using some characteristic of the units that is known in advance of the sample survey. Counties, for example, may be stratified by population and whether they are primarily urban, suburban, or rural. Opinion polls and the Census Bureau surveys that collect national economic and social data employ multistage designs with stratification in the early stages. Another common use of stratification is in sampling economic units. Because the largest corporations, or farms, or bills payable are especially important, economic surveys usually stratify by size and sample a higher proportion of the larger units. Sometimes *all* the large units are chosen and only a sample of the smaller ones. This is a stratified sample where a census is taken in one stratum.*

Stratified samples have two advantages. First, they allow us to gather separate information about each stratum. Second, if the units in each stratum are more alike in the variable measured than is the population as a whole, estimates from a stratified sample will be more precise than from an SRS of the same size. To grasp this, think about the extreme case in which all the units in each stratum are exactly alike. Then a stratified sample of only one unit from each stratum would completely describe the population, but an SRS of the same size would have very low precision. Because of these advantages, stratified samples are widely used.

It may surprise you to notice that stratified samples can violate one of the most appealing properties of the SRS: Stratified samples need not give all units in the population the same chance to be chosen. Some strata may be deliberately overrepresented in the sample. For example, if a poll of student opinion at a Big Ten university used an SRS of moderate size, there would probably be too few blacks in the sample to draw separate conclusions about black-student opinion. By stratifying, we can take an SRS of black students and a separate SRS of other students and then combine them to make up the overall sample. If the university has 30,000 students, of whom 3000 are black, we would expect an SRS of 500 students to contain only about 50 blacks. (Because 10% of the population is black, we expect about 10% of an SRS to be black. Remember those beads in Section 3?) So we might instead take a stratified random sample of 200 blacks and 300 other students. The 200 blacks allow us to study black opinion with fair precision.

You know how to choose this stratified sample if you remember how to take an SRS: Use Table A to select an SRS of 200 of the 3000 blacks; then use Table A a second time to select an SRS of 300 of the other 27,000 students. Because 200 of the 3000 black students are selected, the chance that any one black student is chosen is

$$\frac{200}{3000}, \text{ or } \frac{1}{15}$$

The chance that any one nonblack student is selected is

$$\frac{300}{27,000}, \text{ or } \frac{1}{90}$$

---

*See John Neter, "How Accountants Save Money by Sampling," in *Statistics: A Guide to the Unknown*, for an example of the use of such a sampling design by the Chesapeake and Ohio Railroad.

Each student has a known chance to be chosen, but that chance is different for blacks and others. This is a randomly chosen sample, but not an SRS.

Simple, systematic, and stratified random samples all use chance to select units from the population, as do multistage samples constructed from these building blocks. The details of complex sampling designs should be left to experts. (Strike a blow against poverty: hire a statistician.) But all fit the general statistical framework of *probability sampling*.

> A *probability sample* is a sample chosen in such a way that we know what samples are possible (not all need be possible) and what chance, or probability, each possible sample has to be chosen (not all need be equally probable).

In an SRS of size 500, every group of 500 units from the population is a possible sample, and all are equally likely to be chosen. But in our stratified sample of 500 students, the only possible samples have exactly 200 black students and 300 nonblack students. The laws of probability apply to any probability sample. But haven't we defeated the first aim of an SRS, which was to eliminate bias in sampling? A stratified sample, as we saw, may deliberately overrepresent blacks in a survey of student opinion, or larger corporations in a survey of business practices. True. Fortunately, the fact that probabilities of selection are *known* allows us to correct for overrepresentation when we analyze the data.

> **Example 19.** A university asks a stratified random sample of 200 black students (out of 3000) and 300 other students (out of 27,000) "Do you favor creation of a new degree program in African studies?" In the sample, 162 of the blacks and 174 of the others say "Yes."
> We estimate that
>
> $$\frac{162}{200} = 0.81$$
>
> of all black students favor the new program, as do
>
> $$\frac{174}{300} = 0.58$$
>
> of all other students. To estimate the proportion of all students who

favor the program, first estimate *how many* students are in favor as follows:

$$0.81 \times 3000 = 2430 \text{ blacks}$$
$$0.58 \times 27,000 = 15,660 \text{ others}$$

In all, we estimate that $2430 + 15,660 = 18,090$ of the 30,000 students are in favor. That's

$$\frac{18,090}{30,000} = 0.60$$

or 60%. The stratified sample allows us to say that 81% of black students, 58% of nonblacks, and 60% of all students favor creation of an African studies degree program. What is more, a statistician can make confidence statements about each of these conclusions.

The essential point is not the detail of the calculations, but the fact that statistics from any probability sample share the essential characteristics of statistics based on an SRS. The sampling distribution is known, and confidence statements can be made without bias and with increasing precision as the size of the sample increases. Nonprobability samples such as voluntary response samples do not share these advantages and cannot give trustworthy information about the population.

It may be that you will never have the good fortune to participate in the design of a sample survey. But if you work in advertising, politics, or marketing or if you use government economic and social data, you will surely have to use the results of surveys. We can summarize our study in an outline of the steps in designing a sample survey.

Step 1. *Determine the population*, both its extent and the basic unit. If you are interested in buyers of new cars, your unit could be new-car registrations, new-car owners (individuals), or households that purchased new cars. You must also be specific about the geographic area and date of purchase needed to quality a unit for this population.

Step 2. *Specify the variables to be measured*, and prepare the questionnaire or other instrument you will use to measure them. You will want to test your questionnaire on a pilot group of subjects to be certain it is clear and complete.

Step 3. *Set up the sampling frame.* This is related to Step 1. If you use a list of new-car registrations as a sampling frame (because this list is easy to obtain), households who bought several new cars will appear several times on the list. So an SRS of registrations will not be an SRS of households. (I bought a new car in 1974. The retiring president of General Motors, to express his confidence in the industry in a year of poor sales, bought *five* new GM cars that year. His household is five times as likely as mine to appear in an SRS of 1974 new-car registrations.)

Step 4. *Do the statistical design* of the sample: specify how large the sample will be and how it will be chosen from the sampling frame. You will often want to consult a statistician for technical advice.

Step 5. *Attend to details*, such as training interviewers and arranging the timing of the survey.

Much more might be said about each of these steps. A good deal is known about how to word questions, how to train interviewers, how to increase response in a mail survey, and so on. Much of this is interesting, and some is slightly amusing; for example, colorful commemorative stamps on the outer and return envelopes greatly increase the response rate in a mail survey. But a sample survey would show that you already know more about sampling than 99% of U.S. residents aged 18 or over. Enough is enough.

Section 6 exercises

1.51. A student at a large state university wants to ask the opinion of students about whether the university should advertise itself primarily as a technical school. Students in the School of Engineering and the School of Liberal Arts will probably differ on this issue. How should this fact influence the design of the sample?

1.52. About 20% of the engineering students at a large university are women. The school plans to poll a sample of 200 engineering students about the quality of student life.
    (a) If an SRS of size 200 is selected, about how many women do you expect to find in the sample?
    (b) If the poll wants to be able to report separately the opinions of male and female students, what type of sampling design would you suggest? Why?

1.53. A university employs 2000 male and 500 female faculty members. The equal employment opportunity officer polls a stratified random sample of 200 male and 200 female faculty members.
    (a) What is the chance that a particular female faculty member will be polled?
    (b) What is the chance that a particular male faculty member will be polled?
    (c) Explain why this is a probability sample.
    (d) Each member of the sample is asked, "In your opinion, are female faculty members in general paid less than males with similar positions and qualifications?"

180 of the 200 females (90%) say "Yes."
60 of the 200 males (30%) say "Yes."

So 240 of the sample of 400 (60%) answered "Yes," and the officer therefore reports that "based on a sample, we can conclude that 60% of the total faculty feel that female members are underpaid relative to males." Explain why this conclusion is wrong.

(e) If we took a stratified random sample of 200 male and 50 female faculty members at this university, each member of the faculty would have the same chance of being chosen. What is that chance? Is this an SRS? Explain.

1.54. A club contains 25 students, named

Abel	Fisher	Huber	Moran	Reinmann
Carson	Golomb	Jack	Moskowitz	Silvers
Cryer	Griswold	Jones	Neyman	Sobar
David	Hwang	Kiefer	O'Brien	Thompson
Elashoff	Holland	Lamb	Potter	Valenzuela

and 10 faculty members, named

Andrews	Fischang	Hernandez	Moore	Rabinowitz
Besicovitch	Gupta	Lightman	Phillips	Vincent

The club can send four students and two faculty members to a convention and decides to choose those who will go by random selection. Use Table A to choose a stratified random sample of four students and two faculty members. (There are two strata here, faculty and students. A stratified random sample can be taken only when the strata are chosen in advance and you can identify the members of each stratum.)

1.55. You want to interview 10 male scholarship athletes at length about their attitude toward school and their future plans. Because you believe there may be large differences among athletes in different sports, you decide to interview a stratified random sample of 7 basketball players and 3 golfers. Use the table of random digits, beginning at line 101, to select your sample from the team rosters below.

BASKETBALL

Abdul-Jabbar	Ewing	Robertson
Aguirre	Frazier	Robinson
Baylor	Johnson	Russell
Bird	Jordan	Thomas
Chamberlain	Malone	West
Cousy	McHale	Worthy
Erving	Pettit	

GOLF

Ballesteros	Kite	Palmer
Faldo	Nicklaus	Wadkins
Floyd	Pete	Watson

1.56. You have alphabetized lists of the 2000 male faculty members and of the 500 female faculty members at the university described in Exercise 1.53. Explain how you would assign labels and use Table A to choose a stratified random sample of 200 female and 200 male faculty members. What are the labels of the first five males and the first five females in your sample?

1.57. Using the information in part (d) of Exercise 1.53, give an unbiased estimate of the proportion of the total faculty who feel that females are underpaid.

1.58. A city contains 33 supermarkets. A health inspector wants to check compliance with a new city ordinance on meat storage. Because of the time required, he can inspect only 10 markets. He decides to choose a stratified random sample and stratifies the markets by sales volume. Stratum A consists of 3 large chain stores; the inspector decides to inspect all 3. Stratum B consists of 10 smaller chain stores; 4 out of the 10 will be inspected. Stratum C consists of 20 locally owned small stores; 3 of these 20 will be inspected.

Let "Yes" mean that the store is in compliance and "No" mean that it is not. The population is as follows (unknown to the inspector, of course):

Stratum A		Stratum B		Stratum C			
Store 1	Yes	Store 1	No	Store 1	Yes	11	Yes
2	Yes	2	Yes	2	Yes	12	Yes
3	No	3	No	3	No	13	No
		4	No	4	Yes	14	Yes
		5	Yes	5	No	15	Yes
		6	No	6	No	16	No
		7	Yes	7	No	17	No
		8	No	8	Yes	18	No
		9	No	9	No	19	Yes
		10	Yes	10	No	20	Yes

(a) Use Table A to choose a stratified random sample of size 10 allotted among the strata as described above.

(b) Use your sample results to estimate the proportion of the entire population of stores in compliance with the ordinance.

(c) Use the description of the population given above to find the true proportion of stores in compliance. How accurate is the estimate from part (b)?

1.59. Suppose that the final area chosen in a multistage sampling design contains 500 addresses, of which 5 must be selected. A systematic random sample is chosen as follows:

Step 1. Choose one of the first 100 addresses on the list at random. (Label them 00, 01, . . ., 99 and use a pair of digits from Table A to make the choice.)

Step 2. The sample consists of the address from Step 1 and the addresses 100, 200, 300, and 400 positions down the list from it.

For example, if 71 is chosen at random in Step 1, the systematic random sample consists of the addresses numbered 71, 171, 271, 371, and 471.

(a) Use Table A to choose a systematic random sample of 5 from a list of 500 addresses. Enter the table at line 130.

(b) What is the chance that any specific address will be chosen? Explain your answer.

(c) Is your sample an SRS? Explain your answer.

1.60. A group of librarians once wanted to estimate what fraction of books in large libraries falls in each of several size (height and width) categories. This information would help them plan shelving. To obtain it, they measured all of the several hundred thousand books in a library. Describe a sampling design that would have saved them time and money. That is, outline steps 1 to 4 at the end of Section 6 for a sampling design (simple or complicated) you would suggest to the librarians.

1.61. A labor organization wishes to study the attitudes of college faculty members toward collective bargaining. These attitudes appear to be different at different types of colleges. The American Association of University Professors classifies colleges as follows:

Class I   Offer the doctorate, and award at least 15 per year.
Class IIA   Award degrees above the bachelor's, but are not in Class I.
Class IIB   Award only bachelor's degrees.
Class III   Two-year colleges.

Describe a sampling design that would gather information for each class separately as well as overall information about faculty attitudes.

1.62. An important government sample survey is the monthly Current Population Survey (CPS), from which employment and unemployment data are produced. The sampling design of the CPS is described in the *BLS Handbook of Methods*, published by the Bureau of Labor Statistics. Obtain the handbook from the library, and write a brief description of the multistage sample design used in the CPS.

1.63. Many nationwide surveys "sample the map"; that is, the sampling frame consists of identifiable geographic units rather than of a list of people or places. The Statistical Reporting Service of the U.S. Department of Agriculture makes extensive use of such "area frames" in its surveys of crops, farm economics, and so forth.

The Service finds that an area frame based on maps of rural areas is preferred for surveys of the acreage planted in each crop, but that a frame that lists farms or farmers is superior for surveys of farm income. Can you explain why?

7.   Opinion polls and the political process

Public opinion polls, especially preelection "For whom would you vote?" polls, are the most visible example of survey sampling. They are also one of the most controversial. Most people are happy that sampling methods make employment and unemployment information rapidly available, and few people are upset when marketers survey consumer buying intentions. But the sampling of opinion on candidates or issues is sometimes strongly attacked as well as strongly praised. We will briefly explore three aspects of polls and politics: first, polls of opinion on public issues; second, polls as a tool used by candidates seeking nomination or election; and third, election predictions for public consumption, designed to satisfy our curiosity as to who's going to win.

Polls on public issues (e.g., defense spending, gun control, and legalization of marijuana) are praised as the only way our representatives can know what the people think. You now have the background to understand why other means (such as mail for and against) are unreliable, and why surveys such as the Gallup Poll give accurate information about public opinion. Legislators are constantly under pressure by special-interest groups who back their interests with lobbyists and campaign contributions. Opinion polls give the general public a chance to offset this pressure. The argument in favor of opinion polling is that public opinion is an essential part of democratic government. Polls express this opinion accurately; the alternative is vague impressions and the loud voices of special interests.

Intelligent arguments against polling do not dispute that modern sampling methods guarantee that the polls will give results close to the

"Seventy-three percent are in favor of one through five, forty-one percent find six unfair, thirteen percent are opposed to seven, sixty-two percent applauded eight, thirty-seven percent . . ."

results we would get if we put the poll questions to the entire population. Some would argue against opinion polls on the ground that we elect representatives to use their best judgment, not to slavishly follow public opinion. This seems to be no argument against polls; they only inform our governors what the opinion of the governed is and cannot force them to follow it.

More thoughtful critics ask what the opinions revealed by the polls are worth. Leo Bogart has written a provocative book that raises this very question.[9] He points out that many citizens have not thought about an issue until a poll taker questions them. Unwilling to appear ignorant or uncaring, they often give hasty and uninformed answers. A question put by a telephone interviewer who calls as you were planning supper, a question with no responsibility attached to answering it, will get a low-quality opinion. As Bogart says, "We are likely to answer questions differently when we know the decision is really up to us." He doubts that the 62% of Americans who favored using atomic artillery shells against the Chinese in a Gallup Poll taken during the Korean War would give the same answer if seriously faced with starting a nuclear war. If not all opinions are of equal weight, because some are uninformed and some are flippant, then "public opinion" is not the sum of individual opinions reported by the polls.

What is more, public opinions and attitudes are complex, not easily gauged by a few questions. As Example 16 noted, small changes in the wording of questions can greatly change the response. Because of this, and because some of our answers to a poll lack serious thought, polls sometimes produce contradictory answers to related questions. Polls on proposed treaties limiting nuclear weapons, for example, often find strong support for such a treaty when general questions are asked. More specific questions about the terms of this particular treaty show much less support. The public likes nuclear arms reduction in principle, but tends to be suspicious of any specific agreement. What *is* public opinion, anyway? That's a question worth pondering.

Opinion polls conducted privately by candidates are now a universal tool of campaign strategy. This is the second area of impact of polls on politics. The purpose of these polls is information for more effective campaigning. In what areas and with what groups of voters is the candidate weak? Where are large numbers of uncommitted voters to be found? Which of the opponent's views are liabilities to be exploited? What arguments are most effective in advocating the candidate's views? Campaigners have always sought such information; sampling methods only replace vague impressions and intuition by reliable estimates.

Yet polls are sometimes viewed as part of the transformation of campaigns into exercises in marketing—selling the candidate to the consumers. By market research (sample surveys), the campaign manager discovers what the voters want and then, using all the devices of advertising, cleverly presents the candidate as satisfying those wants.

Most political professionals feel that attempting to present the candidate in a false light to fit voter preferences will fail. It is better politics to use poll results as guides in presenting the candidate's real views and concerns most effectively. If this is true, we as voters need not be alarmed by survey sampling as a campaign tool. As with any tool, it can be used unethically, but the ethical problem is the user rather than the tool. With attention, we should be able to accurately judge the candidate's programs and intentions. The most serious ethical problems in election campaigns, such as negative advertising that slings mud at the opponent, are largely unrelated to the use of statistical sampling.

Polls as election predictors are the third—and most dubious—political use of polls. I speak here of the results that fill the news before each election, informing us that Senator So-and-so is the choice of 58% of Ohio voters. Such polls are certainly popular, as they speak to our wish to know the future. The public is entitled to have its wants satisfied (within reasonable limits), so preelection polls will probably always be with us. Notice also that election polls do not have the drawbacks of surveys of public opinion. In an election poll, as in the voting booth, we are presented with a clear choice in an area where the decision is really up to us.

But election forecasts are somewhat shaky statistically, and some people think they have undesirable political effects. Let us examine both of these problems.

The key question asked in preelection polls takes the form, "If the election were held today, would you vote for X or Y?" Here is the exact question from the Gallup Poll presidential election questionnaire:

> *Suppose you were voting TODAY for President and Vice President of the United States. Here is a Gallup Poll Secret Ballot listing the candidates for these offices. Will you please MARK that secret ballot for the candidates you favor today—and then drop the folded ballot into the box.*

"Suppose you were voting today. . . ." Modern sampling methods give us great confidence that the sample result of a Gallup or Harris poll taken two weeks before the election is close to the truth about the population on that date. But the election is not being held today, and minds may change between the poll and the election. Some voters say that they are undecided, so the polling organization must decide what to do with the undecided vote. After all, these voters cannot choose "undecided" for president on election day. What is more, some of those who said for whom they would vote today will not take the trouble to vote for *anyone* on election day. The polling organizations make great efforts to determine how strongly their respondents hold their preferences and how likely they are to vote. But the problems of changing opinions and low voter turnout cannot be entirely avoided, especially in primary elections. Election forecasting is one of the less satisfactory uses of sampling.

Changing opinions were a major cause of the polls' famous failure in the 1948 Dewey-Truman election. The last poll was conducted three weeks before the election. It is likely (as the polls indicated) that Dewey was leading at the time, but Truman was gaining fast, continued to gain, and won an extremely close election. The other major cause of error in 1948 was the sampling method. Interviewers were given quotas of voters by age, race, sex, and economic status, but selection of individual subjects was left to the interviewer. Such quota samples are far better than "straw polls" that depend on voluntary response, but they are not probability samples. Quota samples favor the well-dressed and prosperous, because the interviewers find it easier to approach such people. In political terms, such a poll has a Republican bias. Gallup and others overestimated the Republican vote in every election from 1936 to 1948, and in the close 1948 election this bias caused an incorrect forecast of a Republican victory.

The opinion polls switched to probability samples after 1948, and computers now enable the final poll to be taken three days rather than three weeks before the election. Despite the problem of failure to vote, election forecasts are now quite accurate. Even more accurate are the *exit polls* that interview voters as they leave their polling places to ask how they actually voted. Exit polls conducted by the television networks often tell the networks who the next president will be by midafternoon of election day.* Under severe political pressure, the networks do not actually announce the winner in each state until the polls have closed in that state. (Did you ever wonder how television can say who won Ohio

---

*For more detail, read Richard F. Link, "Election Night on Television," in *Statistics: A Guide to the Unknown.*

one minute after the polls close? Their exit polls told them hours earlier who the winner was.)

The political effects of election forecasts are much debated. Nobody maintains that they have any major beneficial effects. Here are some of the alleged disadvantages. Voters may decide to stay home if the polls predict a landslide. (Why bother to vote if the results is a foregone conclusion?) Exit polls are particularly worrisome here, because they sometimes allow television to tell the nation who won the presidency while polls on the west coast are still open. Congress is considering uniform voting times for the entire nation to prevent this.

The evidence that voters stay home once the polls declare a winner is not convincing. But there is clear evidence that contributions dry up when the polls show that a candidate is weak. In particular, polls taken far in advance can make it difficult for a little-known candidate to gather resources for primary election campaigns. Such a candidate may do well if, despite the polls, resources are found for an effective campaign that captures the attention of the voters. The charge is that the effect of polls on both candidates and contributors may be to encourage them to act on practical calculations rather than on their convictions.

In reply, note that voters and contributors are likely to react only when the polls show a one-sided contest — a state of affairs that is clear even without polls. Potential contributors surely know that an unknown candidate is unknown. Supporters of Goldwater in 1964 or Mondale in 1984 surely knew that they were the minority. I see no substantial reason to fret over the effects of election forecasts. What is your opinion?

Section 7 exercises

1.64. Never since the beginnings of opinion polls over 50 years ago have less than two-thirds of those sampled favored stronger controls on firearms. Specific gun control proposals have often been favored by 80 to 85% of respondents. Yet little national gun control legislation has passed, and no major national restrictions on firearms exist.

Why do you think this has occurred? Does this mean that opinion polls on issues do not really offer the public a means of offsetting special-interest groups?

1.65. To see whether people often give responses on subjects they are entirely ignorant about, ask several people (we won't ask for an SRS) the following questions:

(a) Have you heard of the Bradley-Nunn bill on veteran's housing?

(b) Have you heard of *Midwestern Life* magazine?

(In a study of a few years ago, 53% said yes to a question like (a) and 25% to (b) even though neither the bill nor the magazine ever existed. The study is cited by Dennis Trewin, "Non-sampling Errors in Sample Surveys," *CSIRO Division of Mathematics and Statistics Newsletter*, June 1977.)

1.66. Bogart (see Sourcenote 9) reports that during the 1968 campaign George Wallace accused the polls of favoring Eastern moneyed interests and neglecting the common people. He would ask crowds at his rallies, "Have any of you-all ever been asked about this here election by Mr. Harris or Mr. Gallup?" and receive shouts of "No" and "Never" in reply (*Silent Politics*, p. 39)

If there were 150 million U.S. residents age 18 and over in 1968 and a poll selected 1500 at random, what was the chance that any one person would be interviewed? Suppose that the major polling organizations conducted 20 surveys during the campaign. What was the chance that a particular person would be interviewed at least once? (Being interviewed by an opinion poll is an unlikely event, and no accusations of bias are needed to explain why you haven't been chosen.)

1.67. A committee of the British Parliament has suggested that no poll results should be published during the three days before an election. Do you feel that this is a good idea? Explain your view.

1.68. Election-night television coverage includes exit polls that often allow quite precise predictions of the outcome before the polls have closed in parts of the country. It is sometimes charged that late voters may stay home if the networks say that the national election is decided. Therefore, predictions based on samples of actual votes should not be allowed until the polls have closed everywhere in the country. Do you agree with this proposal? Explain your opinion.

1.69. An alternative solution to the problem of predicting the winner while the polls are still open is to require all states (except possibly Hawaii) to end voting at the same moment. The polls might close at 10 p.m. in the east and at 7 p.m. in the west, for example. Do you support this proposal? Explain your opinion.

1.70. The Gallup Poll election questionnaire cited in the text used a "secret ballot" arrangement. Why do you think this arrangement was used?

8.   Random selection as public policy

Would you like to be the proud owner of the right to offer cellular
mobile telephone service in your area? So would many other people.
The Federal Communications Commission (FCC) faced the task of
choosing among the many applicants in each area. During 1988 and
1989 the FCC conducted lotteries to award cellular telephone rights in
216 small cities and 428 rural areas. Anyone with enough financial
backing could enter the lottery and hope to emerge with an exclusive
license but also with the obligation to actually offer cellular service.

Again, what should a nation do if there are not enough volunteers for
military service but only a small fraction of eligible youth are needed by
the military? That question last arose during the Vietnam era. Beginning
in 1970, a draft lottery was used to choose draftees by random selection.
Although the draft was ended in 1976, young men must still register at
age 18. The draft lottery may return if the army cannot attract enough
volunteers.

Both the cellular telephone lottery and the draft lottery are examples
of using random selection as public policy. The idea of random selection
is that of drawing lots, or of taking an SRS. The goal is to treat everyone
identically by giving all the same chance to be selected. To give another
example, random selection has been used to allot space in public housing to eligible applicants when there are more applicants than available
housing units.

When should random selection decide public issues? I claim that this
is a policy question, not a statistical question. Random selection does
treat everyone identically; it is fair or unbiased in that sense. If a policy of
identical treatment is desired, random selection is the tool that will
implement that policy. Debate over random selection should concentrate on whether or not distinctions among persons are desirable in a
certain situation. For example, cellular telephone licenses in the 90
largest cities were awarded based on elaborate hearings designed to find
the best candidate. Random selection was used in the smaller areas
because hearings would have been too slow and expensive. In the case of
the military draft, Congress felt that no distinctions should be made
among young men and so requested random selection. As for allotting
public housing, a federal court has ruled that random selection is allowed only when applicants are equally needy. Distinctions *are* to be
made among different degrees of need, so random selection is inappropriate for the entire pool of eligible applicants.

How is random selection carried out? In principle, the same way that an SRS is selected: Label everyone in the pool and use a table of random digits to select at random as many persons from the pool as are needed. In practice, random digits are not used. Instead, physical mixing and drawing of labels is used, for public relations reasons. Because few people understand random digits, a lottery looks fairer if a dignitary chooses capsules from a glass bowl in front of the TV cameras. This also prevents cheating; no one can check the table of random digits in advance to see how cousin Joe will make out in the selection.

Physical mixing *looks* random, but you may recall from Section 2 that it is devilishly hard to achieve a mixing that *is* random. There is no better illustration of this than the first draft lottery, held in 1970. Because an SRS of all eligible men would be hopelessly awkward, the draft lottery aimed to select birth dates in a random order. Men born on the date chosen first would be drafted first, then those born on date number 2, and so on. Because all men ages 19 to 25 were included in the first lottery, 366 birth dates were to be drawn. Here is a news account of the 1970 draft lottery:

They started out with 366 cylindrical capsules, one and half inches long and one inch in diameter. The caps at the end were round.

The men counted out 31 capsules and inserted in them slips of paper with the January dates. The January capsules were then placed in a large square wooden box and pushed to one side with a cardboard divider, leaving part of the box empty.

The 29 February capsules were then poured into the empty portion of the box, counted again, and then scraped with the divider into the January capsules. Thus, according to Captain Pascoe, the January and February capsules were thoroughly mixed.

The same process was followed with each subsequent month, counting the capsules into the empty side of the box and then pushing them with the divider into the capsules of the previous months.

Thus, the January capsules were mixed with the other capsules 11 times, the February capsules 10 times and so on with the November capsules intermingled with others only twice and the December ones only once.

The box was then shut, and Colonel Fox shook it several times. He then carried it up three flights of stairs, a process that Captain Pascoe says further mixed the capsules.

The box was carried down the three flights shortly before the drawing began. In public view, the capsules were poured from the black box into the two-foot deep bowl.[10]

You can guess what happened. Dates in January tended to be on the bottom, while birth dates in December were put in last and tended to be on top. News reporters noticed at once that men born later in the year seemed to receive lower draft numbers, and statisticians soon showed that this trend was so strong that it would occur less than once in a thousand years of truly random lotteries. An inquiry was made, which provided the account quoted.

What's done is done, and off to Vietnam went too many men born in December. But for 1971, the captain and the colonel were given other duties, and statisticians from the National Bureau of Standards were asked to design the lottery. Their design was worthy of Rube Goldberg. The numbers 1 to 365 (no leap year this time) were placed in capsules in a random order determined by a table of random digits. The dates of the year were placed in another set of capsules in a random order deter- mined again by the random-digit table. Then the date capsules were put into a drum in a random order determined by a third use of the table of random digits. And the number capsules went into a second drum, again in random order. The drums were rotated for an hour. The TV cameras

"So you were born in December too, eh?"

were turned on, and the dignitary reached into the date drum: Out came September 16. He reached into the number drum: Out came 139. So men born on September 16 received draft number 139. Back to both drums: Out came April 27 and draft number 235. And so on. It's awful, but it's random.[11] You can now rejoice that in choosing samples we have Table A to do the randomization for us.

Section 8 exercises

1.71. A panel of the National Heart and Lung Institute panel noted that when effective artificial hearts become available they will be too few and too expensive to give to all 50,000 or so patients who need a new heart each year. The panel recommended that hearts be allotted at random among patients of similar medical condition. Discuss this recommendation: Do you favor random selection? If not, how should recipients be chosen? By ability to pay? By value to society (as assessed by whom)? By age and family responsibilities?

1.72. A basketball arena has 8,000 student seats, but 18,000 students would like to watch basketball games. Design a system of allotting tickets at random that seems fair to you. (All students can see some of the 12 home games if you use a rotation system. Will you give upperclassmen some preference? How many tickets may an individual buy? If your school actually does use a random drawing to allot seats, describe the details of the official system and discuss any changes you favor.)

1.73. In 1975, the Dutch government adopted random selection for admissions to university programs in medicine, dentistry, and veterinary medicine. Applicants for these programs are much more numerous than available places. The random selection is stratified so that students with higher grades have a greater chance of being chosen. Do you favor such a system? Why?

1.74. Prior to 1970, young men were selected for military service by local draft boards. There was a complex system of exemptions and quotas that enabled, for example, farmers' sons and married young men with children to avoid the draft. Do you think that random selection among all men of the same age is preferable to making distinctions based on occupation and marital status? Give your reasons.

1.75. Give an example of a situation where you definitely would *approve* random selection. Give an example of a situation where you would definitely *disapprove* random selection.

9. Some ethical questions

Whenever our activities impinge on others (as they usually do), those activities should be carried on with sensitivity to their effects. Sampling of human populations is therefore a possible source of ethical problems. In general, the ethical problems posed by sample surveys are much less severe than those arising in experiments with human subjects. This is because an experiment imposes some treatment (such as a new drug for a medical symptom), while a sample survey only seeks information or opinion from the respondents. Most ethical problems of survey work concern *deceiving respondents*; very rarely is any physical harm possible.

At one extreme of deceit is the use of false pretenses. The telephone caller who says "I'm taking a survey" turns out to be selling cemetery plots or replacement windows. This fraud has little to do with actual survey work and is detested by users of sample surveys because it increases resistance to genuine polling. Another extreme is covert collection of information. Covert operations are not limited to spy rings. Social psychologists, for example, have gathered data by infiltrating small religious groups under the pretense of sincere membership. Spies and researchers both give the same rationale for their violation of privacy: The information they desire cannot be obtained by other methods. The moral problem is to decide when the information sought justifies such deceit.

Some withholding of information from subjects is often essential to avoid bias. A political poll sponsored by a candidate cannot tell subjects who is paying, for knowing that a representative of the Senator X Committee stood before us would influence our answers. Academic survey workers sometimes cannot tell subjects the full purpose of their research for the same reason. This withholding of information does not amount to active deceit, but certain safeguards are called for. The political poll representative should provide subjects with the name, address, and telephone number of the polling organization so that the subject has an avenue of recourse if the poll or poll taker are offensive. The polling organization should have a neutral name to avoid bias. This is easy if the candidate has hired a professional pollster; otherwise it is accomplished by setting up a polling office separate from the campaign office. The academic researcher can inform subjects of the full purpose (and sometimes the results) of the research after the fact.

In any case, potential respondents always must be told how much time the interview will require, what kinds of information are wanted, and how widely this information will circulate with or without personal identification. Only then can the subject make an informed decision to participate or not.

A second area of ethical problems in sampling work is *anonymity and confidentiality*. These are not identical. Anonymity means that the respondent is anonymous; his or her name is not known even to the sampler. Anonymity causes severe problems because it is then not known who responded to a poll and who did not. This means that no follow-up work can be done to increase the response rate. And, of course, anonymity is usually possible only in mail surveys, where responses can be mailed in without any identification.

Confidentiality means that each individual response will be kept confidential, that only sample statistics from the entire sample or parts of it will be made public. *Confidentiality is a basic ethical requirement of sample surveys, and any breach of confidentiality is a serious violation of professional standards.* The best practice is to separate the identity of the respondent from the rest of the data at once and use the identification only to check on who did or did not respond.

"I realize the participants in this study are to be anonymous, but you're going to have to expose your eyes."

Some common practices, however, seem to promise anonymity but deliver only confidentiality. Market researchers often use mail surveys that do not ask the respondent's identity but contain hidden codes on the questionnaire that do identify the respondent. Invisible ink coding and code numbers hidden under the flap of the return envelope are the usual techniques. A false claim of anonymity is clearly unethical; but if only confidentiality is promised, is it not also unethical to hide the identifying code and thus perhaps cause respondents to believe their replies are anonymous? Many market researchers feel that hidden codes are ethical as long as confidentiality is promised and observed. University survey researchers tend to believe that hidden identification is not ethical and that open identification of respondents should be used to make follow-up possible. Some market researchers reply that this reduces the response rate. What do you think?

The *use of information* is the final area I wish to question. A rigorous standard would require public availability of all poll results. Otherwise, the possessors of poll results can use the information gained to their own advantage. This may involve acting on the information, releasing only selected parts of it, or timing the release for best effect. Private polls taken for political candidates are often used in these ways. Is it unrealistic to ask complete disclosure of poll results?

Whatever our response to this question, some aspects of disclosure are agreed upon. The information about the sampling process required by the AAPOR code of ethics (examined at the end of Section 5) should be revealed whenever a sample result is announced. In addition, the AAPOR code requires disclosure of the identity of the sponsor of the survey. After all, we might more carefully inspect a poll paid for by a political party or other interest group than a poll sponsored by a news organization or other neutral party.

Finally, statisticians have an obligation to speak out when their work is misused by those who paid for it. The AAPOR code of ethics recognizes this responsibility in strong terms: "When we become aware of the appearance in public of serious distortions of our research we shall publicly disclose what is required to correct the distortions."

Potential abuses certainly exist in sampling, and we shall meet more difficult ethical problems in studying statistically designed experiments. These abuses and problems should not blind us to the fact that decisions based on incorrect information surely cause far more hardship. Because the statistical ideas we have met reduce the chance of incorrect information being used by decision makers, their overall ethical impact seems to me to be positive.

Section 9 exercises

1.76. In what circumstances is collecting personal information without the subject's consent permitted? Consider the following cases in your discussion:

   (a) A government agency takes a random sample of income tax returns to obtain information on the average income of persons in different occupations. Only the incomes and occupations are recorded from the returns, not the names.

   (b) A psychologist asks a random sample of students to fill out a questionnaire; she explains that their responses will be used to measure several personality traits so that she can study how these traits are related. The psychologist does not inform the students that one trait measured by the test is how prejudiced they are toward other races.

   (c) A social psychologist attends public meetings of a religious sect to study the behavior patterns of members.

   (d) The social psychologist pretends to be converted to membership in a religious sect and attends private meetings to study and report the behavior patterns of members.

1.77. A researcher suspects that orthodox religious beliefs tend to be associated with an authoritarian personality. A questionnaire is prepared to measure authoritarian tendencies and also ask many religious questions. Write a description of the purpose of this research to be read to potential respondents. You must balance the conflicting goals of not deceiving respondents as to what the questionnaire will tell about them and of not biasing the sample by scaring off certain types of people.

1.78. Does having an abortion affect the health of mother or child in any future live births? To study this question, the New York State Health Department traced the reproductive histories of 21,000 women who had abortions in New York in 1970 and 1971 and compared them with those of 27,000 women who gave birth to living children in the same period. The comparison was carried out entirely with Health Department records, beginning in 1970 and tracing any later maternity records for these 48,000 women through 1977.

   Do you consider this an invasion of privacy? If you do, do you think the information to be gained and the difficulties in asking consent justify the study anyway?

1.79. One of the most important nongovernment surveys in the United States is the General Social Survey conducted by the National Opinion Research Center at the University of Chicago. The General

Social Survey regularly monitors public opinion on a wide variety of political and social issues. Interviews are conducted in person in a subject's home. Are a subject's responses to General Social Survey questions anonymous, confidential, or both? Explain your answer.

1.80. The federal government, unlike opinion polls or academic researchers, has the legal power to compel response to survey questions. The census long form, given to 17% of all households in the 1990 census, contained the following question:

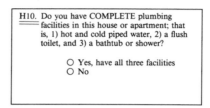

When a question about plumbing was first asked in the 1970 census, some members of Congress felt that this was an invasion of privacy and that the Census Bureau should be prohibited from asking such questions. The bureau replied that as with all census data, no individual information would be released, only averages for various regions. Moreover, lack of plumbing is the best single measure of substandard housing, and the government needs this information to plan housing programs.

Do you feel that this question is proper? If not, when do you think that the government's need for information outweighs a citizen's wish to withhold personal facts? If you do think the plumbing question is proper, where does the citizen's right to withhold information begin?

1.81. If only confidentiality is promised and confidentiality is carefully observed, are hidden codes that identify respondents ethical? Do you feel that this is trickery or do you agree with those who claim that this practice serves a necessary purpose and does not violate the assurances given respondents?

1.82. Discuss how anonymity can be preserved while still recording who did and who did not respond to a survey. (See Exercise 1.70 in Section 7 for one idea.)

1.83. A radical critique of polling is given by Herbert I. Schiller, "Polls Are Prostitutes for the Establishment," *Psychology Today*, July 1972, p. 20. His arguments, which apply to almost any sample survey, are:

(a) The poll taker exercises power over the respondent by wording questions and restricting alternatives in answering.

(b) Polls provide information about the respondent to the poll taker with no reciprocal flow of information in the other direction. They give power to the poll taker by allowing the poll taker to choose the uses of this information.

Thus Schiller feels that polls are an instrument of manipulation that enables the poll taker to control respondents unfairly. Discuss Schiller's position and give your reasons for agreeing or disagreeing.

1.84. Do you favor requiring complete disclosure of the methods, sponsorship, and results (sample statistics only, never individual responses) of all sample surveys? In your discussion, you must balance any benefits of this policy against the cost of carrying it out and the resulting restriction on your ability to ask questions of other people.

1.85. One of the best discussions of the ethics of sampling is by Lester R. Frankel, "Statistics and People — the Statistician's Responsibilities," *Journal of the American Statistical Association*, Volume 71 (1976), pp. 9–16. Mr. Frankel was then president of the American Statistical Association and worked for a firm specializing in sample surveys.

Read this article as an example of the ethical standards that survey statisticians currently hold. Make a list of the areas in which Frankel suggests guidelines for the behavior of survey takers.

NOTES

1. From an article by Michael Kernan of the *Washington Post*, printed in the *Lafayette Journal and Courier*, April 19, 1976.

2. Rice's speech reported in the *New York Times*, February 19, 1974.

3. The finding on women is reported in the *New York Times*, September 12, 1988. The 1984 campaign example is reported in Philip E. Converse and Michael W. Traugott, "Assessing the Accuracy of Polls and Surveys," *Science*, Volume 234 (1986), pp. 1094–1098.

4. See the detailed discussion by Philip L. Martin, "Labor-Intensive Agriculture," *Scientific American*, October 1983, pp. 54–59.

5. Reported in *Public Opinion Quarterly*, Volume 35 (1971–1972), p. 54.

6. From the *New York Times*, August 18, 1980.

7. From a Gannett News Service dispatch appearing in the *Lafayette Journal and Courier*, November 24, 1983.

8. These are the words of John B. Lansing and James N. Morgan, *Economic Survey Sampling* (Ann Arbor, Mich.: Institute for Social Research, University of Michigan, 1971).

9. Leo Bogart, *Silent Politics: Polls and the Awareness of Public Opinion* (New York: Wiley-Interscience, 1972).

10. From the *New York Times*, January 4, 1970. Quoted with extensive discussion by Stephen E. Fienberg, "Randomization and Social Affairs: The 1970 Draft Lottery," *Science*, Volume 171 (1971), pp. 255–261.

11. It's even a little more complicated than I've described. For all the details, see Joan R. Rosenblatt and James J. Filliben, "Randomization and the Draft Lottery," *Science*, Volume 171 (1971), pp. 306–308.

Review exercises

1.86. You are on the staff of a member of Congress who is considering a controversial bill that provides for government-sponsored insurance for nursing home care. You report that 1128 letters have been received on the issue, of which 871 oppose the legislation. "I'm surprised that most of my constituents oppose the bill. I thought it would be quite popular," says the representative. Are you convinced that a majority of the voters oppose the bill? State briefly how you would explain the statistical issue to the representative.

1.87. A newspaper advertisement for *USA Today: The Television Show* once said:

> *Should handgun control be tougher? You call the shots in a special call-in poll tonight. If yes, call 1-900-720-6181. If no, call 1-900-720-6182.* Charge is 50 cents for the first minute.

Explain why this opinion poll is almost certainly biased.

1.88. A magazine for health foods and organic healing wants to establish that large doses of vitamins will improve health. The magazine asks readers who have regularly taken vitamins in large doses to write in and describe their experiences. Of the 2754 readers who reply, 93% report some benefit from taking vitamins.

Is the sample proportion 93% probably higher than, lower than, or about the same as the percent of all adults who would perceive some benefit from large vitamin intake? Why?

1.89. The noted scientist Dr. Iconu wanted to investigate attitudes toward television advertising among American college students. He decided to use a sample of 100 students. Students in freshman psychology (PSY 001) are required to serve as subjects for experimental work. Dr. Iconu obtained a class list for PSY 001 and chose a simple random sample of 100 of the 340 students on the list. He asked each of the 100 students in the sample the following question:

Do you agree or disagree that having commercials on TV is a fair price to pay for being able to watch it?

Of the 100 students in the sample, 82 marked "Agree." Dr. Iconu announced the result of his investigation by saying, "82% of American college students are in favor of TV commercials."

   (a) What is the population in this example?
   (b) What is the sampling frame in this example?
   (c) Explain briefly why the sampling frame is or is not suitable.
   (d) Discuss briefly the question Dr. Iconu asked. Is it a slanted question?
   (e) Discuss briefly why Dr. Iconu's announced result is misleading.
   (f) Dr. Iconu defended himself against criticism by pointing out that he had carefully selected a simple random sample from his sampling frame. Is this defense relevant?

1.90. An advertising agency conducts a sample survey to see how adult women react to various adjectives that might be used to describe an automobile. The firm chooses 400 women from across the country. Each woman is read a list of adjectives, such as "elegant" and "prestigious." For each adjective, she is asked to indicate how desirable a car described this way seems to her. The possible responses are (1) highly desirable, (2) somewhat desirable, (3) neutral, and (4) not desirable. Of the women interviewed, 76% said that a car described as "elegant" was highly desirable.

   (a) What is the population in this sample survey?
   (b) Is the number 76% a parameter or a statistic?
   (c) The sample was a probability sample chosen by using the Gallup Poll's sampling procedure. Use Table 1-1 to make a 95% confidence statement about the reaction of women to the adjective "elegant."

1.91. A simple random sample of 1200 adult residents of California is asked whether they favor restricting imports of foreign cars. The margin of error for 95% confidence in this sample is ±1.4 percentage points. If the survey had selected an SRS of 1200 adults from the city of Chicago (population 3 million) rather than from the state of California (population 29 million), would the margin of error in the confidence statement be wider, the same, or narrower?

1.92. A Gallup Poll survey asked "Do you happen to jog?" Fifteen percent of the sample answered "Yes." A newspaper article reporting the poll said

> *The latest findings are based on in-person interviews with 1540 adults, 18 and older, conducted in more than 300 scientifically selected locations across the country during the period June 7–10, 1985. For results based on samples of this size, one can say . . .*

   (a) Complete the last sentence by making a confidence statement based on the survey results.

   (b) Explain carefully what sources of error are included in the margin of error allowed by your confidence statement. Give two examples of possible sources of error that are not included in the margin of error.

1.93. The New York Times/CBS News Poll conducts regular surveys of public opinion. One of the questions asked in a recent poll was "Do you favor an amendment to the Constitution that would permit organized prayer in public schools?" Sixty-six percent of the sample answered "Yes."

   (a) The article describing the poll says that it "is based on telephone interviews conducted from Sept. 13 to Sept. 18 with 1,664 adults around the United States, excluding Alaska and Hawaii. . . .the telephone numbers were formed by random digits, thus permitting access to both listed and unlisted residential numbers." This sampling method excludes some groups of adults, which may cause some bias in the poll results. What groups of adults are excluded from the sample?

   (b) The article gives the margin of error as three percentage points. Make a confidence statement about the percent of all adults [excluding the groups you named in part (a)] who favor a school prayer amendment.

   (c) The news article goes on to say that "the theoretical errors do not take into account a margin of additional error resulting from the various practical difficulties in taking any survey of public opinion." List some of the "practical difficulties" that may cause errors in addition to the ±3% margin of error.

Chapter **2**

Experimentation

The purpose of sampling is to collect information about some population by selecting and measuring a sample from that population. The goal is a picture of the population, disturbed as little as possible by the act of gathering data. An *experiment*, in contrast, deliberately imposes some treatment on the experimental units or subjects in order to observe the response. A merely observational study, even one based on a sound statistical sample, is a poor way to gauge the effect of an intervention. To see how nature responds to a change, we must actually impose the change. A great advantage of experimentation is that we can study the effects of the specific treatments that interest us, rather than simply observe units as they occur "in nature."

Imagine the frustration of a researcher trying to study the effects of prolonged sleeplessness on reaction time by finding persons who just happen to have been awake for at least 48 hours. Instead, she performs an experiment, keeping volunteer subjects awake for 48 hours and measuring their reaction time at the beginning and end of the sleepless period. She can even keep the subjects awake for 36, 48, 60, and 72 hours and measure their reaction times for each duration of sleeplessness.

The intent of most experiments is to study the effect of changes in some variables on other variables. We call the variables that describe the outcome *response variables*, as distinguished from the *explanatory variables* that the experimenter manipulates. These terms are easy to remember because the explanatory variables explain the response variables. The sleep researcher's explanatory variable is hours without sleep, while reaction time is the response variable. Explanatory variables are often called *independent* variables, and response variables are then called *dependent* variables. We will avoid this terminology because the word "independent" has several other meanings in statistics. Experiments often have several of each kind of variable. We might, for example, wish to study the effect of hours awake (explanatory variable A) and noise level (explanatory variable B) on reaction time (response variable 1) and score on a test of manual dexterity (response variable 2).

A sample survey may also study the effect of some variables (which thus become explanatory variables) on others (the response variables). A survey of natural deaths, for example, might study the relationship between the smoking habits of the deceased and the cause of death. Such a survey may show a relationship between smoking and death from lung cancer, but cannot by itself show that smoking causes lung cancer. People choose whether or not to smoke, and people who choose to smoke may differ in many ways from those who choose not to smoke. Smokers might more often die of lung cancer not because of their smoking but because of some other difference between smokers and nonsmokers.

The greatest advantage of experiments is that, in principle, experiments can establish *causation*: If we change the value of an explanatory variable with no other changes in the experimental conditions, any resulting changes in a response variable must be caused by the changing explanatory variable. This ideal is not often achieved in real experiments. It is hard to arrange an experiment so that nothing affects the response variables except changes in the explanatory variables. Nonetheless, experimentation is far better than observation when we wish to conclude that one variable really does explain another. Sample surveys, on the other hand, are better suited for describing a population. Here is a summary of the vocabulary of experimentation:

Units — the basic objects on which the experiment is done. When the units are human beings, they are called *subjects*.

Variable — a measured characteristic of a unit.

Response variable — a variable whose changes we wish to study; an outcome variable.

Explanatory variable — a variable that explains or causes changes in the response variables. An explanatory variable in an experiment is also called a *factor*.

Treatment—any specific experimental condition applied to the units. A treatment is usually a combination of specific values (called *levels*) of each of the experimental factors.

Here is an example that illustrates our terminology.

> **Example 1.** A fabrics researcher is studying the durability of a fabric under repeated washings. Because the durability may depend on the water temperature and the type of cleansing agent used, the researcher decides to investigate the effect of these two factors on durability. Factor A is water temperature and has three levels: hot (145°F), warm (100°F), and cold (50°F). Factor B is the cleansing agent and also has three levels: regular Tide, low-phosphate Tide, and Ivory Liquid. A treatment consists of washing a piece of the fabric (a unit) 50 times in a home automatic washer with a specific combination of water temperature and cleansing agent. The response variable is strength after 50 washes, measured by a fabric-testing machine that forces a steel ball through the fabric and records the fabric's resistance to breaking.

In this example there are nine possible treatments (combinations of a temperature and a cleansing agent). By using them all, the researcher obtains a wealth of information on how temperature alone, washing agent alone, and the two in combination affect the durability of the fabric. For example, water temperature may have no effect on the strength of the fabric when regular Tide is used, but after 50 washings in low-phosphate Tide, the fabric may be weaker when cold water is used instead of hot water. This kind of combination effect is called an *interaction* between cleansing agent and water temperature. Interactions can be important, as when a drug that ordinarily has no unpleasant side effects interacts with alcohol to knock out the patient who drinks a martini. Because an experiment can combine levels of several factors, interactions between the factors can be observed.

Experiments allow us to study factors of interest to us, either individually or in combination. And an experiment can show that these factors actually cause certain effects. For these reasons, experimentation is the favored method of collecting data whenever our goal is to study the effects of variables rather than simply to describe a population. Experiments are universal in the physical and life sciences. They are carried out whenever possible in the social sciences (that's quite often in psychology, but less often in sociology or economics). Some experiments influence the lives of all of us. For example, the safety of food additives and the safety and effectiveness of drugs must be demonstrated by experiment before public use is allowed.

In this chapter we are concerned with the statistical ideas of experimental design. The *design* of an experiment is the pattern or outline according to which treatments are applied to units. The basic concepts of experimental design apply to experiments in all areas, whether they study agricultural fertilizers, vaccines, or teaching methods.

1. The need for experimental design

Laboratory experiments often have a simple design such as

(1)   Treatment → Observation

in which a treatment is applied (often to several units) and its effect is observed. If before-and-after measurements are made, the design is

(2)   Observation 1 → Treatment → Observation 2

Statistical ideas are not used in the design of such simple experiments. (These experiments are "simple" in their design or pattern, even though the treatment may be quite complex.) When experiments are conducted outside the controlled environment of a laboratory, simple designs such as (1) and (2) often yield *invalid* data; we cannot tell whether the treatment had an effect on the units. The same sad tale must often be told of observational studies, such as sample surveys. Some examples will show what can go wrong.

> **Example 2.** In 1940, a psychologist conducted a study of the effect of propaganda on attitude toward a foreign government. He devised a test of attitude toward the German government and administered it to a group of American students. After reading German propaganda material for several months, the students were tested again to see if their attitudes had changed. This experiment had a design of the form (2), namely,
>
> Test of attitude → Read propaganda → Retest of attitude
>
> Unfortunately, Germany attacked and conquered France while the experiment was in progress. There was a profound

change of attitude toward the German government between the test and the retest, but we shall never know how much of this change was due to the explanatory variable (reading propaganda) and how much to the historical events of that time. The data are invalid; they give no information about the effect of reading propaganda.

Example 3. A high school Latin teacher wished to demonstrate the favorable effect of studying Latin on mastery of English. She therefore obtained from school records the scores of all seniors on a standard English-proficiency examination. The average score for seniors who had studied Latin was much higher than the average score for those who had not. The Latin teacher concluded that "the study of Latin greatly improves one's command of English." But students elect whether or not to study Latin. Those who elect Latin are probably (on the average) both smarter and more interested in language than those who do not. This self-selected group would have a higher average English-proficiency score whether or not they studied Latin. Whether studying Latin raised the English scores yet more we cannot tell; the data are invalid for this purpose. (This is a census of the school's seniors, not an experiment. But it is similar to Example 2 in that the effect of the explanatory variable on the response variable cannot be ascertained.)

No valid conclusion can be drawn in either of these examples because the effect of the explanatory variable cannot be distinguished from the effect of influences outside the study. Variables not of interest in a study that nonetheless influence the response variable are *extraneous variables*. In Example 2, reading propaganda is the experimental factor (the explanatory variable), and the events of current history are an extraneous variable. In Example 3, study of Latin is the explanatory variable and the innate ability of the students is an extraneous variable.

> **The effects of two variables (explanatory variables or extraneous variables) on a response variable are said to be *confounded* when they cannot be distinguished from one another.**

In Example 2, the effect of reading propaganda was confounded with the effect of historical events; the influences of these two variables on attitude toward Germany cannot be separated. In Example 3, the effect of studying Latin was confounded with the ability of the students; both influence English-proficiency scores, and their influences cannot be separated.

Confounding different variables (mixing up their effects) often obscures the true effect of explanatory variables on a response variable. Here are some additional examples.

> **Example 4.** An article in a women's magazine reported that women who nurse their babies feel more receptive toward the infants than mothers who bottle-feed. The author concluded that nursing has desirable effects on the mother's attitude toward the child. But women choose whether to nurse or bottle-feed, and it is possible that those who already feel receptive toward the child choose to nurse, while those to whom the baby is a nuisance choose the bottle. The effect on the mother's attitude of the method of feeding is confounded with the already existing attitude of the mother toward the child. Observational studies of cause and effect, such as this one, rarely lead to clear conclusions because confounding with extraneous variables almost always occurs.

> **Example 5.** A particularly important example of confounding occurs in clinical trials of drugs and other medical treatments. Many patients respond positively to *any* treatment, even a dummy medication. This is no doubt a reaction to personal attention and especially to the authority of the doctor who administers the treatment. Dummy medications are called *placebos*, and the response of patients to any treatment in which they have confidence is called the *placebo effect*. Many studies have shown, for example, that placebos relieve pain in 30 to 40% of patients, even those recovering from surgery. The placebo effect is not confined to the patient's imagination. Not only subjective effects ("My pain is less") but objectively measured responses often occur. In a clinical trial of the effectiveness of vitamin C in preventing colds, patients who were given a placebo that they thought was vitamin C actually had fewer colds than patients given vitamin C who thought it was a placebo! There is no doubt that faith healing works (sometimes).
>
> Experiments of designs (1) and (2) are often useless in testing drugs or other medical treatments. The placebo effect is confounded with the effect of the treatment; the patients might have responded as well to a dummy pill as they did to the drug being tested.

So both observation and simple experiments often yield invalid data owing to confounding with extraneous variables. This situation is difficult to remedy when only observation is possible. Experiments enable us to escape the effects of confounding. The first goal of experimental design is to make possible valid conclusions, to enable us to say how the

*"I want to make one thing perfectly clear, Mr. Smith. The medication
I prescribe will cure that run-down feeling."*

explanatory variables affected the response variables. It is now clear that
some new ideas are needed to reach this goal.

Section 1 exercises

2.1. It has been suggested that there is a "gender gap" in political party
preference in the United States; women seem more likely than
men to prefer Democratic candidates. A political scientist selects a
large sample of registered voters, both men and women, and asks
each voter whether they voted for the Democratic or Republican
candidate in the last Congressional election. Is this study an experiment? Why or why not? What are the explanatory and response
variables?

2.2. Some people believe that exercise raises the body's metabolic rate
for as long as 12 to 24 hours and thus enables us to continue to
burn off fat after we end our workout. In a study of this effect,
subjects were asked to walk briskly on a treadmill for several
hours. Their metabolic rate was measured before, immediately
after, and 12 hours after the exercise. The study was criticized

because eating raises the metabolic rate, and no record was kept of what the subjects ate after exercising. Was this study an experiment? Why or why not? What are the explanatory and response variables?

2.3. The question of whether a radical mastectomy (removal of breast, chest muscles, and lymph nodes) is more effective than simple mastectomy (removal of the breast only) in prolonging the life of women with breast cancer has been debated intensely. To study this question, a medical team examines the records of 25 large hospitals and compares the survival times after surgery of all women who have had either operation.

(a) What are the explanatory and response variables here?

(b) Explain carefully why this study is not an experiment.

2.4. An educator wants to compare the effectiveness of computer software that teaches reading with that of a standard reading curriculum. She tests the reading ability of each student in a class of fourth graders, then divides them into two groups. One group uses the computer regularly, while the other studies a standard curriculum. At the end of the year, all students are retested, and the two groups are compared for increase in reading ability.

(a) Is this an experiment? Why or why not?

(b) What are the explanatory and response variables?

2.5. A study of the effect of living in public housing on family stability and other variables in poverty-level households was carried out as follows: A list of applicants accepted for public housing was obtained, together with a list of families who applied but were rejected by the housing authorities. A random sample was drawn from each list, and the two groups were observed for several years.

(a) Is this an experiment? Why or why not?

(b) What are the explanatory and response variables?

(c) Does this study contain confounding that may prevent valid conclusions on the effects of living in public housing? Explain.

2.6. The article "Safety of Anesthetics" by Lincoln E. Moses and Frederick Mosteller in *Statistics: A Guide to the Unknown* describes a study undertaken when it was observed that the death rates of surgical patients are different for operations in which different anesthetics are used. Here are the death rates for four leading anesthetics:

Anesthetic	Halothane	Pentothal	Cyclopropane	Ether
Death rate	1.7%	1.7%	3.4%	1.9%

This is *not* good evidence that cyclopropane is more dangerous than the other anesthetics. Suggest some variables that may be confounded with the choice of anesthetic in surgery and that could explain the different death rates.

In each of Exercises 2.7 to 2.9 confounding is present. Explain briefly what variables are confounded and why the conclusions drawn about the effect of the explanatory variable on the response variable are not valid.

2.7. Last year only 10% of a group of adult men did not have a cold at some time during the winter. This year all the men in the group took 1 gram of vitamin C each day, and 20% had no colds. A writer claims that this shows that vitamin C helps prevent colds.

2.8. The educator in Exercise 2.4 asked for teachers who would volunteer to use the computer software. The experimental group of students was taught by teachers who volunteered to use the computer, while the control group was taught by teachers who did not volunteer. At the end of the year, the computer group had improved their reading scores somewhat more than the control group.

2.9. A college student thinks that drinking herbal tea will improve the health of nursing home patients. She and some friends visit a large nursing home regularly and serve herbal tea to the residents in one wing. Residents in the other wing are not visited. After six months, the residents who were served tea had fewer days ill than the second.

2.10. A chemical engineer is designing the production process for a new drug. The drug will be manufactured in batches, which may have a higher or lower yield depending on the temperature and pressure at which the batch is processed. The engineer decides to try three temperatures and two pressures in a search for the combination having the highest yield. She processes four batches at each temperature-pressure combination and measures the percent yield of the drug in each batch.

   (a) What are the experimental units and the response variable in this experiment?
   (b) How many factors are there? How many treatments?
   (c) How many experimental units are required for the experiment?

2.11. New varieties of corn with altered amino-acid patterns may have higher nutritive value than standard corn, which is low in the amino acid lysine. An experiment is conducted to compare two new varieties, called opaque-2 and floury-2, with normal corn. Corn-soybean meal diets using each type of corn are formulated at each of three protein levels, 12% protein, 16% protein, and 20% protein. Ten 1-day-old male chicks are assigned to each diet, and their weight gains after 21 days are recorded. The weight gain of the chicks is a measure of the nutritive value of their diet.

   (a) What are the experimental units and the response variable in this experiment?

   (b) How many factors are there? How many treatments?

   (c) How many experimental units are required for the experiment?

2.12. A survey of physicians found that some doctors give a placebo to a patient who complains of pain for which the physician can find no cause. If the patient's pain improves, these doctors conclude that it had no physical basis. The medical school researchers who conducted the survey claimed that these doctors do not understand the placebo effect. Why?

2.13. After the July 1976 convention of the American Legion in Philadelphia, 29 convention goers died of a mysterious "Legionnaires' disease." It was at first suspected that the disease might be a variety of influenza. To check this theory, laboratory experimenters injected material from dead and sick Legionnaires into chick embryos and watched to see if influenza virus grew in the embryo. (The presence of small numbers of virus in the material injected would go undetected. But if they grew into many, they could be detected.) No influenza virus was found. A Yale Medical School virologist commented that these results were not quite convincing. He said that a known dose of influenza virus should have been injected into other chick embryos as a "positive control."

   Can you see the virologist's point and explain his objection in more detail?

2. First steps in statistical design of experiments

The central idea in avoiding confounding of experimental with extraneous variables is *comparative experimentation*. If we can set up two equivalent groups of units, then give the treatment to only one group (the experimental group) while treating the other group (the control group) exactly the same in every way except not giving it the treatment, then any differences between the groups at the end of the experiment must be due to the effect of the treatment. Any extraneous variables influence both groups, while the experimental treatment influences only one; so by comparing the two groups, the effect of the treatment can be discovered. If two treatments (two drugs, or two fertilizers, for example) are to be compared, we can give one to each of two equivalent groups, and no control group is needed.

Comparative studies need not be experiments. We might, for example, assess the safety of various surgical anesthetics by analyzing hospital records to compare death rates during surgery when the different anesthetics are used. But the groups being compared are not equivalent, for some anesthetics tend to be used in serious operations or on patients who are old or in poor physical condition, while others are used in less risky situations. A high death rate may mean only that this anesthetic is used in high-risk operations.* If comparison is to eliminate such confounding, we must have equivalent groups of subjects. Arranging for equivalent groups to receive the treatments is the kind of active intervention that distinguishes experiments from other types of studies.

Comparative experimentation was first used in earnest in agricultural research, beginning in the nineteenth century. Agronomists tried to obtain equivalent groups of units (small plots of land) by carefully matching plots in fertility, soil type, and other variables. It is difficult to match units in all important extraneous variables, especially because the experimenter may not think of them all ahead of time. What is more, experimenter judgment in assigning units to groups opens the door to bias. A medical researcher may unconsciously tend to assign more seriously ill patients to the standard treatment and leave less serious cases to the new and untried treatment. The moral is clear: If comparative experimentation is to be effective, we need a better way of assigning units to groups.

Randomized comparative experiments

The better way was provided in the 1920s by R. A. Fisher, a statistician working for an English agricultural experiment station.† Fisher realized that equivalent groups for experimental use can be obtained by *randomly assigning* units to groups. Just as a simple random sample is likely

---

*An account of an important study of this problem appears in Lincoln E. Moses and Frederick Mosteller, "Safety of Anesthetics," in *Statistics: A Guide to the Unknown*. Because the hospital records contained information on the kind of surgery and type of patient, it was possible to adjust for the effects of these extraneous variables.
†Sir Ronald Aylmer Fisher (1890–1962) was one of the century's greatest scientists in two fields, genetics and statistics. He invented the statistical design of experiments and many other statistical techniques.

to be representative of the population, so a random selection of half the available units is likely to create two groups (the one selected and the one left behind) similar in every respect. We have now reached the simplest statistically designed experiment:

(3)

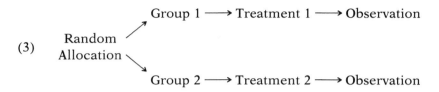

This is the replacement for design (1), which involved no comparison. If before-and-after observations are made, as in design (2), we simply make them on each group:

(4)

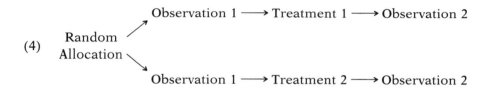

Let us study some examples of simple designed experiments.

Example 6. Suppose that we wish to give an experimental design for the propaganda study (Example 2) that will allow conclusions to be drawn. There are 100 subjects available. We first choose an SRS of 50 subjects in the usual way: Label the subjects 00, 01, . . ., 99 and use the table of random digits to choose 50. If we enter the table in line 136, we obtain

Group 1

08	42	14	47	53	77	37	72	87	44
75	59	20	85	63	79	09	24	54	64
56	68	12	61	78	36	60	91	73	98
48	11	45	92	66	83	16	89	40	22
15	58	13	35	86	17	70	69	28	55

Group 2 consists of the 50 remaining subjects. All 100 subjects are tested for attitude toward the German government. The 50 subjects in Group 1 then read German propaganda regularly for several months, while the 50 subjects in Group 2 are instructed not to read German propaganda; otherwise, both groups go about their normal lives. Then all 100 subjects are retested. This experiment has design (4):

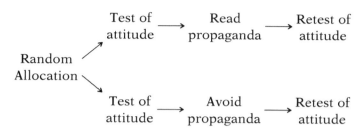

If the first group shows a more positive (or less negative) change in attitude toward Germany between test and retest, we can attribute this difference to the effect of reading propaganda.

Some comments on Example 6 are in order. First, notice that random allocation to groups 1 and 2 is done by selecting an SRS from the available units to be group 1. Nothing new here. The purpose of randomization is to create groups that are equivalent prior to the experiment. Many variables (such as sex, age, race, religion, and political opinion) may influence a subject's reaction to German propaganda. Random allocation should "average out" the effects of these extraneous variables by dividing them evenly between the groups. Moreover, random allocation will average out the effects of extraneous variables we have not thought of, as well as those we have listed.

Second, notice how randomization and comparison ensure that valid conclusions can be drawn. The control group shares the experiences of the experimental group except for propaganda reading. The fall of France, for example, influences the attitudes of both groups. The groups were equivalent prior to the experiment (we can check this by comparing their average scores on the first test), and the groups were identically treated except that group 1 read propaganda while group 2 did not. So any difference in the average change in the attitudes of the two groups must be due to the effect of the propaganda. It might happen, for example, that the attitude of both groups becomes much more negative after the fall of France, but the attitude of group 1 changes less in the negative direction than does the attitude of group 2. This would show the effect of the propaganda read by group 1 but not by group 2.

Example 7. It has been claimed, most notably by the two-time Nobel laureate Linus Pauling, that large doses of vitamin C will prevent colds. An experiment to test this claim was performed in Toronto in the winter of 1971–1972. About 500 volunteer subjects were randomly allocated to each of two groups. Group 1 received 1 gram per day of vitamin C and 4 grams per day at the first sign of a cold. (This is a large amount of vitamin C; the recommended daily allowance of this vitamin for adults is only 60 milligrams, or 60/1000 of a gram.) Group 2 served as a control group and received a placebo pill identical in appearance to the vitamin C capsules. Both groups were regularly checked for illness during the winter. The experimental design is (3):

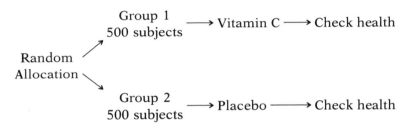

Some of the subjects dropped out of the experiment for various reasons, but 818 completed at least two months. Groups 1 and 2 remained well-matched in age, occupation, smoking habits, and other extraneous variables. At the end of the winter, 26% of the subjects in group 1 had not had a cold, compared with 18% in group 2. Thus vitamin C did appear to prevent colds better than a placebo, but not much better.[1]

The randomized comparative experiment in Example 7 avoided confounding, so the effect of the treatment can be separated from the effects of extraneous variables, including the placebo effect. If group 2 had received nothing, the effect of vitamin C would still have been confounded with the placebo effect, because group 1 might be responding to being given any treatment at all. As it is, *both* groups receive pills, and both were equivalent before the experiment began, so the chemical effect of vitamin C is the only difference between the groups. Thus we should be able to draw valid conclusions as to whether vitamin C prevents colds better than a placebo does.

Examples 6 and 7 show how valid data can be obtained in the settings of Examples 2 and 5 by a randomzied comparative experiment. Examples 3 and 4 are observational studies. Simple experiments also would yield valid data in those settings. But such experiments would mean randomly assigning students to study or avoid Latin, and women to nurse or bottle-feed their children. These treatments cannot be imposed, for practical and moral reasons. It is only when experimentation is not possible that we try (often without success) to study cause and effect by observation alone.

Why randomize?

Randomization in an experiment is the analog of probability sampling in a sample survey and serves the same purposes. First, randomization is a fair or unbiased way of assigning units to experimental groups, just as an SRS is a fair or unbiased way of selecting units from a population. As in sampling, randomization in experimental design has a second equally important function. Recall that probability sampling produces sample statistics having a known sampling distribution, so we know how likely the statistic is to fall within any given distance of the population parameter. Similarly, if experimental units are randomly allocated to treatments, we know how likely any degree of difference among the groups is. For example, in the Toronto vitamin C experiment, a difference between the groups of 26% versus 18% illness-free was observed. Because of the use of randomization, we can say that if the two treatments (vitamin C and placebo) have the same effect, a difference at least this large would occur in less than 1% of a large number of experiments. We therefore can be confident that the difference is due to the differing effects of the treatments. Randomization as the basis for drawing conclusions will receive much attention in Chapter 8. At this point, we concentrate on randomization as a means of eliminating possible bias by averaging extraneous variables and creating equivalent groups that allow valid comparison of treatments.

Increasing the number of units assigned to each group has the same effect as increasing the size of a sample. In both cases, a greater number of observations produces less variable (more precise) results. That is, we can be more confident that a repetition of the experiment would give results that differ little from those we obtained. Thus we can be more confident that our experimental results reflect the effect of the treatments and are not just an accident arising from bad luck in the random assignment of units to groups. Let us sum up the analogy between random sampling and randomized experiments.

1. Random sampling has no bias; it favors no part of the population in choosing a sample. Random allocation of experimental units to groups has no bias; units with special properties are not favored in choosing any group.

2. The sample obtained by random sampling varies in repeated sampling and may by chance be unrepresentative of the population. The experimental groups obtained by randomization vary in repeated trials and may by chance fail to be closely equivalent.

3. Using larger random samples decreases the variability of the result and increases our confidence that the sample is representative of the population. Using larger randomly chosen groups of experimental units decreases the variability of the result and increases our confidence that the groups are equivalent before treatments are applied.

How well does randomization work in practice? Here is an example in which this question was investigated:

Example 8. When the University Group Diabetes Program, a major medical experiment on treatments for diabetes, produced evidence that the drug tolbutamide was ineffective and perhaps unsafe, the design of the experiment was questioned. In response, statistician Jerome Cornfield wrote a detailed account of the conduct of the study that includes a look at the effectiveness of the randomization.[2]

Subjects were randomly assigned to four treatments. The tolbutamide group showed higher cardiovascular mortality (i.e., more subjects died of heart attacks) than the other groups. Could this have happened because the tolbutamide group contained higher-risk subjects? Cornfield examined how the patients assigned to each group compared in eight risk factors that increase the chance of a heart attack. These risk factors were age 55 or older, high blood pressure, history of digitalis use, history of chest pains, abnormal electrocardiogram, high cholesterol level, overweight, and calcification of the arteries. The results appear in Table 2-1. The groups do appear quite equivalent in these risk factors. For example, 84.9% of the patients assigned to placebo had one or more risk factors, compared with 86.8% of those assigned to tolbutamide. "All in all," Cornfield commented, "the luck of the draw does not seem to have been too bad." He reminded his medical readers that randomization produces groups that are also comparable in other extraneous variables, such as a patient's smoking history, for which information was not available.

The simplest kind of randomization allocates experimental units at random among *all* treatments. Such experimental designs are called *completely randomized designs*. They correspond to the SRS in sampling. Designs (3) and (4) are completely randomized designs with two groups. If more treatments are to be compared, we can randomly divide the units into more groups. It is not necessary to allot the same number of units to each group, but the data are easier to use when this is done. The University Group Diabetes Study (Example 8) was a completely randomized design with four groups. This design is illustrated in the following diagram:

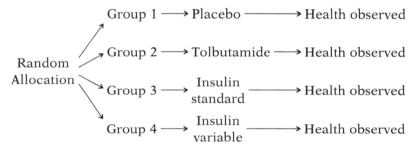

Table 2-1. Number of subjects in the University Group Diabetes Program having each number of risk factors*

Number of risk factors	Group 1 (placebo)	Group 2 (tolbutamide)	Group 3 (insulin standard)	Group 4 (insulin variable)
0	28	25	22	15
1	60	50	62	76
2	59	58	60	57
3	26	34	34	30
4	10	17	8	4
5	2	4	8	4
6	0	1	1	1
Total number of subjects	185	189	195	187

SOURCE: Jerome Comfield, "The University Group Diabetes Program," *Journal of the American Medical Association*, September 20, 1971, pp. 1676–1687. Copyright 1971, American Medical Association.

*This table covers the 756 subjects (out of a total of 823) for which information on all eight risk factors was available.

Section 2 exercises

2.14. A vaccine has been developed for use against a dangerous virus. You have available 10 rats (named below), which will be exposed to the virus. Unprotected rats usually die when infected. Design an experiment to test the effectiveness of the vaccine. Use line 140 of Table A to carry out the random assignment.

Alfie	Bernie	Chuck	David	Frank
Harry	Lyman	Mercedes	Polyphemus	Zaffo

2.15. Some medical researchers suspect that added calcium in the diet reduces blood pressure. You have available 40 men with high blood pressure who are willing to serve as subjects.
  - **(a)** Outline an appropriate design for the experiment, taking the placebo effect into account.
  - **(b)** The names of the subjects are given below. Use Table A beginning at line 111 to do the randomization required by your design, and list the subjects to whom you will give the drug.

Abrams	Danielson	Duttman	Lippman	Rosen
Adamson	Durr	Hipp	Marsden	Solomon
Afifi	Edwards	Hruska	McNeill	Thompson
Bikalis	Fluharty	Iselin	Morse	Travers
Burkett	Fratianna	Janle	O'Brian	Turing
Clemente	Gerson	Kaplan	Obramowitz	Ullmann
Cranston	Green	Krushchev	Plochman	Wang
Curzakis	Gutierrez	Lattimore	Rodriguez	Wise

2.16. A group of 200 first graders is available to compare the effectiveness of method A and method B for teaching reading.
  - **(a)** Outline the design of an experiment to make this comparison. Include a brief description of your response variable.
  - **(b)** Explain carefully how you would carry out the random assignment called for in your design. Use Table A, beginning at line 137, to assign five students as a demonstration of your randomization.

2.17. You wish to compare three treatments for effectiveness in preventing flu: (1) a flu vaccine, (2) 1 gram of vitamin C per day, and (3) a placebo taken daily. Describe how you would use 600 volunteer subjects in a designed experiment to compare these treatments. (Do not actually do any randomization, but do include a diagram showing your design. Be sure to state carefully what your response variable is.)

2.18. Turkeys raised commercially for food are often fed the antibiotic salinomycin to prevent infections from spreading among the birds.

Salinomycin can damage the birds' internal organs, especially the pancreas. A researcher believes that adding vitamin E to the diet may prevent injury. He wants to explore the effects of three levels of vitamin E added to the diet of turkeys along with the usual dose of salinomycin. There are 30 turkeys available for the study. At the end of the study, the birds will be killed and each pancreas examined under a microscope.

Give a careful outline of the design of a randomized comparative experiment in this setting. Then label the 30 turkeys 00 to 29 and use Table A beginning at line 125 to carry out the random assignment required by your design.

2.19. To demonstrate how randomization reduces confounding, consider the following situation: A nutrition experimenter intends to compare the weight gain of newly weaned male rats fed diet A with that of rats fed diet B. To do this she will feed each diet to 10 rats. She has available 10 rats of genetic strain 1 and 10 of strain 2. Strain 1 is more vigorous, so if the 10 rats of strain 1 were fed diet A, the effects of strain and diet would be confounded, and the experiment would be biased in favor of diet A.

Label the rats 00, 01, . . ., 19 (use repeated labels if you wish). Use Table A to assign 10 rats to diet A. Do this four times, using different parts of the table, and write down the four experimental groups you obtained.

Suppose that the strain 1 rats are numbers 00, 02, 04, 06, 08, 10, 12, 14, 16, and 18. How many of these rats were in each of the four diet A groups that you generated? What was the average number of strain 1 rats assigned to diet A?

2.20. Comparison alone does not make a study an experiment. Examples 3 and 4 in Section 1 are observational studies that compare two groups, but they are not experiments. Describe clearly the difference between a comparative observational study and a comparative experiment. What advantages do comparative experiments have over comparative observational studies?

2.21. Read "The Biggest Public Health Experiment Ever," by Paul Meier, in *Statistics: A Guide to the Unknown*. Describe the experimental design used in the Salk polio vaccine trials. What extraneous variables would have been confounded with the treatment if the "observed control" approach had been used?

2.22. Makers of competing pain relievers sometimes claim that their pills dissolve faster and illustrate this claim with pictures of pills dissolving in a glass of water. You are to compare the dissolving time of Anacin and aspirin.

(a) Outline the statistical design of an experiment to make this comparison.

Heart Treatment's Value Doubted

By Michael Specter
Washington Post Staff Writer

In a study with far-reaching implications for the routine treatment of heart attack patients, researchers have found that the immediate use of special balloons to force open clogged arteries is unnecessary in the vast majority of cases if the patient is treated with a clot-dissolving drug.

Most heart specialists have assumed that the balloon treatment, an expensive and increasingly popular procedure called balloon angioplasty, should routinely follow the use of drugs, such as TPA, that dissolve the blood clots that cause most heart attacks.

But in a study of 3,262 heart attack patients that is expected to transform the standards for treatment of the nation's leading killer, researchers at medical centers across the country found the extra measures were rarely needed. Half of the randomly selected patients were treated solely with the clot-dissolving drug TPA, while the other half were given TPA followed by angioplasty.

The results, reported in today's issue of The New England Journal of Medicine, showed that for most people adding angioplasty was no better than relying on the less complicated and less costly drug treatment.

Only 10 percent of the group assigned solely to drug treatment died or had another heart attack in the six weeks following treatment. By comparison, 11 percent of the group assigned to receive the combination drug and angioplasty treatment suffered the same fate. The difference could have occurred by chance. But the fact that angioplasty did not prove to be the better option was a shock even to some of the nation's most renowned heart experts.

"There is no question that it bucks the trend," said Eugene Braunwald, professor of medicine at Harvard Medical School and the study chairman for the research project. "It will spare many patients unnecessary surgery and reduce the cost of medical care greatly."

(b) Discuss any other aspects of the experiment that may be important to your conclusion. (For example, what do you mean by "aspirin"?)

2.23. Think of a simple question of interest to you that might be settled by an experiment. (For example, in which gear can you ride a bicycle up a long hill fastest?) Discuss in detail the design of an appropriate experiment.

2.24. The article from which the excerpts at left were taken appeared in the *Washington Post* of March 9, 1989. Describe the design of the experiment in as much detail as the article allows. Include the treatments, the response variable, and the design outline. Does the design of the experiment give you some confidence in its result?

2.25. The following Associated Press dispatch appeared in the *Lafayette Journal and Courier* of November 30, 1974:

(a) What important information about the experimental design is given in the article?

(b) Is any important information omitted?

(c) Do you consider the results of the experiment convincing? Why or why not?

Praying to soybeans aids in higher yields

By GEORGE CORNELL

With county officials measuring the results, experimenters on an Ohio farm say they found that portions of a field that had been the object of loving prayers yielded the biggest crop.

The case offered an unusual instance of recent stepped-up interest in psychic phenomena, viewed by many with keen skepticism.

"Somehow God's creative energy of growth can be channeled through us even to plants," says Gus Alexander of Wright State University, who holds a doctorate in communications research and who set up the project.

It was carried out on a soybean field near Jamestown, Ohio, east of Dayton, with daily prayerful attention of a church group focused on six designated plots, but not on six adjoining control plots.

Alexander says the yield of soybeans receiving the special attention was increased by 4 percent over the comparable control plot, even though the experiment had extended over only a third of the growing season.

In checking results of the experiment late in October, the Greene county agent's technical assistant, Donald H. Tate, was on hand to weigh the yields from the six experimental and six control plots.

According to the figures, five of the experimental strips had produced heavier yields than had adjacent control strips, while in the sixth case the control strip had a slightly greater yield.

In the experiment, a group of 10 people at Dayton's Church of the Golden Key, supplied with diagrams of the soybean plots, took on the task of "sending love" to the experimental areas each night at 11:30 p.m. for about 40 days.

Reprinted by permission of The Associated Press.

3.   Difficulties in experimentation

Just as probability sampling does not ensure accurate results from a sample survey, so statistically designed experiments are not a total solution to the problem of obtaining valid experimental data. In this section, we increase our understanding of experimentation by looking at some common difficulties in drawing valid conclusions.

Difficulty 1: To what population do the conclusions apply?

If you've seen one alpha particle, you've seen them all. So experimental conclusions about alpha particles reached in Denmark in 1923 remain true in Berkeley in 1993. The physical sciences often work with essentially identical experimental units. This has two heavenly consequences: Random assignment is not needed to create equivalent groups when all units are identical, and conclusions about one set of units generalize to the population of all such units. In other areas of study, the units are often highly variable. This may be due to variation in the units themselves or to the effects of extraneous variables. We know that random assignment can, on the average, create equivalent groups from variable units. Now we must ask how widely one can apply the conclusions from such an experiment. An example will convey the difficulty.

> **Example 9.** A psychologist selects as experimental subjects students enrolled in freshman psychology, PSY 101, at Upper Wabash University. Because two treatments are being compared, the students are assigned at random to two groups; that is, design (3) of Section 2 is used. There is a large difference in the average response of the two groups, too large to be reasonably due to the accidents of random assignment. The psychologist therefore concludes that the effects of the two treatments on the units differ.

For what population is this conclusion valid? In a strict sense, only for the group of students who served as subjects. If they are not representative of some larger population, the experimental conclusions are not valid for any larger population. The situation is quite similar to taking an SRS from a sampling frame that is not representative of the population. To take an extreme example, an experiment on psychological reactions to a pain-relieving drug, however well designed, will not give valid conclusions about adults in general if the subjects are mental patients.

What shall we do? The ideal solution is clear: Take an SRS from the population of interest to obtain the experimental subjects, and then apply the random assignment needed for the experiment. This ensures (or at least makes it very likely) that the experimental subjects are representative of the larger population. Such an ideal solution is almost never possible in practice. What is possible is to use an SRS from the pool of available subjects. The psychologist, for instance, should draw an SRS of PSY 101 students rather than allow students to choose which of a list of experiments they will participate in. If, for example, the experiment concerns the effect of watching pornographic movies, the reactions of students who choose to watch may differ sharply from the reactions of randomly selected students. (Of course, telling randomly selected students they must watch pornography is morally objectionable, so student choice of experiments rather than random selection is often necessary. It's enough to make psychologists envy physicists.)

If the experimental subjects are an SRS of a pool of available units, the experimental results are surely valid for that pool. Our psychologist can now draw conclusions about the population of students in PSY 101 this semester. No doubt she had grander things in mind. So she must persuade us that the experimental subjects are representative of some larger population. Here she draws on her understanding of her subject matter, not on statistics. If the experiment involved eye movements in response to visual stimuli, the experimenter can probably persuade us that the conclusions are valid for all persons with normal vision. If the experiment involved response to pornographic films, it is not clear that conclusions drawn from student subjects can be generalized to executives. If the subjects were volunteers, this limits still more the population to which the conclusions apply.

Difficulty 2: Lack of realism

Our psychologist, let us say, is interested in response to frustration. Her experiment requires the subjects to play a game rigged against them. The game demands team cooperation, and the experimenter will observe the relations among team members as they repeatedly lose. This is an artificial situation, and the PSY 101 students know it's "only an experiment," even if they don't know that the game is rigged. Do their reactions in this situation give information about their response to genuine frustration outside the laboratory? As with Difficulty 1, the question involves how far the experimental conclusions can be generalized. But the barrier to general conclusions here is the lack of realism of the experimental treatment. Once again the experimenter must try to persuade us. And once again, the eye-movements expert has a better case than the researcher studying pornography or frustration. Inability to apply realistic treatments is a serious barrier to effective experimentation in many areas of the social sciences.

Lack of realism in experiments is an issue that touches not just the social sciences, but our health and prosperity. Food additives are required, not unreasonably, to be safe. The Food and Drug Administration (FDA) is charged with seeing that they are safe. The FDA's exercise of judgment is restricted by laws, in particular by the famous Delaney Amendment, an all-or-nothing statement that outlaws additives found to cause cancer in humans *or animals*. A typical trial of a food additive involves adding large amounts of the substance to the diet of laboratory rats. If significantly more tumors appear in this experimental group than in the control group of rats fed an additive-free diet, the additive is found guilty and must be banned.

Now, such an experiment can provide strong evidence that large doses of an additive over a short period do (or don't) induce cancer in rats. We would really like to know whether small doses over a long period induce cancer in people. As the list of condemned additives grows (cyclamates, the food color Red 2, saccharin, . . .), so does the grumbling. Part of the grumbling is about the all-or-nothing character of the law. Surely, some say, if so useful a substance as saccharin causes just a little cancer, we ought to allow it in our colas, cookies, and cakes. That's a policy question, not a scientific question, and you must choose your own poison. But part of the grumbling concerns the realism of rat experiments as indicators of human health hazards.

THE DELANEY AMENDMENT

Sec. 409 (c) (3) (A). **No additive shall be deemed to be safe if it is found to induce cancer when ingested by man or animal, or if it is found, after tests which are appropriate for the evaluation of the safety of food additives, to induce cancer in man or animal.**
—*Federal Food, Drug and Cosmetic Act, 1958*

Informed opinion holds that the rats can be trusted in general. But every species has its peculiarities, and the case of the widely used artificial sweetener saccharin may ride on a peculiarity of rats. Saccharin caused bladder tumors in rats, and the bladders of rats are special. Rats concentrate their urine very highly before excreting it. Saccharin is not metabolized but is excreted unchanged. So the saccharin sits there in the rat's bladder for a long time, in raw form and in high concentration, waiting for the rat to get around to urinating. Some respectable scientists think that all that saccharin could cause tumors by physically irritating the bladder—only in rats, of course, not in people.

Can we check this hypothesis that saccharin does not cause bladder cancer in humans? Not easily. Such experiments on people are frowned on. We might try to see if bladder-cancer patients tend to be heavy users of saccharin. Several such observational studies have shown no link, but these studies are not sensitive enough to detect the small increase in cancer predicted by the animal experiments. Saccharin remains on the market by special decree of Congress, despite the scientific consensus that it probably does cause just a little cancer.[3]

This—and many another—example points to a clear principle: Rarely does a single experiment definitely establish that A causes B; there is almost always some flaw. (What flaws there might be is the subject of this section.) Repeated experiments, perhaps combined with other kinds of studies, are usually needed to found a conclusion on rock rather than sand.

Difficulty 3: Avoiding hidden bias

Many of our warnings about sources of bias in sample surveys concerned the need for care in areas other than statistical design. There is an art to wording questions; there are special techniques to increase the response rate. In experimentation, there are also special techniques and precautions that go beyond statistical design. We will mention two common precautions: the double-blind technique and randomization for purposes other than assigning units to treatments.

> **Example 10.** *Double-blind experiments.* In the vitamin C trial of Example 7, the health of the subjects was checked regularly and a judgment was made as to whether each was illness free. These judgments are necessarily somewhat subjective, especially because the experiment concerned mild illnesses such as the common cold. The physician making the judgment may be unconsciously in-

"Dr. Burns, are you sure this is what the statisticians call a double blind experiment?"

fluenced by knowledge of what treatment the subject received, especially if the physician knows that the treatment was a placebo and so "ought" to have no effect. (Keep in mind that this is not a deliberate bias but simply an unconscious influence similar to the placebo effect in patients.) Therefore, the diagnosing physician is kept ignorant of which treatment each subject received, so the diagnosis cannot be different for subjects in the two groups. This is called the *double-blind technique*. If only the subjects are ignorant of which treatment they are given, we have a single-blind experiment. When, as here, both the subjects and those who evaluate the outcome are ignorant of which treatment was given, we have a double-blind experiment. Only the director of the experiment knows which patients received vitamin C pills and which received placebo pills.

The double-blind procedure is used whenever possible in medical trials, because experience has shown that even careful investigators can be influenced by knowledge of the treatments used. Such ideas ought to be used in other experimental settings whenever they are appropriate. For example, in a traditional ESP (extrasensory perception) experiment, the experimenter looks at cards printed with various shapes (star, square, circle, etc.) not visible to the subject. The subject guesses the shapes, and the experimenter records whether the guesses are correct. It was noticed that the experimenter often recorded some incorrect guesses as correct. This does not necessarily show deliberate distortion but may be a result of the experimenter's desire to discover ESP. The remedy is to have the subject press a computer key to record guesses. The computer's record of the guesses is later compared with a written record (made before the experiment) of the shape of each card in the shuffled deck used. Failure to use blind experimentation often invalidates "experiments" set up to prove a point.

> **Example 11.** *More randomization.* In experiments on nutrition, it is common to use newly weaned male white rats as experimental units. Rats are randomly assigned to the diets to be compared; this assures that the stronger rats are not somehow assigned to one diet. Weight gain over a several-week period is observed. Now it turns out that rats placed in top cages gain weight somewhat faster than rats housed in bottom cages.* If rats fed diet A were placed in the bottom row, those fed diet B in the row above, and those fed diet C in the top row, the effects of diet and cage location would be confounded. The remedy is to assign rats to cages at random. Note that this is not the same as the random assignment to diets used in the experimental design. It is quite common to use additional randomization somewhere in carrying out an experiment.

The design of the ESP experiment also can be improved by an additional randomization. Instead of using a deck of cards, we might place cards bearing different shapes on a table in front of the sender, with a light next to each. The lights are lit in an order determined by a table of random digits. The sender concentrates on the card whose light is on.

---

*See Example 15 in this chapter for a more detailed study of this example, which is based on Elisabeth Street and Mavis B. Carroll, "Preliminary Evaluation of a New Food Product," in *Statistics: A Guide to the Unknown.* This fact about rats appears on p. 163 of that book.

The subject being tested for ESP sits in another room with the shapes in front of him and pushes a button under the shape he thinks the sender is concentrating on. Score is kept automatically by a computer that checks whether the button pushed by the subject matches the sender's light. The use of a table of random digits prevents the subject from picking up patterns in a reshuffled deck and keeps the sender "blind." The entire format is designed to prevent communication between sender and subject.

No precaution is too elaborate in such an experiment, for many professional psychics are clever frauds, and many researchers on psychic phenomena want so badly to find paranormal effects that they unconsciously collaborate with the psychic. It is not surprising that the most publicized "scientific experiments" on psychic phenomena have been arranged by physicists, who are naive where human subjects are concerned. As one skeptic put it, "Electrons and rats don't cheat. Professional psychics do."[4] Anyone planning to test the powers of psychics should have at least a psychologist, and preferably also a professional magician who knows the tricks of the trade, present. (Houdini made a regular practice of exposing phony spiritualist mediums, and contemporary magicians perform like services.)

Difficulty 4: Attention to detail

Even when your subjects are not so slippery as psychics, no data-collection design can survive lack of attention to detail. It is surprising how often sloppiness infects even important experiments. In 1976, the FDA banned Red 2, the food color most widely used in the United States, as a potential cause of cancer. Here is a quote from an account of the experiment that led to the ban:

> **Example 12.** The study involved feeding Red 2 to four different groups of rats, each at a different dosage level, and then comparing the health of these treated groups with the health of a control group. There were 500 rats in all—seemingly enough for a solid evaluation. But the study was left unsupervised for a long period of time after a scientist was transferred, and it developed two serious flaws. To begin with, the animal handlers managed to put some of the rats back in the wrong cages part way through the experiment, so that an undetermined number of rats were shifted among the control group and the four treated groups. Second, the animal handlers were slow in retrieving dead rats from their cages and

rushing them off to the pathologist for examination. As a result, virtually all the rats that died during the course of the experiment were so badly decomposed as to be of little use for evaluation. Only those rats that survived to the end of the experiment and were killed—some 96 in all—were available for detailed histopathological examination, "It was the lousiest experiment I've seen in my life," commented one scientist who reviewed the data.[5]

A decision had to be made, and a clever statistician managed to rescue some information. But careless supervision prevented the kind of clear conclusion that properly designed experiments are intended to produce. The detailed statement of exactly how an experiment is to be conducted is called (especially in clinical trials) the *protocol* of the experiment. Writing a careful protocol and making certain that it is followed is part of an experimenter's job.

Section 3 exercises

2.26. For what population do you think the experimental conclusions are valid in each of the following cases? (We cannot be definite about this question without expert knowledge, so our answers will be partly guesses.)

    (a) The vitamin C study of Example 7. (The subjects were adult men.)

    (b) The propaganda study of Example 6. (The subjects were college students.)

    (c) The dissolving-time study of Exercise 2.22 in Section 2.

2.27. Fizz Laboratories, a pharmaceutical company, has developed a new pain-relief medication. Sixty patients suffering from arthritis and needing pain relief are available. Each patient will be treated and asked an hour later, "About what percentage of pain relief did you experience?"

    (a) Why should Fizz not simply administer the new drug and record the patients' responses?

    (b) Design an experiment to compare the drug's effectiveness with that of aspirin and of a placebo.

    (c) Should patients be told which drug they are receiving? How would this knowledge probably affect their reactions?

    (d) If patients are not told which treatment they are receiving, the experiment is single-blind. Should this experiment be double-blind also? Explain.

2.28. An experiment that was publicized as showing that a meditation technique lowered the anxiety level of subjects was conducted as follows: The experimenter interviewed the subjects and assessed their levels of anxiety. The subjects then learned how to meditate and did so regularly for a month. The experimenter reinterviewed them at the end of the month and assessed whether their anxiety levels had decreased or not.

    (a) There was no control group in this experiment. Why is this a blunder? What extraneous variables may be confounded with the effect of meditation?

    (b) The experimenter who diagnosed the effect of the treatment knew that the subjects had been meditating. Explain how this knowledge could bias the experimental conclusions.

    (c) Briefly discuss a proper experimental design, with controls and blind diagnosis, to assess the effect of meditation on anxiety level.

2.29. There is some evidence that nursing home residents are happier and more active when they can make more decisions for themselves. You are planning an experiment in which the treatment of interest will be a talk given to residents by the nursing home manager describing what decisions they can and should make for themselves. Briefly discuss the design of such an experiment. Should there be a control group? What treatment will you give the control group? It is not practical to randomly assign residents living next to each other different groups. How will you form the treatment and control groups in this situation?

2.30. Taste tests ask subjects to compare the taste of two food products, such as Pepsi and Coke, and say which they prefer. There is no separate control group; each subject serves as his or her own control by tasting both products.

    (a) Randomization remains important in a taste test. How should randomization be used?

    (b) How does the idea of blindness apply in a taste test?

2.31. A chemist wants to compare the variability of a new and simpler assay method with the standard method. She prepares a batch of solution, divides it into 40 specimens, and then selects 20 specimens at random. She asks her technician to analyze these 20 with the new method and the remaining 20 specimens with the standard method. Since all of the specimens come from the same solution, all assays should give the same answer except for variation in carrying out the chemical analysis. The technician performs the analyses with the new method. Then the technician falls ill, so the chemist performs the 20 old method assays herself. The results using the new method are much more variable than those using the old, and the chemist therefore rejects the new assay technique. This experiment is fatally flawed despite a randomized comparative design. Explain why the results cannot be trusted.

2.32. Experiments with human subjects that continue over a long period of time face the problem of dropouts (subjects who quit the experiment) and nonadherers (subjects who stay in but don't follow the experimental protocol). In his account of the University Group Diabetes Program (Example 8 in Section 2), Dr. Cornfield makes the following statement:

> For this complex problem, the UGDP has followed the generally accepted practice of comparing the mortality experience of the originally randomized groups, and of not eliminating dropouts or nonadherers from the analysis. This practice is conservative in that it dilutes whatever treatment effects, beneficial or averse, are present.

Explain why counting dropouts and nonadherers in the final results reduces any real effect of a treatment relative to a placebo.

2.33. A number of clinical trials have studied whether reducing blood cholesterol using either drugs or diet reduces heart attacks. These experiments have usually followed their subjects for five to seven years. In order to see results in this relatively short time, the subjects have been chosen from the group at greatest risk: Middle-aged men with high cholesterol levels or existing heart disease. The experiments have generally shown that reducing blood cholesterol does decrease the risk of a heart attack. But some doctors doubt the usefulness of these experimental results for many of their patients. Why?

2.34. One of the clinical trials that have demonstrated that lowering the level of cholesterol in the blood can reduce the risk of heart attacks is described in Gina Kolata, "Lowered Cholesterol Decreases Heart Disease," *Science*, Volume 223 (1984), pp. 381–382. Read this article as an example of the size and complexity of a major medical experiment. Describe the statistical design of the experiment (which was quite simple). Describe the serious practical difficulties that had to be overcome.

4. More on experimental design

The basic ideas of statistical design of experiments are *randomization* and *control*.

Randomization is the random allocation of experimental units among treatments, most simply by assigning an **SRS** of units to each treatment.

Control is taking account of extraneous variables in the experimental design, most simply by the use of equivalent groups for comparison.

Completely randomized designs use both randomization and control in their simplest form, allocating the units at random among all the treatments. The examples in Section 2 were particularly simple in that the treatments were levels of a single factor, such as "drug administered" in the University Group Diabetes Program. Many experiments have more than one factor, so interactions among the factors can be studied. Here is an example with two factors and six treatments:

Example 13. A food products company is preparing a new cake mix for marketing. It is important that the taste of the cake not be changed by small variations in baking time or temperature. An experiment is done in which batches of batter are baked at 300, 320, and 340°F for 1 hour and for 1 hour and 15 minutes. All possible combinations are used, resulting in six treatments that can be outlined as follows:

		Factor A (temperature)		
		300°F	320°	340°F
Factor B	1 h	1	2	3
(time)	$1\frac{1}{4}$ h	4	5	6

Sixty batches of batter — ten for each treatment — will be prepared from the mix and baked. The batches should be assigned at random to the treatments and the cakes baked in the random order that results. Draw an SRS of size 10 from the numbers 01 to 60 to obtain the positions in which cakes will receive treatment 1 on baking, then an SRS of the remaining 50 numbers to obtain the baking positions for treatment 2, and so on. If, for example, 01 is assigned to treatment 4, the first cake prepared is baked at 300° for $1\frac{1}{4}$ hours. After baking, each cake will be scored for taste and texture by a panel of tasters. These scores are the response variables.

It would be a serious mistake in this example to prepare first 10 batches of batter to bake under treatment 1, then 10 under treatment 2, and so on. Any variable changing systematically over time, as room humidity might, would be confounded with the experimental treatments. Notice that there are two experimental factors, or explanatory variables, namely baking time and oven temperature. Each of the six treatments is a combination of a particular level of each factor. This design is somewhat more complex than those with only one experimental factor, but it is completely randomized because the 60 units (batches of batter) are assigned completely at random to the six treatments.

Matched-pairs designs

Control appears in completely randomized designs only in the basic idea of comparative experimentation. Much of the advanced study of experimental design concentrates on more elaborate ways of controlling extraneous variables. We will introduce one additional principle of experimental design to illustrate how control can be improved.

Example 14. Is the right hand of right-handed people generally stronger than the left? A student designs an experiment to study this question. She fastens an ordinary bathroom scale to a shelf five feet from the floor, with the end of the scale projecting out from the shelf. Subjects squeeze the scale between their thumb beneath it and their fingers on top: The scale reading in pounds measures hand strength.

A completely randomized design would require two groups of subjects, one for each hand. Surely it is more natural to have each subject use both hands, so that a direct right-left comparison is obtained. This is a *matched-pairs* design. The name reminds us that instead of comparing two groups, we make comparisons within matched pairs of observations, in this case the two hands of the same person.

What about randomization in this experiment? We can't assign subjects at random to groups. Instead, choose at random which hand each subject tries first. A subject may gain confidence or learn better how to grasp the scale after the first try; randomizing the order prevents this learning from being confounded with right versus left hand.

In the matched pairs design of Example 14, each subject serves as his or her own control. Some subjects are stronger than others, and will have higher scale readings for both hands. The matched-pairs design takes account of this extraneous variable. A direct right versus left comparison for each subject gives better information than comparing right hands in one group versus left hands in a separate group. Comparing two treatments on the same subjects is not the only situation in which the added control provided by a matched-pairs design is helpful. Suppose that before comparing two treatments for high blood pressure, we match the subjects in pairs that are as similar as possible. Each pair of subjects shares the same age, sex, race, occupation, and current blood pressure. The two treatments are assigned at random within each matched pair, say by tossing a coin. Comparisons within such matched pairs should be more informative than comparisons between two randomly chosen groups of subjects, because we have controlled some important extraneous variables by matching rather than relying completely on randomization.

Block designs

Matched-pairs designs are one kind of *block design*. A *block* is a group of experimental units that are similar in some way that affects the outcome of the experiment. In a block design, the random assignment of treatments to units is done separately within each block. Here is an example.

> **Example 15.** It is common in nutrition studies to compare diets by feeding them to newly weaned male rats and measuring the weight gained by the rats over a 28-day period.* If 30 such rats are available and three diets are to be compared, each diet will be fed to 10 rats. Random assignment of rats to diets will average the effect of extraneous variables, such as the health of the rats. It is nonetheless wise to take additional steps to produce equivalent groups of rats for each diet. For example, standard strains of laboratory rats are available; this minimizes hereditary differences among the rats.
>
> We can come still closer to equivalent groups by forming blocks. The initial weight of the rats is an extraneous variable that is especially important as an influence on weight gain. Rather than ran-

---

*For more information on this example, see Elisabeth Street and Mavis B. Carroll, "Preliminary Evaluation of a New Food Product," in *Statistics: A Guide to the Unknown.*

domly assign 10 rats to each diet, we therefore first divide the animals into 10 groups (blocks) of three rats each based on weight. The three lightest rats form the first block, the next lightest three form the second block, and so on, with the tenth group consisting of the three heaviest rats. Now we randomly assign one rat from each block to each diet. The blocks help to create equivalent groups of 10 rats to be fed each diet, because the rats in each block are approximately equal in weight and one of them is assigned to each diet.

In Example 15, we did *not* use the completely randomized design described by

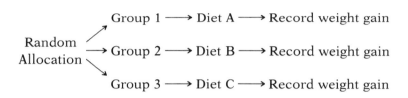

A completely randomized design handles all extraneous variables by randomization. If the weight gained by individual rats varies a lot from rat to rat, the groups fed diets A, B, and C may have quite different average weight gains just because of this individual variation. Thus it will be hard to detect the systematic effect of the diets. One remedy is to add more rats to each group to obtain less variable average results. That's expensive. Another remedy is to take account of the worrisome extraneous variable in the layout of the experiment, that is, to control for the effect of initial weight directly rather than simply relying on randomization to average the effect. This was done in Example 15. First divide the rats into blocks consisting of rats of about the same weight. This is *not* random but is based on a particular extraneous variable: the initial weight of the rats. Then randomize, but only within the blocks, by randomly assigning one of the three rats in each block to each diet. This is called a *randomized complete block design*. It has the pattern shown in Figure 2-1.

There are still three experimental groups of 10 rats each, one fed each of diets A, B, and C. Each group contains one rat from each of the 10 blocks, so the initial weights of the rats in the three groups closely match. This matching greatly reduces the variation in average weight gain among the groups owing to different initial weights. It is now much easier to detect any difference in weight gain resulting from the different

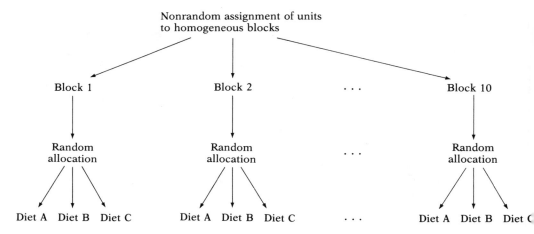

Figure 2-1. Randomized complete block design. In such a design, judgment is used to form blocks of similar units, and randomization is applied within each block separately.

diets. Each block in a randomized complete block design contains exactly as many experimental units as there are treatments. In a matched-pairs design, for example, there are just two treatments. Each treatment is then applied once within each block. Other block designs may use larger blocks so that each treatment is assigned to several units within each block. Still other block designs use incomplete blocks having fewer units than there are treatments.

Blocking in experimental design is similar to stratification in sampling. Both require judgment to classify units into groups (blocks or strata) in such a way that units within a group show less variability than the entire population of units. Randomization is then used only within each block or stratum; that is, randomization no longer is "complete" as in a completely randomized design but is restricted to each block separately.

The definition of *control* as taking account of extraneous variables in the experimental design is now clearer. The blocks in a randomized block design are groups of units that agree in some extraneous variable important to the outcome of the experiment. The term "block" originated in agricultural experimentation, where it refers to a compact plot of land with little variation in soil type, fertility, and so forth. In experiments on fertilizers or cultivation methods, we are not interested in such extraneous variables as soil type and fertility. Random assignment of treatments (fertilizers or whatever) to small plots of ground (experimental units) within the same block gives equivalent units for comparison of treatments. Several blocks give a larger number of observations and also provide data on more than one soil type and fertility.

In Example 15, a block consisted of animals of approximately equal weight. We are interested in the effect of diet, not of initial weight. Because animals within a block have the same initial weight, it should be easier to detect the effect of diet when these animals are fed different diets. We use several blocks to obtain more observations; data on three rats would not be precise even if all had the same initial weight. In summary, blocks are a way of holding fixed an extraneous variable that would otherwise cause large variations in the experimental results. A randomized block design therefore will give more precise (more repeatable) results than a completely randomized design with the same number of units.

Extraneous variables can be dealt with by either randomization or control. Completely randomized experiments rely exclusively on the averaging effects of randomization, while randomized block experiments include an extraneous variable directly in forming blocks. Elaborate experimental designs often deal with many variables by control. In a general sense, variables with a large influence on the outcome should be controlled. Others can be randomized. Using a completely randomized design in place of a randomized block design is not wrong in the sense of producing invalid results. But it is inefficient; that is, the more elaborate design requires fewer experimental units to give equally precise results. In Section 1, I stated that the first goal of experimental design is to obtain valid results, to discover the true effects of the treatments. The second goal is to do this as efficiently as possible, to use as few units as possible for a given degree of precision. The randomized block design hints at how this second goal is pursued.

Section 4 exercises

2.35. Most motor vehicles are equipped with catalytic converters to reduce harmful emissions. The ceramic used to make the converters must be baked to a certain hardness. The manufacturer must decide which of three temperatures (500, 750, and 1000°F) is best. The position of the converter in the oven (front, middle, or back) also affects the hardness. So there are two experimental factors: temperature and placement.

    (a) List the treatments in this experiment if all combinations of levels of the two factors are used.

(b) Design a completely randomized experiment with five units in each group.

(c) Using Table A, beginning with line 101, do the randomization required by your design.

2.36. A drug is suspected of affecting the coordination of subjects. The drug can be administered in three ways: orally, by injection under the skin, or by injection into a vein. The potency of the drug probably depends on the method of administration as well as on the dosage administered. A researcher therefore wishes to study the effects of the two factors, dosage at two levels and method of administration by the three methods mentioned. The response variable is the score of the subjects on a standard test of coordination. Ninety subjects are available.

(a) List the treatments that can be formed from the two factors.

(b) Describe an appropriate completely randomized design. (Just outline the design; don't do any randomization.)

(c) The researcher could study the effect of dosage in an experiment comparing two dosage levels for one method of administration. He then could separately study the effect of administration by comparing the three methods for one dosage level. What advantages does the two-factor experiment you designed in (a) have over these two one-factor experiments taken together?

2.37. A psychologist is interested in the effect of room temperature on the performance of tasks requiring manual dexterity. She chooses as treatments temperatures of 70 and 90°F. The response variable is the number of correct insertions in a half-hour period in an elaborate peg-and-hole apparatus that requires the use of both hands simultaneously. Each subject is trained on the apparatus and then asked to make as many insertions as possible in 30 minutes of continuous effort.

(a) Describe a matched-pairs design for this experiment. Be sure to state how you will randomize.

(b) Why is a matched-pairs design preferable to a completely randomized design that assigns a separate group of subjects to each temperature?

2.38. Do consumers prefer the taste of Pepsi or Coke in a blind test in which neither cola is identified? Describe briefly the design of a matched-pairs experiment to investigate this question.

2.39. There is good evidence that physical stress — even someone stroking the leaves of a plant for a minute a day — inhibits plant growth. Some claim (without good evidence) that speaking kindly to plants encourages growth. We are going to investigate the effects of physical contact, talking to plants, or both, on growth. Our experimental units are tomato seedlings that have just developed their first pairs of true leaves. Discuss the design of such an experiment. You must carefully describe the treatments and other aspects of the protocol as well as the statistical design.

2.40. The batches to be processed in the drug-making experiment of Exercise 2.10 in Section 1 are mixed and processed one at a time. Label the first batch processed 01, the second 02, and so on. Use Table A to do the randomization required for a completely randomized design. Which batches did you assign to each treatment?

2.41. Twenty overweight females agree to participate in a study of the effectiveness of four reducing regimens. The researcher first compares each subject's weight with her "ideal weight" and then calculates how many pounds overweight each is. The subjects and their excess poundage are:

Birnbaum	35	Hernandez	25	Moses	25	Smith	29
Brown	34	Jackson	33	Nevesky	39	Stall	33
Brunk	30	Kendall	28	Obrach	30	Suggs	35
Dixon	34	Loren	32	Rakov	30	Tran	42
Festinger	24	Mann	28	Siegel	27	Williams	22

The subjects are grouped into five blocks of four subjects each by excess weight, and the four regimens A, B, C, and D are assigned at random, each to one subject in each block. This is a randomized complete block design. After eight weeks the researcher measures each subject's weight loss.

(a) How many experimental units are there? How many factors? How many treatments?

(b) Arrange the subjects in order of increasing excess weight, then group the four least overweight, the next four, and so on. These groups are blocks.

(c) Use Table A to do the required random assignment of subjects to treatments separately within each block. Explain carefully how you used the table.

2.42. In Exercise 2.19 of Section 2, a nutritionist had 10 rats of each of two genetic strains. The effect of genetic strain can be controlled by treating the strains as blocks and randomly assigning five rats of each strain to diet A. The remaining five rats of each strain receive diet B. Use Table A, beginning at line 111, to do the randomization. This is a randomized block design but not a randomized complete block design. Why not?

2.43. An agronomist wishes to compare the yield of five corn varieties. The field in which the experiment will be carried out increases in fertility from north to south. The agronomist therefore divides the field into 30 plots of equal size, arranged in six east-west rows of five plots each, and employs a randomized complete block design.

Identify the experimental units, the treatments, and the blocks. Describe the arrangement of the randomized complete block design, but do not do the randomization.

2.44. Hearing loss is more common among premature infants than among full-term babies. It has long been thought that this is a physical effect of premature birth. Recently it has been suggested that hearing loss develops in infants who spend long periods in incubators, because of the high noise level in standard incubators. (You see that the effect of being premature is confounded with the effect of spending a long period in an incubator, because usually only premature babies spend time in incubators.)

Design an experiment to decide which explanation is true. (Ignore practical and moral problems. You can get more or less information out of a correct experiment here, depending on how elaborate you want to make it.)

2.45. Does the use or omission of the zip code affect the number of days a letter takes to reach another city? Describe briefly the design of an experiment to investigate this question. Be sure to specify the treatments exactly and to tell how you will handle extraneous variables such as the day of the week on which the letter is mailed.

5.   Social experiments

The effectiveness of a new fertilizer or a new drug is always tested by a controlled and randomized experiment—and for good reason: A well-designed experiment can provide clearer answers than any other method of study. What about testing the effectiveness of a new welfare program or health insurance system or preschool education program? Public policy decisions in these areas have usually been based on much supposition and little knowledge. It is tempting to try an experiment in the hope of finding clear answers. Many such *social experiments* have been done in recent years.

Changes in the welfare system that supports poor families have been the subject of several social experiments. These illustrate nicely the pitfalls as well as the advantages of the method. The basic issues have remained the same: How can welfare best support poor families, especially those with children, while at the same time giving poor people an incentive to become self-supporting and costing no more tax dollars than necessary?

> **Example 16.** The New Jersey Income-Maintenance Experiment was conducted in four urban areas in New Jersey and Pennsylvania, beginning in 1968.[6] The new welfare policy tested in this experiment replaced fixed welfare payments that end entirely as soon as the family earns more than a small amount of other income with a sliding scale of payments that decrease gradually as the family earns more. Supporters of the new policy argued that it gives more incentive to work because welfare payments don't disappear as soon as the recipient finds a minimum-wage job. Opponents said the new program would be too expensive.
>
> The population from which units for the experiment were selected consisted of low-income households containing at least two persons, one of them a healthy male between 18 and 58 years old. Most poverty-stricken families have no employable male present, and so are not in the target population. This population was chosen because of special interest in the effects of the new welfare system on working-age males, an interest aroused not by sexism but by the fact that working-age males generally cannot collect welfare under the present system and so might have less incentive to work under a new system that would provide welfare benefits to them.
>
> As recommended in Section 3, experimental units were chosen from the population by sampling. Poverty areas in four cities were selected from census data and on-site inspection. Then blocks were selected at random within these areas, and field workers were sent out to make lists of all dwelling units in each block selected. From these, a sample of households was chosen. Interviewers visited these households to see if they belonged to the target population and to explain the experiment. In the end, 1357 families were chosen, and agreed, to participate.
>
> The experimental design was a bit complicated. There were two factors, the guaranteed minimum income level and the rate at which welfare payments decreased as earnings went up. Eight combinations of these factors were tried, with the current welfare system as a control, giving nine treatments in all. The many response variables included measures of family income, employment, and family stability. The 1357 families that participated were first divided into three blocks of 400 to 500 families each, depending on

their recent income level. Blocks were used because recent income helps predict future income. There was another reason for blocking as well: The experiment could not afford to assign too many of the poorest families to the treatments that promised the highest income. So the randomization did not make all treatments equally likely. The chance of being assigned to a specific treatment was different in each of the three blocks. But each family had a known chance of being assigned to each treatment, and these chances were the same for every family in a block. It was a complicated bit of randomization, much akin to moving from an SRS to a general probability sample.

The experimental design of the New Jersey Income-Maintenance Experiment is more involved than any we met earlier, in three ways: First, it made use of blocks in a two-factor experiment; second, not all the possible combinations of the two factors were used; and third, different randomization patterns, none of them giving a household the same chance of landing in all nine groups, were employed in each block. The real complexity, however, lay not in the statistical design but in the practical difficulties of social experiments. The researchers had to contend with dark stairwells, vicious dogs, hostile local politicians, and militant community groups. Moreover, 10% of the dwelling units in the initial sample were empty, 19% of the remaining households were never at home (in five tries), and 18% refused to speak to the interviewer in four visits. An initial sample of 48,000 households yielded 27,000 that were interviewed. Of these, only 3124 were eligible to participate in the experiment; low income families with able-bodied men present are not common in urban areas. Another 425 families had vanished by the time interviewers returned to invite them to join the experiment and others refused, leaving the final 1357. The life of an experimenter out there in the real world has its frustrations.

The New Jersey experiment showed that the new welfare payment system succeeded in increasing the income of poverty-level families without reducing the employment rate of the husbands. But the new system did lower the employment of their wives; the wives in the experimental groups worked 23% fewer hours per week than the controls. (Because these families had an average of four children, the wives were working hard at home.) On the whole, the new system was not as successful as its supporters had hoped, and the experimental results did not influence legislation. This lack of influence was also due to the fact that final results were not available until 1977, nine years after the study began.

Example 17. The Baltimore Options Program was part of another round of experiments with new welfare programs that began in 1982.[7] The goal this time was to compare the existing welfare system with a new program that did not change welfare payments but offered basic education, job training, unpaid work experience, and job-search help. The population of interest was welfare recipients who were able-bodied single parents (90% of them women) with no children less than six years old. The experimental design was much simpler than that of the New Jersey study; subjects were simply assigned at random to one of the two programs. The practical difficulties were also reduced because participation in the program was required as a condition of receiving welfare payments. This was possible because all subjects received their regular welfare payments, with or without the extra preparation for employment. There were 1362 subjects in the experimental group and 1395 subjects in the control group.

Even the simpler Baltimore Options Program suffered its share of practical problems. Many subjects had legitimate reasons not to participate in the training and work programs, and many left the program through marriage or the birth of another child. As a result, only 45% of the subjects in the experimental group actually participated in one of the new activities within a year of the start of the study. Nonetheless, the results were quite clear: The new program did increase employment and earnings. By the third year, the earnings of the experimental group were almost 25% higher than in the control group. The costs of the new training programs were balanced by reduced welfare payments to those who found work and by the taxes they paid. Partly as a result of evidence produced by the Baltimore Options Program and other welfare experiments, requirements for participation in job training and placement programs were included in new welfare legislation passed by Congress in 1988.

The advantages of social experimentation are clear: There is a real possibility that experiments will give a better picture of the effects of a new welfare policy than any amount of debate. The disadvantages of social experiments are also clear: They are hard to do well because of practical problems, and they take a long time and are expensive. Conditions may so change between the planning of an experiment and the release of the results that the answers given by the experiment no longer respond to the right questions. As the contrast between the two welfare experiments suggests, the chances of success are higher for a simple experiment designed to test the effect of a modest change in policy.

Issues such as police response to domestic violence calls or the most effective treatment for persons convicted of drunk driving are easier to handle than welfare reform.

Public policy decisions affect us all, either directly or indirectly by changing our society and spending our taxes. Experimentation is a powerful tool for gathering information on the effects of treatments. Such a tool should certainly help to guide policy decisions. Social experimentation, used carefully, is an idea whose time has come.

Section 5 exercises

2.46. Would requiring cars to keep their headlights on during daylight hours reduce traffic accidents? Discuss the design of an experiment to investigate this question.

2.47. Once a person has been convicted of drunken driving, the purpose of court-mandated treatment or punishment is to prevent future offenses of the same kind. Suggest three different treatments that a court might require. Then outline the design of an experiment to compare their effectiveness. Be sure to specify the response variables you will measure.

2.48. The usual police response to domestic violence cases has long been to calm the situation and to warn the offender, but not to arrest him unless use of a weapon or other circumstances require an arrest. The first experiments on police response to domestic violence suggested that arrest reduces future incidents. This evidence, along with pressure from women's groups, has changed police policy in many cities, where persons accused of domestic violence are now arrested.

Outline the design of an experiment to compare "warn and release" with "arrest and hold." What are your response variables? How will you help the police on the scene do the randomization required? (These experiments raise ethical problems, but you may ignore them here.)

2.49. You are interested in whether lowering highway speed limits reduces traffic fatalities. Observation cannot answer this question, because areas with lower speed limits differ from high speed limit regions in many ways. Heavily populated urban regions usually have lower speed limits than rural areas, for example. How would you design an experiment to compare 55-, 65-, and 75-mile-per-hour speed limits? In addition to an outline of the statistical design, describe the physical working of the experiment. What kinds of roads would you use? Would you change speed limits on the same roads over time or assign different speed limits on similar roads at the same time?

2.50. Another of the early experimental trials of new welfare systems was conducted in Gary, Indiana. Unlike the New Jersey experiment, households both with and without working-age males present were included. The statistical design of the Gary experiment divided families into two blocks, those with and those without an employable adult male. Families were randomly assigned to treatments separately within each block.

Explain why blocking was used in this social experiment.

2.51. The article "A Health Insurance Experiment" by Joseph P. Newhouse in *Statistics: A Guide to the Unknown* describes a social experiment. The purpose of the experiment was to discover if changing the percent of medical costs that are paid by health insurance has an effect on either the amount of medical care that people use or on their health. The treatments were insurance plans that paid all costs above a ceiling but one of 100%, 75%, 50%, or 5% of costs below the ceiling.

(a) Outline the design of a randomized comparative experiment suitable for this study.

(b) Describe briefly the practical and ethical difficulties that might arise in such an experiment.

(c) If possible, read the Newhouse essay. Was the design more complex than yours? In what way? What practical problems are mentioned in the essay?

2.52. Choose an issue of public policy that you feel might be clarified by a social experiment. Briefly discuss the statistical design of the experiment you are recommending. What are the treatments? The response variables? Should blocking be used?

6.   Ethics and experimentation

Carrying out experiments on human subjects raises serious ethical problems. We will consider two situations that carry different risks and raise somewhat different ethical difficulties — medical experiments and experiments in the behavioral and social sciences.

Medical research

Medical experiments (often called *clinical trials*) have been particularly controversial because the giving or withholding of medical treatment can result in direct danger to a patient. Yet comparative experiments are necessary if life-saving medical knowledge is to be obtained. Some treatments—"magic bullets" like penicillin—have effects so dramatic that properly designed experiments are not needed to detect them. But this is rare. Medicine usually advances in smaller steps: The death rate from the new operation is 10% lower than for the old one; patients given the new drug live 5% longer than those given the old. Only a randomized comparative experiment can give clear evidence for or against such moderate gains. The dilemma is that the benefit of medical advances goes mainly to future patients, while the risk falls on the current patients who serve as subjects.

> **Example 18.** Simply avoiding medical experiments is not the answer to the ethical dilemma. Medical treatments that are introduced without the firm evidence of effectiveness that only a randomized comparative experiment can give may be ineffective and dangerous. Consider the case of mammary artery ligation.[8] This surgical treatment for angina, the severe pain caused by inadequate blood supply to the heart, was so popular in the 1950s that it was the subject of an article in *Reader's Digest*. The surgeon opened the patient's chest and tied off the internal mammary arteries in the hope of forcing more blood to flow to the heart by other routes, thus relieving the patient's angina. In 1958 and 1959, some skeptical researchers finally conducted a randomized comparative double-blind experiment. Mammary artery ligation was compared to a "placebo operation" in which the surgeon did not tie off the arteries. Result: Some improvement in angina for both groups, but no difference between them. The claims for mammary artery ligation had been based entirely on the placebo effect. Surgeons abandoned mammary artery ligation at once. Thanks to a properly designed experiment, angina sufferers no longer undergo chest surgery for no good reason.

How can we protect experimental subjects while still allowing medicine to use the unequaled power of statistically designed experiments? Concern for subjects has in recent years placed ever-tighter restraints on the conduct of clinical trials. The regulations now in place are elaborate, but based on a few general principles.

All agree that medical experiments must be guided by the first principle of medical ethics, expressed in the ancient Latin motto *"Primum non nocere,"* or "First of all, do no harm." Exposing patients to treatments *known* to be harmful is clearly unethical. The application of this first principle is not as simple as might at first appear, because it also prohibits the giving of treatments known to be less than the most effective available. A control group, for example, cannot be given a placebo if there is an accepted treatment for the medical condition being studied.

In a comparative trial of a new drug or surgical procedure against a standard treatment, one of the groups of subjects always receives an inferior treatment. The experiment is ethical if we don't know which group this is. A clinical trial requires genuine uncertainty about which of the treatments is superior. The new drug or surgical procedure cannot be used unless there is reason to believe that it may be better than the standard treatment. But there must also be some remaining ignorance to justify giving the standard treatment to a control group. After all, a promising new therapy may be less effective than hoped or may cause unsuspected side effects.

> **Example 19.** The principle that patients cannot receive a treatment known to be inferior affects the statistical design of clinical trials. If a clinical trial begins to produce convincing evidence that one treatment is more effective, the trial is often stopped before all results are in so that all subjects can be given the superior treatment. The Physicians' Health Study examined the claim that aspirin taken regularly can reduce the risk of heart attacks. More than 22,000 male physicians were divided at random into two groups. One group took a dose of aspirin every second day, while the other took a placebo pill that was identical in appearance. When 189 of the 11,034 subjects in the placebo group had suffered heart attacks as against 104 of the 11,037 men in the aspirin group, the researchers stopped the experiment and announced the result. Deciding when the evidence is conclusive is a difficult problem both statistically and ethically.

The second ethical standard for clinical trials requires that the *informed consent* of the subjects be obtained. Subjects are told that they will be part of a research study, that is, that something beyond treating their illness is going on. The nature of the study and the possible risks and benefits are explained, and the patients are asked to agree to participate and to consent to random assignment to a treatment. Those who refuse are not included in the experiment and usually receive the standard therapy. (Can you see why these patients cannot be used as part of the control group, even though they receive the standard therapy?)

Many individuals are willing to participate in random clinical trials for the future benefit of all. But many patients find the idea of being subjects in an experiment hard to accept, and the idea that their treatment will be chosen by the toss of a coin even harder to accept. In studies of serious conditions such as cancer, as many as 80 or 90% of the subjects approached sometimes refuse to participate. The strict rules on disclosure and informed consent now enforced in clinical trials do make clinical research more difficult.

Informed consent can be somewhat difficult to achieve, especially when the subjects are a dependent group rather than paying consumers of medical care. In the past, prisoners were often volunteer subjects for medical experiments. We might doubt whether prisoners requested to "volunteer" are really free to refuse. In early 1976, the Federal Bureau of Prisons announced that medical experimentation on federal prisoners no longer would be permitted. The *New York Times* commented in an editorial (March 8, 1976),

> *Truly voluntary consent is virtually impossible to achieve in prison and there is a large temptation to undervalue prisoners' interests during the course of such research. The new Federal policy is clearly the appropriate response to these problems and it should serve as an example to the states which still permit experiments to be conducted in their prisons.*

What all of this adds up to is that medical experimentation in the United States is subject to tight rules for the protection of subjects, rules drawn up by the federal agencies that pay for (and hence can control) almost all medical studies. Yet the appearance of AIDS has raised the conflict between the welfare of current patients and the search for knowledge to benefit future patients to new heights.[9] Because AIDS is fatal, activists urged that experimental drugs be made widely available to patients in the hope that lives would be saved. In 1989, the federal government agreed to allow patients for whom there is no satisfactory alternative treatment to receive experimental drugs at the same time as clinical trials of these drugs are underway. Some such drugs will be dangerously toxic and some will be ineffective. More disturbing to defenders of the traditional rules, some AIDS patients who are enrolled in clinical trials try any new drug in place of or in addition to those given in the clinical trial. Making untested drugs widely available runs the risk of preventing researchers from producing reliable information that will lead to effective treatments for AIDS. AIDS is of course not the only fatal disease, so that the pressure from AIDS activists may lead to widespread changes in procedures for testing drugs. The standards for medical research are continuing to change under the impact of ethical questions.

Behavioral and social science research

When we move from medicine to the social and behavioral sciences, the direct risks to experimental subjects are less acute. But so is the value to the subjects of the knowledge gained. Consider, for example, the experiments conducted by psychologists in their study of human behavior.

> **Example 20.** Stanley Milgram of Yale conducted a famous experiment "to see if a person will perform acts under group pressure that he would not have performed in the absence of social inducement."[10] The subject arrives with three others who (unknown to the subject) are confederates of the experimenter. The experimenter explains that he is studying the effects of punishment on learning. The Learner (one of the confederates) is strapped into an electric chair, mentioning in passing that he has a mild heart condition. The subject and the other two confederates are Teachers. They sit in front of a panel with switches with labels ranging from "Slight Shock" to "Danger: Severe Shock" and are told that they must shock the Learner whenever he fails in a memory learning task. How badly the Learner is shocked is up to the Teachers and will be the *lowest* level suggested by any Teacher on that trial. All is rigged. The Learner answers incorrectly on 30 of the 40 trials. The two Teacher confederates call for a higher shock level at each failure. As the shocks increase, the Learner protests, asks to be let out, shouts, complains of heart trouble, and finally screams in pain. (The "shocks" are phony and the Learner is an actor, but the subject doesn't know this.) What will the subject do? He can keep the shock at the lowest level, but the two Teacher stooges are pressuring him to torture the Learner more for each failure.
>
> What the subject most often does is give in to the pressure. "While the experiment yields wide variation in performance, a substantial number of subjects submitted readily to pressure applied to them by the confederates." Milgram noted that in questioning after the experiment, the subjects often admit that they acted against their own principles and are upset by what they did. "The subject was then dehoaxed carefully and had a friendly reconciliation with the victim."[11]

While this experiment offends many people, we must agree that the potential harm to the subjects is not so clear as in some medical experiments. That sick feeling of knowing that you were tricked into acts you consider despicable doesn't put Dr. Milgram in the class of the medical researchers of the early 1960s who injected live cancer cells into patients

"I'm doing a little study on the effects of emotional stress. Now, just take the axe from my assistant."

to study the possibility that cancer is infectious. But then neither is knowledge of the behavior of individuals under group pressure so clearly valuable as knowledge of whether or not cancer can be infectious.

Social scientists, like medical researchers, have steadily tightened their standards of conduct. A report on a 1979 conference on ethics in social research said of Milgram's work: "A decade ago it was hailed as a brilliant if disturbing experiment. Now it is regarded as raising serious ethical questions; the dominant view is that to conduct such a study is wrong."[12] Universities, hospitals, and other research institutions now have review committees that screen all proposed research with human subjects with a view to protecting subjects. Experiments with possible emotional effects much less severe than those in Milgram's studies are regularly vetoed by review committees.

The committee system enforces the principles of no harm to subjects and informed consent. This enforcement mechanism is so important that the institutional review committee deserves to be called a third accepted ethical principle. The review committee's job is not easy. Experiments in the social sciences include treatments in the gray area where "possible emotional harm" shades into "attacks human dignity" or merely into the slippery category of "in bad taste." Solomon himself could not make decisions acceptable to all in so foggy a situation.

In addition to the question of what constitutes "possible harm to the subject," Milgram's work illustrates another problem common to studies of human behavior. The degree of informed consent required by medical ethics is simply not possible. If the subjects were aware of the true purpose of the experiment, the experiment would not be valid. So subjects are asked to consent to participation without being given a detailed description of the experiment and its purpose. Do such experiments amount to unjustified manipulation of the subjects, bringing out aspects of their behavior that they would prefer not to reveal to scrutiny? Or are the experiments justified by the knowledge gained, by uninformed consent, and by the fact that the treatments are not harmful? Some of the exercises afford an opportunity for you to make this judgment in typical cases. Here is an extreme case.

> **Example 21.** Police departments in major cities receive large numbers of domestic violence calls, reporting violence or threats of violence within a household. In the past, police officers calmed the situation and often told the offender to leave the house for the night, but rarely made an arrest unless a weapon or other circumstances demanded it. Women's groups argued that beating a wife or girl friend should lead to arrest, but police doubted the value of making an arrest when the victim was unwilling to press charges. Will arrest reduce future offenses? That's a question that experiments have tried to answer.
>
> A typical domestic violence experiment compares two treatments: arrest the suspect and hold him overnight, or warn the suspect and release him. When police officers reach the scene of a domestic violence call, they calm the participants and investigate. Weapons or death threats require an arrest. If an arrest is warranted by the facts but not legally required, an officer radios headquarters for direction. The person on duty opens the next in a file of envelopes, prepared in advance by a statistician, that contain the treatments in random order. If the card in the envelope says "arrest," the suspect is arrested. If it says "warn and release," that is done. Then police records are watched and the victim is visited by case workers to see if the domestic violence has reoccurred.

Such experiments appear to show that arresting domestic violence suspects does indeed reduce their future violent behavior. As a result of this evidence, arrest has become the standard police response to domestic violence in many cities.

Domestic violence experiments address an important issue and have helped change long-standing policy. But they are on the edge of what is ethically acceptable. The abuser and the abused do not know that they are subjects in an experiment. The police response to their situation is decided at random and without their knowledge. The ethical rules that govern medical trials and most social science experiments would prevent the domestic violence experiments and with them the knowledge that arrest is effective. Are these studies ethically justified?

Experimentation is necessary for the advance of knowledge, and, at least with medical trials, few would deny that the knowledge gained serves the general good. Nevertheless, experimenters are often tempted to place their search for knowledge ahead of the welfare of their subjects. It is fortunate that interest in the welfare of subjects has greatly increased in recent years. The three principles I have emphasized (no deliberate harm, informed consent, and review in advance by an independent committee), although often difficult to apply, provide a framework for protecting subjects while allowing the power of statistically designed experiments to aid the growth of knowledge.

Section 6 exercises

There are no "right" answers to these exercises. Thoughtful persons are found on both sides of the dilemmas presented. I wish mainly to invite you to think about the ethics of experimentation.

2.53. *(Placebos)* There is at present no vaccine for a serious viral disease. A vaccine is developed and appears effective in animal trials. Only a comparative experiment with human subjects in which a control group receives a placebo can determine the true worth of the vaccine. Is it ethical to give some subjects the placebo, which cannot protect them against the disease? (See Paul Meier, "The Biggest Public Health Experiment Ever" in *Statistics: A Guide to the Unknown* for a discussion of the experiment that established the effectiveness of the Salk polio vaccine, a case that fits this exercise.)

2.54. *(Only the best treatment?)* The standard surgical treatment for breast cancer for many years was radical mastectomy, an operation to remove the infected breast, underlying chest muscles, and lymph nodes in the armpits. This operation is disfiguring, so that many women prefer a simple mastectomy (removal of the breast only) or even removal of the tumor only. These operations cannot be more effective than a radical mastectomy and may be less effective, because in removing less tissue they may leave cancer cells behind. The effectiveness of these operations can be compared only by a clinical trial. But is it ethical to assign women randomly to a treatment that cannot be more effective than the radical surgery? (In fact, randomized trials were carried out and found no significant differences in survival a decade after either a radical mastectomy or a simple mastectomy.)

2.55. *(Deception of subjects)* A psychologist conducts the following experiment: A team of subjects plays a game of skill against a computer for money rewards. Unknown to the subjects, one team member is a stooge whose stupidity causes the team to lose regularly. The experimenter observes the subjects through one-way glass. Her intent is to study the behavior of the subjects toward the stupid team member.

This experiment involves no risk to the subjects and is intended simply to create the kind of situation that might occur in any pickup basketball game. To create the situation, the subjects are deceived. Is this deception morally objectionable? Explain your position.

2.56. *(Enticement of subjects)* A psychologist conducted the following experiment: She measured the attitude of subjects toward cheating, then had them play a game rigged so that winning without cheating was impossible. Unknown to them, the subjects were watched through one-way glass, and a record was kept of who cheated and who did not. Then a second test of attitude was given.

Subjects who cheated tended to change their attitudes to find cheating more acceptable. Those who resisted the temptation to cheat tended to condemn cheating more strongly on the second test of attitude. These results confirmed the psychologist's theory.

Unlike the experiment of Exercise 2.55, this experiment entices subjects to engage in behavior (cheating) that probably contradicts their own standards of behavior. And the subjects are led to believe that they can cheat secretly when in fact they are observed. Is this experiment morally objectionable? Explain your position.

2.57. *(Morality, good taste, and public money)* In 1976 the House of Representatives deleted from an appropriations bill funding for an experiment to study the effect of marijuana on sexual response.

Like all government-supported research, the proposed study had been reviewed by a panel of scientists, both for scientific value and for risk to the subjects. *Science* [Volume 192 (1976), p. 1086] reported,

> *Dr. Harris B. Rubin and his colleagues at the Southern Illinois Medical School proposed to exhibit pornographic films to people who had smoked marihuana and to measure the response with sensors attached to the penis.*
>
> *Marihuana, sex, pornographic films — all in one package, priced at $120,000. The senators smothered the hot potato with a ketchup of colorful oratory and mixed metaphors.*
>
> *"I am firmly convinced we can do without this combination of red ink, 'blue' movies, and Acapulco 'gold,'" Senator John McClellan of Arkansas opined in a persiflage of purple prose. . . .*
>
> *The research community is up in arms because of political interference with the integrity of the peer review process.*

Two questions arise here:
- **(a)** Assume that no physical or psychological harm can come to the volunteer subjects. I might still object to the experiment on grounds of decency or good taste. If you were a member of a review panel, would you veto the experiment on such grounds? Explain.
- **(b)** Suppose we concede that any legal experiment with volunteer subjects should be permitted in a free society. It is a further step to say that any such experiment is entitled to government funding if the usual review procedure finds it scientifically worthwhile. If you were a member of Congress, would you ever refuse to pay for an experiment on grounds of decency or good taste?

2.58. (*Informed consent*) The information given to potential subjects in a clinical trial before asking them to decide whether or not to participate might include
- **(a)** The basic statement that an experiment is being conducted; that is, something beyond simply treating your medical problem occurs in your therapy.
- **(b)** A statement of any potential risks from any of the experimental treatments.
- **(c)** An explanation that random assignment will be used to decide which treatment you get.
- **(d)** An explanation that one "treatment" is a placebo and a

statement of the probability that you will receive the placebo.

Do you feel that all of this information is ethically required? Discuss.

2.59. (*Informed consent*) The subjects in the New Jersey Income-Maintenance Experiment (Example 16 in Section 5) were given a complete explanation of the purpose of the study and of the workings of the treatment to which they were assigned. They were not told that there were other treatments that would have paid them more (or less), nor that the luck of randomization had determined the income they would receive. Do you agree or disagree that the information given is adequate for informed consent?

2.60. (*Prison experiments*) The decision to ban medical experiments on federal prisoners followed the uncovering of experiments in the 1960s that exposed prisoners to serious harm. But experiments such as the vitamin C test of Example 7 in Section 2 are also banned from federal prisons. Is it necessary to ban experiments in which all treatments appear harmless because of the difficulty of obtaining truly voluntary consent in a prison? What is your overall opinion of this ban on experimentation?

2.61. (*Dependent subjects*) Students in PSY 101 are required to serve as experimental subjects. Students in PSY 102 are not required to serve, but they are given extra credit if they do so. Students in PSY 103 are required either to sign up as subjects or to write a term paper.

Do you object to any of these policies? Which ones, and why?

2.62. (*Popular pressure*) The substance Laetrile received abundant publicity in 1977 as a treatment for cancer. Like hundreds of thousands of other chemical compounds, Laetrile had earlier been tested to see if it showed antitumor activity in animals. It didn't. Because no known drug fights cancer in people but not in animals, the medical community branded Laetrile as worthless. Advocates of Laetrile wanted the FDA to conduct a clinical trial on human cancer patients.

It is usually considered unethical to use a drug on people without some promise based on animal trials that it is safe and effective. What is more, Laetrile may have toxic side effects. Do the popular interest in Laetrile and the fervor of its advocates justify a clinical trial?

2.63. (*Men versus women*) Almost all clinical trials that have studied the effects of such factors as blood cholesterol, taking aspirin, or exercise on heart attacks have used middle-aged male subjects.

Women's groups have complained that this leads to better health information about men than about women. The researchers reply that in order to get clear results in the five years or so that such a study lasts, they must choose their subjects from the groups that are most likely to have heart attacks. That points to middle-aged men. What would you suggest?

2.64. (*Equal treatment*) A group of researchers on aging proposed to investigate the effect of supplemental health services on the quality of life of older persons. Eligible patients on the rolls of a large medical clinic were to be randomly assigned to treatment and control groups. The treatment group would be offered hearing aids, dentures, transportation, and other services not available without charge to the control group. A review committee felt that providing these services to some but not other persons in the same institution raised ethical questions. Do you agree?

2.65. (*Animal welfare*) Many people are concerned about the ethics of experimentation with living animals. Some go so far as to regard any animal experiments as unethical, regardless of the benefits to human beings. Briefly discuss each of the following examples:

(a) Military doctors use goats that have been deliberately shot (while completely anesthetized) to study and teach the treatment of combat wounds. Assume that there is no equally effective way to prepare doctors to treat human wounds.

(b) Several states are considering legislation that would end the practice of using cats and dogs from pounds in medical research. Instead, the animals will be killed at the pounds.

(c) The cancer-causing potential of chemicals is assessed by exposing lab rats to high concentrations. The rats are bred for this specific purpose. (Would your opinion differ if dogs or monkeys were used?)

NOTES

1. Read "Is Vitamin C Really Good for Colds?" in *Consumer Reports*, February 1976, pp. 68–70, for a discussion of this conclusion. The article also reviews the need for controlled experiments and the Toronto study.

2. Jerome Cornfield, "The University Group Diabetes Program," *Journal of the American Medical Association*, Volume 217 (1971), pp. 1676–1687.

3. The saccharin question is discussed in detail in the "News and Comment" section of *Science*, Volume 196 (1977), pp. 1179–1183, and Volume 208 (1980), pp. 154–156. Details of the data from the rat experiment that led to FDA action appear in a letter to the editor in *Science*, Volume 197 (1977), p. 320.

4. Martin Gardner, "Supergull," *New York Review of Books*, March 17, 1977.

This review of two books on psychic researchers illustrates nicely the naive experiments and vague and deceptive accounts that pervade this field.

5. From Philip M. Boffey, "Color Additives: Botched Experiment Leads to Banning of Red Dye No. 2," *Science*, Volume 191 (1976), p. 450.

6. For a full account of this social experiment, see David Kershaw and Jerilyn Fair, *The New Jersey Income-Maintenance Experiment, Volume I* (New York: Academic Press, 1976).

7. See Daniel Friedlander et al., *Maryland: Final Report on the Employment Initiatives Evaluation* (New York: Manpower Demonstration Research Corporation, 1985) and Daniel Friedlander, *Supplemental Report on the Baltimore Options Program* (New York: Manpower Demonstration Research Corporation, 1987).

8. See Earnest M. Barsamian, "The Rise and Fall of Internal Mammary Artery Ligation," in J. P. Bunker, B. A. Barnes, and F. Mosteller (eds.), *Costs, Risks, and Benefits of Surgery* (New York: Oxford Univ. Press, 1977), pp. 212–220.

9. More detailed accounts of the AIDS drug trials issues through the fall of 1989 appear in Eliot Marshall, "Quick Release of AIDS Drugs," *Science*, Volume 245 (1989), pp. 345–347 and Joseph Palca, "AIDS Drug Trials Enter New Age," *Science*, Volume 246 (1989), pp. 19–21.

10. Stanley Milgram, "Group Pressure and Action Against a Person," *Journal of Abnormal and Social Psychology*, Volume 69 (1964), pp. 137–143.

11. Postexperimental attitudes and the final quotation are in Stanley Milgram, "Liberating Effects of Group Pressure," *Journal of Personality and Social Psychology*, Volume 1 (1965), pp. 127–134.

12. Quoted from Constance Holden, "Ethics in Social Science Research," *Science*, Volume 206 (1979), pp. 537–540. A book-length discussion is Arthur G. Miller, *The Obedience Experiments: A Case Study of Controversy in Social Science* (New York: Praeger, 1986).

Review exercises

2.66. To compare surgical and nonsurgical treatments of intestinal cancer, researchers examined the records of a large number of patients. Patients who received surgery survived much longer (on the average) than patients who were treated without surgery. The study concludes that surgery is more effective than nonsurgical treatment.

    (a) What are the explanatory and response variables in this study?

    (b) Is this study an experiment? Why or why not?

    (c) Discussion with medical experts reveals that surgery is reserved for relatively healthy patients; patients who are too ill to tolerate surgery receive nonsurgical treatment. Explain why the conclusion of the study is not supported by the data.

    (d) Outline the design of a randomized comparative experiment to compare the effectiveness of surgical and nonsurgical treatment for intestinal cancer.

2.67. You are in charge of a writing course for college freshmen with 1200 students enrolled. You wish to learn if the students write better essays when required to use computer word processing than when essays are written and revised by hand. Describe the design of an experiment to determine whether word processing results in better essays. Be sure to include a description of the treatments and the response variable or variables.

2.68. In a study of the relationship between physical fitness and personality, middle-aged college faculty who have volunteered for an exercise program are divided into low-fitness and high-fitness groups on the basis of a physical examination. All subjects then take the Cattell Sixteen Personality Factors Questionnaire, and the results for the two groups are compared. Is this study an experiment? Explain your answer.

2.69. A study of the personality effects of running involved 231 male runners who ran about 20 miles each week. The runners were given the Cattell Sixteen Personality Factors Questionnaire, a 187-item multiple-choice test often used by psychologists. A news report (*New York Times*, February 15, 1988) said, "The researchers found statistically significant personality differences between the runners and the 30-year-old male population as a whole." A headline on the article said, "Research has shown that running can alter one's moods." Explain carefully, to someone who knows no statistics, why the headline is misleading.

2.70. The Associated Press article on the following page appeared in the *New York Times* of October 16, 1975. Describe in detail how you think the experiment was designed. Does the news report omit any important information about the design? Are there any inconsistencies in the details given?

2.71. You are participating in the design of a medical experiment to investigate whether a calcium supplement in the diet will reduce the blood pressure of middle-aged men. Preliminary work suggests that calcium may be effective and that the effect may be greater for black men than for white men.

    (a) A block design should be used in this experiment. How are the blocks formed?

    (b) Outline in graphic form the design of an appropriate experiment.

2.72. The design of controls and instruments has a large effect on how easily people can use them. A student investigates this effect by

Marijuana Is Called An Effective Relief In Cancer Therapy

BOSTON, Oct. 15 (AP)—Marijuana is far more effective than any other drug in relieving the vomiting and nausea that plagues thousands of cancer patients undergoing chemical therapy, researchers say.

Clinical trials with cancer patients using a marijuana derivative have been so successful that the drug should be considered seriously as a treatment for the chemical therapy side effects, they add.

In a report to be published tomorrow in the New England Journal of Medicine, Harvard Medical School researchers at the Sidney Farber Cancer Center say they tested the effectiveness of the marijuana drug against a dummy drug in 22 patients with a variety of cancers.

For patients who completed the study, 12 of 15 cases involving marijuana drug treatments resulted in at least a 50 per cent reduction in vomiting and nausea after cancer therapy. And in five of these treatments, the patients suffered no nausea at all, the report added.

There was no decrease in nausea or vomiting in 14 cases in which placebo, or dummy treatment, was used, the researchers said. In the "double-blind" experiment, neither patients nor doctors knew in advance who got the real or dummy drugs.

Dr. Stephen E. Salan said in an interview that about 75 per cent of the thousands of patients getting chemotheraphy for cancer suffered moderate to extreme nausea and vomiting.

And of this group, 90 per cent get no relief from conventional antinausea drugs, he added.

Dr. Salan said he and his colleagues in the study, Dr. Norman E. Zinberg and Dr. Emil Frei 30, did not know specifically why marijauna worked to decrease nausea.

asking right-handed students to turn a knob (with their right hands) that moves an indicator by screw action. There are two identical instruments, one with a right-hand thread (the knob turns clockwise) and the other with a left-hand tread (the knob must be turned counterclockwise). The response variable is the time required (in seconds) to move the indicator a fixed distance. Thirty right-handed students are available to serve as subjects. You have been asked to lay out the statistical design of this experiment. Describe your design, and carry out any randomization that is required.

Chapter 3

Measurement

Professor Nous is interested in how much the intelligence of a child is influenced by the environment in which the child is raised. Children raised in the same environment still differ in intelligence because of heredity and other factors, so a comparative study is needed.

Professor Nous daydreams about randomized comparative experiments in which newborn babies are snatched from their mothers and assigned to various environments. Alas, in this instance she must be content with an observational study. A brilliant idea strikes: Why not look at the variation in intelligence between identical twins who for some reason have been raised in separate homes since birth? Because identical twins have the same heredity, differences in their intelligence must reflect the effect of environment. A careful search turns up 20 pairs of separated twins. Professor Nous feels that her troubles are over.

Not so. Once the units for study (the twins) are chosen, it remains to collect data by measuring the properties of each unit that interest the good professor. In particular, she must measure the intelligence of each child. Because intelligence, unlike sex or age, is not directly observable, Professor Nous is faced with a serious problem.

In Chapters 1 and 2, we thought about designing a study by choosing or assigning the units of study. Professor Nous has successfully completed this task. Now we must think about measurement, for the final step in collecting data is to measure some property of the units.

To *measure* a property means to assign numbers to units as a way of representing that property.

It is measurement that actually produces data. You always thought that statistics concerns numbers, but we have not paid much attention to numbers until this chapter. The real point of thinking about measurement is to help you become more comfortable with numbers. We will ask when numbers are meaningful (Section 1), inquire about their accuracy (Section 2), and learn that not all numbers are equally informative (Section 3). All you have learned will be applied, in Section 4, to the art of looking at numbers intelligently.

1. First steps in measurement: Validity

Measurement requires three things: An *object* to be measured, a well-defined *property* of the object to measure, and a *measuring instrument* that actually does the job of assigning numbers to represent the property. If you wish to measure the length of your bed, the object (the bed) is at hand, the property (length) is clearly understood, and you might use a tape measure as your measuring instrument. The basic ideas of measurement are clearest for measurements of physical properties such as length because we understand and agree on what these properties are. The conceptual difficulties of measurement are greatest when the property to be measured is vague and disputed. To measure intelligence, we must have a clear concept of what we mean by "intelligence." It is in fact not at all clear what intelligence is, so that any attempt to measure a person's intelligence will be controversial.

In practice, we cannot neatly separate discussion of the property to be measured and the instrument that measures it. To measure the length of a bed, we stretch out a tape measure marked off in inches or centimeters; the length of the bed is the length of the tape stretched out beside it. This somewhat circular reasoning doesn't bother us because we can see length and are happy to use lengths along the tape measure as a standard for measuring other lengths. We can't see intelligence. The instrument in this case is an IQ test, a long multiple-choice test designed by psychologists. A discussion of measuring intelligence involves asking just what IQ tests measure.

In Chapters 1 and 2, we defined a *variable* as a measured characteristic of a unit. To be useful, a variable must be exactly defined. Both length and IQ are exactly defined because we know what instruments were used to measure these properties. To use data intelligently, you must know the definitions of the variables whose values are reported. What, for example, does it mean when the Bureau of Labor Statistics (BLS) announces that last month's unemployment rate was 7.3%? Here is the official BLS definition of what it means to be unemployed, quoted from the Bureau's monthly news release.

> *People are classified as* employed *if they did any work at all as paid civilians; worked in their own business or profession or on their own farm; or worked 15 hours or more in an enterprise operated by a member of their family, whether they were paid or not. People are also counted as employed if they were on unpaid leave because of illness, bad weather, disputes between labor and management, or personal reasons. Members of the Armed Forces stationed in the United States are also included in the employed total.*

> *People are classified as* unemployed, *regardless of their eligibility for unemployment benefits or public assistance, if they meet all of the following criteria: They had no employment during the survey week; they were available for work at that time; and they made special efforts to find employment sometime during the prior 4 weeks. Persons laid off from their former jobs and awaiting recall and those expecting to report to a job within 30 days need not be looking for work to be counted as unemployed.*

"Unemployed? Not me, I'm out of the labor force."

The labor force, says the BLS, consists of all people at least 16 years of age and not in an institution who are either employed or unemployed. If you aren't working and also are not available for work and looking for work, then you aren't in the labor force. The monthly unemployment rate is the percent of the labor force that is unemployed. It is of course a sample statistic, based on the Current Population Survey, a monthly probability sample of about 55,800 households conducted by the Census Bureau. The measuring instrument that enables the BLS to classify each person as employed, unemployed, or not in the labor force consists of a lengthy set of questions asked in person by a trained interviewer.

Definitions, such as the definition of what it means to be unemployed, are important to our understanding of data. Having read the BLS definition, we now know more precisely what a 7.3% employment rate means. Changing the definition would change the numbers. The Conservative government of Prime Minister Margaret Thatcher in Britain changed the definition of the unemployment rate 18 times between 1979 and 1989. Seventeen of these 18 changes had the effect of lowering the reported unemployment rate, which of course is just what any government would like to see.

What about our insistence that measurement is always an operation that produces a number? My height, measured by a tape measure, is 180 centimeters. Dr. Nous might decide to measure intelligence by the Wechsler Adult Intelligence Scale, the most common IQ test. I won't tell you my IQ, but it's a number determined by that specific process. My employment status is measured by classifying me according to the BLS definitions. I'm employed. That's not a number, you say. True enough, but when I run my employment status through a computer, the computer assigns a numerical code to each employment status. We, too, can do that. Let's use 0 for "not in the labor force," 1 for "unemployed," and 2 for "employed." Now my employment status is 2. Any measurement *must* result in a number.

It is possible to talk about classifying or categorizing separately from measurement because classifying persons by sex or race or employment status does not naturally produce numbers. I prefer to lump classifying in with measurement by insisting that we assign numbers to the classes. As we saw with employment status, that's no hardship. In Chapters 1 and 2, variables could have values such as "agree" or "disagree" (as when an opinion poll asks your reaction to a statement). Now we insist that variables have numerical values.

Validity

Statistics deals with variables, that is, with numbers resulting from measurement. Beware of the easy passage from a property of units to a variable that claims to represent the property. The variable must be clearly and exactly defined. The property may be vague, inexact, and not directly observable. In this situation, the variable is rarely a complete representation of the property.

> **Example 1.** The BLS definition of "unemployed" may not agree in detail with your vague idea of what it means to be unemployed. Labor says this variable understates unemployment by insisting that only persons actively seeking work can be unemployed. If you are without a job and so discouraged that you stop looking for one, you are out of the labor force and therefore no longer unemployed! On the other hand, management argues that the BLS overstates unemployment because some persons who are looking for work may refuse to accept a job unless it is exactly right for them. Many teenagers and people whose spouses have steady jobs, secure because the household already has one income, shop around for the right job. By the BLS definition, these persons are unemployed. (I pass over in silence the claim of the National Association of Manufacturers, recorded on the front page of the *Wall Street Journal* on September 13, 1976, that the official unemployment rate is inflated because "criminal elements" are included in the labor force.)

Though both management and labor would prefer slightly different definitions of "unemployed," neither accuses the BLS of having a completely inappropriate definition. The official unemployment rate is a useful indicator of the state of the economy. That is, the BLS produces a *valid* measure of employment and unemployment.

A variable is a *valid* measure of a property if it is relevant and appropriate as a representation of that property.

Validity of a measurement process is a simple but slippery idea. Does the process measure what you want it to? That is the question of validity. If I measure your height in inches and record you as employed if your height exceeds 65 inches, that is an invalid measure of your employment status. The BLS uses a valid measure — not a perfect measure of employment (whatever that might mean), not the whole story about employment status, not the only variable that might measure employment status, but an appropriate and relevant variable.

It is easy for persons who are not experienced with numbers to fall into the trap of using an invalid measure. A common case is the use of absolute numbers when *rates or proportions* are appropriate. Here is an example.

> **Example 2.** If customers returned 36 coats to Sears and only 12 to La Boutique Classique next door, this does not mean that Sears' customers were less satisfied. Sears sold 1100 coats that season, while La Boutique sold 200. So Sears' return rate was
>
> $$\frac{36}{1100} = 0.033, \text{ or } 3.3\%,$$
>
> while the return rate at La Boutique was
>
> $$\frac{12}{200} = 0.06, \text{ or } 6\%.$$
>
> This return *rate*, or percentage of coats returned, is a more valid measure of dissatisfaction than the *number* of returns.

Validity for what?

It is easy to decide whether a variable is a valid measure of a property when we understand the property well. That's true of physical properties, such as length. Employment status and customer satisfaction (Examples 1 and 2) are a bit less clear, but we still have a good idea of what we want to measure. But it sometimes happens that the property to be measured is so fuzzy that reasonable people can disagree on the validity of a variable as a measure of that property. This situation is not uncommon in the social and behavioral sciences. Psychologists wish to measure such things as intelligence or authoritarian personality or mathematical aptitude. The variables are typically scores on a test—an IQ test for intelligence, the mathematics part of the Scholastic Aptitude Test (the "college boards") for mathematical aptitude, and so on. The validity of these variables is controversial.

> **Example 3.** The Scholastic Aptitude Tests (SATs) are taken each year by about a million high school seniors seeking admission to colleges and universities. There are two SATs, testing verbal and

mathematical aptitude. Scores range from 200 to 800. These tests have been attacked because black and Hispanic students score considerably lower, on the average, than do whites. Moreover, women do somewhat more poorly than men. Here are the average SAT scores by sex and racial/ethnic group for 1988.[1]

Group	Verbal score	Math score
Men	435	498
Women	422	455
American Indian	393	435
Asian	408	522
Black	353	384
Mexican-American	382	428
Puerto Rican	355	402
Other Hispanic	387	433
White	445	490

As these data suggest, decisions that rely heavily on SAT scores will place women and minorities at a disadvantage. Critics find this objectionable, and some courts have agreed. For example, a federal judge ruled in 1989 that awarding New York state scholarships based on SAT scores discriminates against women. But it is possible that the tests are simply reporting real differences among the groups. Minorities (on the average) come from poorer families and attend weaker schools than whites, and women (on the average) take fewer science and math courses than men. These and other factors might cause differences in average scores on a valid test, that is, a test that really does measure scholastic aptitude. What can we say about the validity of SAT scores?

To discuss the issue sensibly, we must ask validity *for what purpose?* and validity *for what population?* Critics of the SATs, for example, sometimes complain that these tests measure only cognitive ability, ignoring emotional and physical abilities that are just as important. That is much like criticizing a yardstick for not giving the time of day, because the SATs do not claim to measure such other abilities. On the other hand, it is difficult to claim that SATs measure some innate cognitive ability. Scholastic aptitude probably has many components and cannot be measured by two multiple-choice exams. SAT scores can be improved by intensive coaching courses, which surely could not change a student's innate ability. In short, scholastic aptitude is so poorly understood in itself that we cannot say that SAT scores are a valid measure of this vague concept.

What can be said is that SAT scores are valid for predicting academic success in college. Success in college is a clear concept, and there is an observed connection between high SAT scores now and high grades in college next year. This is predictive validity.

A measurement of a property has *predictive validity* if it can be used to predict success on tasks that are related to the property to be measured.

Predictive validity is much less vague than our original definition of validity. It is often the most useful answer to "validity for what?"

There remains the question of "validity for whom?" The most common criticism of SATs is that they are *culturally biased*, that they are valid for the white middle class but not for other groups.

> **Example 4.** Almost all measurements of human behavior have some cultural bias. Even tests of physical ability and perception may not be valid across cultures. A test of spatial perception that asks children to put rectangular pegs through holes may not be valid for children in rural Mali. Houses in this west African nation are round, and rectangular objects such as books and boxes are uncommon in rural areas. The shape of the pegs is therefore unfamiliar to the children.

The charge that the SATs are culturally biased is plausible if the tests are taken to measure scholastic ability. Asians, for example, have the highest math SAT scores of any group, but their verbal scores are less than the white average. This may reflect the fact that many Asians are not native speakers of English. Similarly, the family income of blacks taking the SATs is less than half that of whites. The lower SAT scores of blacks are certainly influenced by the great disparity between the social and economic status of blacks and whites in America.

But when the SATs are considered as measures not of ability but of future academic success, they appear to be innocent of bias. SAT scores predict future grades for minorities at least as well as for whites. Success in college depends on exposure to middle-class education and habits as well as on innate ability. White, middle-class students have an advantage on the SATs, but they have the same advantage in college.

Much of the controversy over standardized tests such as the SATs arises because some users ignore the limitations on the validity of the tests. The SATs are valid predictors of future academic success, but they are quite imperfect predictors. Many factors not measured by such tests also influence a student's success in college. It is not appropriate to base college admissions decisions on test scores alone. It *is* appropriate to regard low SAT scores as a danger signal, to look for other abilities that will help the student overcome them, or even to plan special compensatory training for such students. Like a sharp knife, a standardized test can be used wisely by the wise or foolishly by the foolish. Those who call for an end to such tests finally base their case on the judgment that the foolish so outnumber the wise that sharp knives ought to be confiscated.[2]

The example of standardized tests illustrates the care needed in discussing the validity of measures of fuzzy concepts. It also illustrates the way statisticians deal with this problem: They try to show a connection between the measuring variable and other variables that ought to be connected with the original fuzzy idea. Persons of high mental ability ought to do better in school than those of low ability. Persons with high IQ or SAT scores do tend to do better in school than those with low scores. The final step is to replace the vague idea with the precise variable that seems to be connected with the same behavior. That is, for statistical purposes, *the best way to define a property is to give a rule for measuring it*. The saying that "intelligence is whatever it is that IQ measures" is an example of such a definition. IQ measures some combination of innate ability, learned knowledge, test-taking skills, and exposure to mainstream culture. This is a clearly defined property of people because it is established by the Wechsler Adult Intelligence Scale. It is an interesting variable because it is related to success in school and other interesting response variables. But IQ is not the same as our everyday idea of intelligence. Our old friend Professor Nous should probably decide to use her sample of separated twins to study the effect of environment on IQ. To study intelligence is beyond her reach.

Statistics can deal only with measured properties. Intelligence and maturity cannot be studied statistically, though variables such as IQ can be. Beware of the arrogance that says that everything can be measured or that only things we can measure are important. The world contains much that is beyond the grasp of statistics.

Section 1 exercises

3.1. You are studying the relationship between political attitudes and length of hair among male students. You will measure political attitudes with a standard questionnaire. How will you measure length of hair? Give precise instructions that an assistant could follow. Include a description of the measuring instrument that your assistant is to use.

3.2. One of the preliminary difficulties facing the New Jersey Income-Maintenance Experiment comparing alternative welfare systems (Example 16 in Section 5 of Chapter 2) was to give a definition of the income of a family. This was important because initial family income was used to form blocks and because the amount of welfare payments depended on the family's income. Write an exact definition of income for this purpose. A short essay may be needed. For example, will you include nonmoney income, such as the value of food stamps or of subsidized housing? Will you allow deductions for the cost of child care needed to permit the parent to work?

3.3. You want to measure the physical fitness of college students. Give an example of a clearly invalid way to measure fitness. Then briefly describe a measurement process that you think is valid.

3.4. We understand "intelligence" to mean something like general problem-solving ability. Why is it *not* valid to measure intelligence by a test that asks questions such as

Who wrote *Romeo and Juliet*?

or

Who won the last soccer World Cup?

3.5. The number of persons killed in bicycle accidents rose from about 600 in 1967 to over 1100 in 1973. Does this indicate that bicycle riding was becoming much less safe during these years? Let us put this question more specifically.

    (a) Is the total number of fatalities per year a valid measure of the danger of bicycle riding? Why or why not?

    (b) If you question the validity of total deaths as a measure of danger, suggest a variable that is a more valid measure. Explain your suggestion.

3.6. A teenager argues that young people are safer drivers than the elderly. He cites government data showing that in 1988, 5375 drivers at least 65 years of age were involved in fatal accidents. In contrast, only 3225 drivers aged 16 and 17 had a fatal accident. You suspect that there are more elderly drivers than teenage drivers. In fact, the *Statistical Abstract of the United States* tells you that there were about 4,304,000 drivers aged 16 and 17 in 1988, and about 20,417,000 drivers aged 65 and over. Write a short rebuttal of the teenager's claim, using this information to calculate a more valid measure of involvement in fatal accidents.

3.7. Congress wants the medical establishment to show that progress is being made in fighting cancer. Some variables that might be used are:

    (a) Total deaths from cancer. (These have risen sharply over time, from 331,000 in 1970 to 488,000 in 1988.)

    (b) Death rates—the percent of all Americans who die from cancer. (These death rates are rising steadily, from 17.2% in 1970 to 22.5% in 1988.)

    (c) Survival rates among cancer patients—the percent of cancer patients who survive for five years from the time the disease was discovered. (These rates are rising slowly for all cancer, though the survival rate has improved greatly for a few kinds of cancer.)

Discuss the validity of each of these variables as a measure of the effectiveness of cancer treatment.

3.8. "Domestic automakers led foreign manufacturers in the number of vehicles recalled in the United States because of safety problems in the first half of 1983, the National Highway Traffic Safety Administration says." That is the lead sentence in an Associated Press dispatch, which my local paper (*Lafayette Journal and Courier*, July 3, 1983) headlined "U.S. AUTOMAKERS RECALL MORE THAN COMPETITION." The article goes on to state that domestic companies recalled 2,047,400 cars and trucks, while foreign manufacturers recalled 903,100. Both the article and the headline suggest that domestic vehicles had a more serious safety-recall problem than imports. This is quite wrong: The imports suffered worse. Explain how an invalid measure led to this incorrect conclusion.

3.9. You wish to study the effect of violent television programs on antisocial behavior in children.

    (a) Define (that is, tell how to measure) the explanatory and response variables. You have many possible choices of variables; just be sure that the ones you choose are valid and clearly defined.

    (b) Do you think that an experiment is practically and morally possible? If so, briefly describe the design of an experiment for this study.

    (c) If you were unable or unwilling to do an experiment, briefly discuss the design of a sample study. Will confounding with other variables threaten the validity of your conclusions about the effect of TV violence on child behavior?

3.10. The law requires that tests given to job applicants must be shown to be directly job-related. The Labor Department believes that an employment test called the General Aptitude Test Battery (GATB) is valid for a broad range of jobs. As in the case of the SATs, blacks and Hispanics get lower average scores on the GATB than do whites. Describe briefly what must be done to establish that the GATB has predictive validity as a measure of future performance on the job.

2.   Accuracy in measurement

Even the most exact laboratory measurements are not perfectly accurate. Physicists and chemists, who make little use of statistical ideas in data collection, use statistics heavily in analyzing errors in their measurements. When a measurement is made as accurately as the instrument allows, repeated measurements do not always give the same result. (Try Exercises 3.11 and 3.12 to see that for yourself.) Accuracy in measurement has two aspects: small bias and high reliability.

A measurement process is *unbiased* if it does not systematically overstate or understate the true value of the variable.

A measurement process is *reliable* if repeated measurements on the same unit give the same (or approximately the same) results.

Unbiasness and reliability in measurement carry the same meanings as do unbiasedness and precision in data collection. (In fact, it is common in the physical sciences and engineering to use the word "precision" instead of "reliability" to describe the repeatability of the results of measurement.) "Unbiased" means correct on the average; "reliable" means repeatable. The difference is that these meanings now apply to the process of measuring a property of a unit, not to the process of choosing units to be measured. If an egg scale always weighs 10 grams heavy, it is biased. If the scale has dust in its pivot and gives widely different weights when the same egg is weighed several times, it is unreliable.

The analogy between repeated measurement and repeated sampling extends further. The results of repeated measurement are *random* in the same sense that the results of repeated sampling are random. That is, individual results vary, but there is a definite *distribution* of results when many trials are made. The average of several measurements is less variable (more reliable) than a single measurement, just as a sample statistic becomes more precise in larger samples. That is why laboratory instructors in physics or chemistry suggest that you repeat your measurements several times and use the average value. The National Geodetic Survey requires field crews gathering data for its very accurate maps to repeat angle measurements 32 times. It is a pleasing instance of the harmony of thought that measurement error and sampling error can be studied and overcome by the same ideas.

> **Example 5.** If you want to know how much something weighs, you use a scale. If an analytical chemist wants to know how much a specimen weighs, he uses a better scale. His scale is less biased and more reliable than yours, and he calibrates it regularly by weighing a standard weight whose mass is accurately known. The mass of standard weights is accurately known because they have been compared with superstandard weights kept by the National Institute of Standards and Technology (NIST) in Washington, D.C. And the mass of *these* weights is known very accurately because they have been compared with the International Prototype Kilogram, which lives in a guarded vault in Paris and indirectly determines all weights in the world.
>
> The NIST standard weight called NB 10, for example, weighs about 9.9996 grams. (It is supposed to weigh 10 grams but is light by about the mass of a grain of salt.) Because the NIST knows that the results of repeated measurements are random, it repeatedly weighs standards such as NB 10 to determine the reliability of the weighing process. Then, when the analytical chemist sends in his standard weight for calibration, the NIST can tell him the reliability of its answer, just as the Gallup Poll can state the precision of its conclusions. Here are the results of 11 determinations of the mass (in grams) of NB 10 made with all the care the NIST could apply:[3]

9.9995992	9.9995985
9.9995947	9.9996008
9.9995978	9.9996027
9.9995925	9.9995929
9.9996006	9.9995988
9.9996014	

> You see that the 11 measurements do vary. There is no such thing as an absolutely accurate measurement. The average (mean) of

These National Institute of Standards and Technology mass standards provide the basis for measuring weight in the United States, whether by a laboratory balance or your bathroom scale. [Photo courtesy of Mass and Volume Section, National Institute of Standards and Technology.]

these measurements is 9.9995982 grams, which is a more reliable estimate of the true mass than a single measurement. In fact, the NIST says it is 95% confident that this average is within ± 0.0000023 gram of the truth. Such reliability statements also apply to unknown masses, such as that of the analytical chemist's standard weight.

I hope that this excursion into the world ruled by the International Prototype Kilogram has reminded you that even physicists and chemists don't get "the right answer" on every measurement. If you have spent part of your youth in laboratories, you need no reminder. I well remember the dark midnight when, after hours of failing to get the same answer twice in an optics lab, I faced the alternative of smashing the wretched interferometer or choosing a major other than physics. Because the interferometer was expensive, I got out of physics.

As usual, things are tougher yet outside the laboratory. Measuring unemployment is also measurement, and the concepts of bias and reliability apply to it just as they do to measuring the mass of NB 10. The sampling process from which the unemployment rate is obtained has a high but not perfect precision. The precision can be clearly stated because probability sampling is used. The measurement of employment status is also not perfectly reliable. The BLS checks the reliability of its measurements by having supervisors reinterview about 5% of the sample. This is repeated measurement on the same unit, just as when NB 10 is weighed several times. It turns out that interviewers and supervisors almost always agree on who is not in the labor force and who has a full-time job. These measurements are extremely reliable. But supervisors and interviewers disagree on the status of about 10% of the "unemployed." The distinctions between unemployed, temporarily laid off, and underemployed are a bit subjective. The measurement of unemployment is therefore somewhat unreliable. To sum up: The precision of sampling refers to the repeatability of sample statistics in *different* samples. The reliability of measurement refers to the repeatability of measured values on the *same* units.

Bias has the same meaning in measuring unemployment as in measuring mass: systematic deviation of the measured result from the "true value" that perfect measurement would produce. The "true value" isn't exactly known even for a mass, but the idea of a biased scale that systematically weighs high or low is clear enough. For unemployment, the "true employment status" of a person might be his status by the BLS definitions when classified by the Commissioner of Labor Statistics herself. Other interviewers will sometimes disagree with the commissioner (the reliability of measurement is not perfect), but there is no bias unless, for example, the interviewers systematically call "unemployed" persons the commissioner would say are "not in the labor force." There appears to be little bias in measuring unemployment.

Take note of one detail: If the definition of unemployment used by the BLS understates the "unemployment problem," this is not bias in measurement. It is instead a question of the validity of the BLS definition as a measure of unemployment. Bias in a measurement process means that the process gives measurements on a variable that are systematically higher or lower than the true value of that variable. Whether the variable is a valid measure of a property such as intelligence or employment status is a different, and harder, question. You might also note that there is a small bias due to nonsampling error in the Current Population Survey that collects the unemployment data. The survey can't locate perhaps 5 or 6% of the population. We guess that these people are more likely to be unemployed than the rest of us, so that the survey slightly underestimates the true unemployment rate. This is again not a question of bias in the measurement process.

Example 6. Data on college enrollments are collected by the federal government's National Center for Education Statistics. The preliminary data for the fall of 1976 showed that the number of first-time students in public universities had grown by 4% over the previous year. But the questionnaire for collecting preliminary data contained confused wording that led many colleges to include new graduate students in the category of first-time students, a category meant to include only students with no previous higher education. The final figures showed that the number of new students in public institutions was *down* 9%, not up 4%.

This is an example of measurement bias. The badly worded form produced values of the variable "number of first-time students" that were systematically higher than the true value.

Reliability of weighings or of employment status classifications is checked by making repeated measurements on the same units. Reliability of IQ tests or the SATs cannot be checked this way because subjects taking a test the second time have an advantage over their first try. Even with different but (one hopes) equivalent forms of the test, this learning effect rules out assessing reliability from many repeated measurements. Behavioral scientists must fall back on less direct and more complicated ways of checking reliability, though the basic ideas we have covered still apply. Here is yet another way in which psychology is a more complicated subject than physics.

Section 2 exercises

3.11. Use a ruler to mark off a piece of stiff paper in inches (mark only full inches, no fractions) about as shown here:

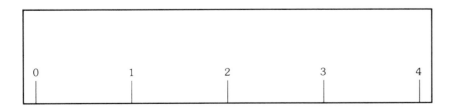

Measure the line below with your instrument and record your answer to a hundredth of an inch (such as 2.23 or 2.39 inches).

To make this measurement, you must estimate what portion of the distance between the 2- and 3-inch marks the line extends. Careful measurements usually involve an uncertainty such as this; we have merely magnified it by using an instrument divided into inches only.

(a) What is the result of your measurement?

(b) Ask four other people to measure the line with your instrument and record their results. Average the five measurements. What margin of error do you think a measurement with your instrument has? (That is, how reliable is it?)

(c) Suppose that someone measures the line by placing its left end at the end of the instrument instead of at the 0 mark. This causes bias. Explain why. Is the reliability of the measurement also affected, or not?

Comment: If you collect the results of the entire class for part (a) and display them in a graph like Figure 1-2 in Chapter 1, you will have a picture of the reliability of this measurement similar to the picture used to present the precision of a sample statistic.

3.12. Take a 1-foot ruler and measure the length of one wall of a room to the nearest inch. Do this five times and record your answers in feet and inches. What is your average result? If possible, now use a tape measure at least as long as the wall to get a more accurate measurement. Did your ruler measurements show bias? (For example, were they almost all too long?)

3.13. All the members of a health class are asked to measure their pulse rate as they sit in the classroom. The students use a variety of methods. Method 1: Count heart beats for 6 seconds and multiply by 10 to get beats per minute. Method 2: Count heart beats for 30 seconds and multiply by 2 to get beats per minute. Which method is more reliable? Why? Is either method clearly more biased than the other? Why?

3.14. One student in the class of the previous exercise proposes a third method: Starting exactly on a heart beat, measure the time needed for 50 beats and convert this time into beats per minute. This method is more accurate than either of the two methods mentioned in the previous exercise. Why?

3.15. A psychologist claims that a standard psychological test for "authoritarian personality" does not really measure an aspect of the subject's personality, but instead measures other factors such as religious beliefs. Is she attacking the *accuracy* or the *validity* of the test? Explain your answer.

3.16. A news article reported a study of preemployment job performance tests and subsequent job performance for 1400 government technicians. Such tests are often accused of being biased against minority groups. A psychologist for the Educational Testing Service commenting on the results of the study said, "Six years later, we found that belief wrong, if you define bias as meaning the scores are unrealistically low in relation to performance on the job." (*New York Times*, July 27, 1973.) Is "bias" in the everyday sense used in this news article the same as "measurement bias" in the technical sense of this chapter? Why or why not?

3.17. Give an example of a measurement process that is valid but has large bias. Then give an example of a measurement process that is invalid but highly reliable.

3.18. In the mid-nineteenth century, craniometry was a respectable scientific endeavor. It was thought that measuring the volume of a human skull would measure the intelligence of the skull's owner. It was difficult to measure a skull's volume accurately, but at last Paul Broca, a professor of surgery, perfected a method of filling the skull with lead shot that gave nearly the same answer in repeated measurements. Alas, cranial capacity has turned out to have no visible relation to achievement or intelligence. Were Broca's measurements of intelligence by skull volume reliable? Were they valid? Explain your answer.

3.19. A friend reads that "Reliability is to the process of measurement as precision is to a sample statistic." She asks you to explain this statement. Give a brief explanation.

3. Scales of measurement

Measurement of a property means assigning a number to represent it. Having designed our data-collection process and stated what measurements are to be made on the units, we can cheerfully amass our data — numbers resulting from the measurements we made. The next step is usually to find averages or prepare some other summary of these data. Before plunging ahead, it is wise to ask how much information our numbers carry. Consider, for example, my employment status. We agreed to represent this by the variable having value 0 if I'm not in the labor force, value 1 if I'm unemployed, and value 2 if I'm employed. Now, 2 is twice as much as 1. And 2 inches is twice as much as 1 inch.

But an employment status of 2 is *not* twice an employment status of 1. That's obvious, but sneaky. The numbers used to code employment status are just category labels disguised as numbers. We could have used the labels A, B, and C, except that to include categorization as part of measurement, we insisted on numbers. So not all numbers resulting from measurement carry information, such as "twice as much," that we naturally associate with numbers. What we can do with data depends on how much information the numbers carry.

We speak of the kind of information a measurement carries by saying in what kind of *scale* the measurement is made. Here are the kinds of scales:

> A measurement of a property has a *nominal scale* if the measurement tells only *what class* a unit falls in with respect to the property.
>
> The measurement has an *ordinal scale* if it also tells when one unit has *more of* the property than does another unit.
>
> The measurement has an *interval scale* if the numbers tell us that one unit *differs by a certain amount* of the property from another unit.
>
> The measurement has a *ratio scale* if in addition the numbers tell us that one unit has *so many times as much* of the property as does another unit.

Measurements in a *nominal scale* place units in categories, nothing more. Such properties as race, sex, and employment status are measured in a nominal scale. We can code the sex of a subject by

$$0 - \text{female}$$
$$1 - \text{male}$$

or by

$$0 - \text{male}$$
$$1 - \text{female}.$$

Which numbers we assign makes no difference; the value of this variable indicates only what the sex of the subject is.

In an *ordinal scale*, the order of numbers is meaningful. If a committee ranks 10 fellowship candidates from 1 (weakest) to 10 (strongest), the candidate ranked 8 is better than the candidate ranked 6 — not just different (as a nominal scale would tell us), but better. But the usual arithmetic is not meaningful: 8 is not twice as good as 4, and the difference in quality between 8 and 6 need not be the same as between 6 and 4. Only the order of the values is meaningful. Ordinal scales are important when social scientists measure properties such as authoritarian personality by giving a test on which a subject can score, say, between 0 and 100 points. If the test is valid as a measure of this property, then Esther who scores 80 is more authoritarian than Lydia who scores 60. But if Jane scores 40, we can probably not conclude that Esther is twice as authoritarian as Jane. Nor can we say that "the difference in authoritarianism between Esther and Lydia is the same as between Lydia and Jane" just because their scores differ by 20 in each case. Whether a particular test has an ordinal scale or actually does carry information about differences and ratios we leave for psychologists to discuss. Many tests have ordinal scales.

With *interval and ratio scales* we reach the kind of measurement familiar to us. These are *measurements made on a scale of equal units,* such as height in centimeters, reaction time in seconds, or temperature in degrees Celsius. Arithmetic such as finding differences is meaningful when these scales are used. A cockroach 4 centimeters long is 2 centimeters longer than one 2 centimeters long. There is a rather fine distinction between interval and ratio scales. A cockroach 4 centimeters long is twice as long as one 2 centimeters long; length in centimeters has a ratio scale. But when the temperature is 40°C it is not twice as hot as when it is 20°C. Temperature in degrees Celsius has an interval scale, not a ratio scale. Another way of expressing the difference is that ratio scales have a meaningful zero. A length of 0 centimeters is no length, a time of 0 seconds is no time. But a temperature of 0°C is just the freezing point of water, not no heat. (There is a temperature scale, the absolute or Kelvin scale, with 0° at absolute zero, the temperature at which molecules stop moving, and there is literally no heat. This is a ratio scale.)

We will not pay attention to the distinction between interval and ratio scales. But it is important (and usually easy) to notice whether a variable has a nominal scale (objects are put into categories), an ordinal scale (objects are ordered in some way), or an interval/ratio sale (measurements are made on a scale marked off in units).

One concluding fine point: The scale of a measurement depends mainly on the method of measurement, not on the property measured. The weight of a carton of eggs measured in grams has an interval/ratio scale. But if I label the carton as small, medium, large, or extra large, I

have measured the weight in an ordinal scale. If a standard test of authoritarian personality has an ordinal scale, this does not mean that it is impossible to measure authoritarian personality on an interval/ratio scale, only that this test does not do so.

Section 3 exercises

3.20. Identify the scale of each of the following variables as nominal, ordinal, or interval/ratio:

 (a) The concentration of DDT in a sample of milk, in milligrams per liter.

 (b) The species of each insect found in a sample plot of cropland.

 (c) A subject's response to the following personality test question: "It is natural for people of one race to want to live away from people of other races.

> Strongly agree
> Agree
> Undecided
> Disagree
> Strongly disagree"

(d) The pressure in pounds per square inch required to crack a specimen of copper tubing.

3.21. Identify the scale of each of the following variables as nominal, ordinal, or interval/ratio:

(a) The position of the Chicago Cubs in the National League Eastern Division Standings (1st, 2nd, 3rd, 4th, 5th, or 6th).

(b) The reaction time of a subject, in milliseconds, after exposure to a stimulus.

(c) The score of a student on an examination in this statistics course.

(d) A person's occupation as classified by the Bureau of Labor Statistics (managerial and professional, technical, sales, and so on).

3.22. The Gallup Poll from time to time asks its sample how highly they regard political candidates. The respondents are asked to rate each candidate on a 10-point scale that runs from "highly favorable" down to "highly unfavorable." Dwight Eisenhower received one of the highest ratings, with 65% of the voters giving him a highly favorable rating during the 1956 campaign. Barry Goldwater, in contrast, was rated highly favorable by only 15% of the voters in 1964. What type of scale does Gallup use to measure the personal enthusiasm of a voter toward a candidate? Explain your answer.

3.23. What type of scale is illustrated by the numbers on the shirts of a basketball team?

3.24. What type of scale is illustrated by house address numbers along a typical city street?

3.25. The 1990 census long form, given to a sample of 17% of all households, asks, "In what U.S. State or foreign country was this person born?" Another question asks "If this person is a female — How many babies has she ever had, not counting stillbirths?" What type of measurement scale is used in each of these two questions?

3.26. An advertising firm conducts a sample survey to see how adult women react to various adjectives that might be used to describe an automobile. The firm chooses 300 women from across the country. Each woman is read a list of adjectives, such as "elegant" and "prestigious." For each adjective she must indicate how desirable a car described this way seems to her. The possible answers she can give are (1) highly desirable, (2) somewhat desirable, (3) neutral, and (4) not desirable.

(a) What is the population in this sample survey?

(b) What is the response variable?

(c) What type of measurement scale does the response variable have?

4. Looking at data intelligently

Political rhetoric, advertising claims, debate on public issues—we are assailed daily by numbers employed to prove a point or to buttress an argument. Asking a few preliminary questions of such data will help us to distinguish sense from nonsense.

What is the source of the data?

Knowledge of the source helps us decide whether to trust the data. Knowing the source of data also allows us to check whether it was quoted correctly and to use the knowledge gained in Chapters 1 and 2 to assess the quality of the data. Here are some examples.

> **Example 7.** In a climate of increased concern over the risks of exposure to radiation, veterans' groups collected instances of multiple myeloma (a form of cancer) among veterans of the Hiroshima and Nagasaki occupation forces. They claimed that the number of cases was unusual and called for government study and compensation. This is *anecdotal evidence*, based on a few cases without systematic comparison or data collection. A committee of the National Research Council found no evidence that the rate of multiple myeloma for the atomic veterans was higher than that in other similar populations. Anecdotal data may be an incentive to more careful investigation, but in themselves they reflect rumor and emotion as often as fact.

> **Example 8.** "Charles D. Masters of the United States Geological Survey told oil industry officials that the world's supply of crude oil would run out in about the year 2046 if used at the current rate." Here we have an *expert opinion*. We cannot sample yet-to-be-discovered oil supplies, so any conclusion must be based on past data and informed judgment about the future. We must ask ourselves if we trust the expert. Is his statement a careful estimate or a shot in the dark? Is he speaking under political pressure or in support of a pet project? Does his judgment agree with that of other experts? Is it supported by a reasonable reading of past data?

> **Example 9.** "So we went to 21 major cities and asked 550 drinkers to compare white rum with the leading brands of gin and vodka. 24.2% preferred gin. 34.4% preferred vodka. And 41.1% preferred white rum." This statement in a rum advertisement by the Commonwealth of Puerto Rico describes a comparative taste test, but it

says little about the design of the experiment. Government regulations require that claims such as "more drinkers prefer rum than either gin or vodka" be based on actual studies and that the details of such studies be available on request. We can write the advertiser if we wish to assess the claim.

Anecdotal evidence (Example 7) should be given little credence until substantiated by systematically collected data. We are well-equipped to assess data from sample surveys or experiments (Example 9) if only we are given the details we need. Example 8 lies in the murky middle ground between these cases. Informed opinion is a common source of data in public issues concerning future trends. While potentially more trustworthy than anecdotal accounts, no informed judgment is as reliable as a properly designed survey or experiment. Whenever surveys or experiments are possible, informed judgment is a second-rate source of data. The rapid adoption of random sampling by experienced political campaign managers is one of many examples of the replacement of expert judgment by statistical methods.

Because we have no data about the future, predicting future trends requires *extrapolation*. Extrapolation means projecting the trend of past data to future results. Extrapolation is risky, because the future can differ from past trends. Figure 3-1 shows the price of crude oil, the world's

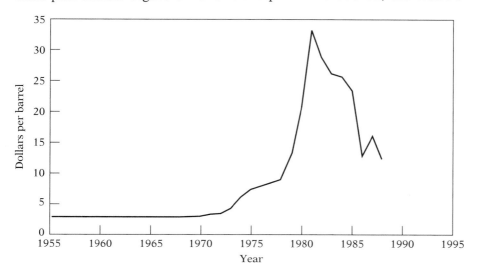

Figure 3-1. Average annual price of crude oil at the point of production, in dollars per barrel. [Energy Information Administration, reported in 1990 *Statistical Abstract of the United States.*]

most important source of energy. After years of stability, oil prices were suddenly driven upward by the 1973 oil embargo and the 1978 Iranian revolution. Oil cost eight times more in 1981 than in 1973. This unexpected increase caused economic disruption throughout the world. Experts, who had not foreseen the sudden rise, now expected oil prices to continue upward. The graph shows another surprise: Crude oil prices dropped by half in the next few years. Prices then jumped sharply once more in 1990 following the Iraqi invasion of Kuwait. As this example shows, experts are not exempt from the risks of extrapolation. Nor are they immune to outside pressure or without private prejudices. Reaching a clear conclusion when extrapolation is necessary and political and economic special interests are strong is extraordinarily difficult. Unfortunately, crucial issues such as future energy supply fall in this category.

Do the data make sense?

You would not accept verbal nonsense in a discussion. Don't accept numerical nonsense either.

> **Example 10.** "The mayor said that 90% of the police force had never taken a bribe. These honest men should not be tarnished by the misdeeds of a few." That "90%" is a *meaningless number*. The mayor has no idea what percent of the police have never taken a bribe. He wants to say that "a great majority" or "all but a few" have not. We can (and do) ignore his "90%."

> **Example 11.** "True cigarettes have 5 milligrams less tar." This is a *meaningless comparison*. Five milligrams less tar than what?

A more serious aspect of asking whether data make sense is to examine them for *internal consistency*. Do the numbers fit together in a way that makes sense? A little thought here will do wonders. Here is part of an article dealing with a cancer researcher at the Sloan-Kettering Institute who was accused of committing the ultimate scientific sin, falsifying data:

> **Example 12.** "One thing he did manage to finish was a summary paper dealing with the Minnesota mouse experiments . . . That paper, cleared at SKI and accepted by the *Journal of Experimental Medicine*, contains a statistical table that is erroneous in such an

elementary way that a bright grammar school pupil could catch the flaw. It lists 6 sets of 20 animals each, with the percentages of successful takes. Although any percentage of 20 has to be a multiple of 5, the percentages that Summerlin recorded were 53, 58, 63, 46, 48, and 67."[4]

In Example 12, lack of internal consistency led to the suspicion that the data were phony. *Too much precision or regularity* can lead to the same suspicion, as when a student's lab report contains data that are exactly as the theory predicts. The laboratory instructor knows that the accuracy of the equipment and the student's laboratory technique are not good enough to give such perfect results. He suspects that the student made them up. Here is an example drawn from another account of fraud in medical research.

> **Example 13.** ". . . Lasker had been asked to write a letter of support. But in reading two of Slutsky's papers side by side, he suspected that the same 'control' animals had been used in both without mention of the fact in either. Identical data points appeared in both articles, but . . . the actual number of animals cited in each case was different. This suggested at best a sloppy approach to the facts. Almost immediately after being asked about the statistical discrepancies, Slutsky resigned and left San Diego."[5]

In this case, suspicious regularity (identical data points) combined with inconsistency (different numbers of animals) led a careful reader to suspect fraud.

Careful statistical writers avoid the appearance of more precision than the data warrant. In technical writing, a confidence statement can be made. In nontechnical writing, the degree of precision is indicated by rounding off. When the Alabama Development Office reports that the state has attracted 422,657 new industrial jobs in the past 25 years, we suspect at once that the office doesn't know this number down to the last job. The Alabama Department of Industrial Relations, in cooperation with the Bureau of Labor Statistics, counts industrial workers in the state. They recorded a gain of 103,000 instead of the Development Office's 422,657. We can surmise that the gain of 103,000 is accurate to about the nearest thousand. Aside from spurious precision, the great difference between the two numbers is a matter of definition of variables. The Development Office does not bother to subtract jobs eliminated, so the actual number of jobs might even be decreasing. The

Department of Industrial Relations reports the actual change in the number of industrial jobs. The latter is surely a more valid measure of employment trends in Alabama.

The final part of asking if the data make sense is to *ask if they are plausible*. Numbers are easily misquoted, and the result is often wildly too high or too low. A little knowledge and common sense will detect many unbelievable numbers. For example, *Organic Gardening* magazine (July 1983) says that "the U.S. Interstate Highway System spans 3.9 million miles and is wearing out 50% faster than it can be fixed. Continuous road deterioration adds $7 billion yearly in fuel costs to motorists." Now, 3.9 million miles of pavement would build 1300 separate highways across the 3000 miles separating the east and west coasts. Common sense says that's wrong. Sure enough, the *Statistical Abstract of the United States* gives a more plausible 41,000 miles of interstate highways. *Organic Gardening* probably meant 39,000 rather than 3.9 million. I will leave you to assess that $7 billion per year in extra fuel costs. Here is another example.

> **Example 14.** A writer in *Science* [Volume 192 (1976), p. 1081] stated that "people over 65, now numbering 10 million, will number 30 million by the year 2000, and will constitute an unprecedented 25 percent of the population." Such explosive growth of the elderly—tripling in a quarter century to become a fourth of the population—would profoundly change any society. But wait. Thirty million is 25% of 120 million, and the U.S. population is already more than twice that size. Something is wrong with the writer's figures. Thus alerted, we can check reliable sources such as census reports to learn the truth. A reader of *Science* did so [letter to the editor, Volume 193 (1976)] and noted that in 1975 there were 22.4 million persons over 65, not 10 million. The projection of 30 million by the year 2000 is correct, but that is only 11 or 12% of the projected population for that year. The explosive growth of the elderly vanishes in the light cast by accurate statistics.

I hope that reading this book will help you form the habit of looking at numbers closely. Your reward will be the reputation for brilliance that accrues to those who point out that a number being honored by everyone else is clearly nonsense.

Is the information complete?

A subtle way of using data in support of dubious conclusions is giving only part of the relevant information. This is perhaps the single most common trick employed by those who use numbers to make an impression rather than to tell the whole truth. Here are some typical examples.

> **Example 15.** A television advertisement by the Investment Company Institute (the mutual fund trade association) said that a $10,000 investment made in 1950 in an average common stock mutual fund would have increased to $113,500 by the end of 1972. That's true. The *Wall Street Journal* (June 7, 1972) pointed out that the same investment spread over all the stocks making up the New York Stock Exchange Composite Index would have grown to $151,427; that is, mutual funds performed worse than the stock market as a whole.

> **Example 16.** Anacin was long advertised as containing "more of the ingredient doctors recommend most." The ad did not mention that the ingredient is aspirin. Another over-the-counter pain reliever claimed that "doctors specify Bufferin most" over other "leading brands." Bufferin also consists primarily of aspirin and was specified most only because doctors rarely recommend a particular brand of pure aspirin. Both advertising claims were literally true; the Federal Trade Commission found them both misleading.

"Sure your patients have 50% fewer cavities. That's because they have 50% fewer teeth!"

Example 17. A television commercial for Schick Super Chromium razor blades showed a group of barbers shaving with the same blade, one after another. The twelfth, thirteenth, fifteenth, and seventeenth men to use the blade were interviewed. All said the shave was satisfactory. Consumers Union repeated this experiment with 18 men. The seventeenth and eighteenth to shave with the same blade were satisfied, but the seventh, eighth, ninth, and tenth all said the blade needed changing. Did all 17 users in the commercial get good shaves, or was the interviewing selective? Were the barbers chosen at random? Was their judgment biased by the knowledge that Schick was sponsoring the test and that they might appear on TV?[6]

Examples 15, 16, and 17 illustrate the misleading effect of giving true information out of context. The truth without the *whole* truth can lie.

Even complete and accurate data may mislead us if we are not aware of changes in the process of measuring and collecting the data. This too is part of the background information needed to interpret data intelligently. The reported size of a university's faculty changed when postdoctoral researchers, who had been listed as faculty members, were dropped from the list. This is a change of definition. The number of petty larcenies reported in Chicago more than doubled between 1959 and 1960 because a new police commissioner had introduced an improved reporting system. The new system gave a much better count of crimes committed, so the number of crimes reported rose. This is a change in data-collection procedure. Almost all series of numbers covering many years are affected by changing definitions and collection methods. Often these changes are pointed to by sudden jumps in the series of numbers (lack of internal consistency), but not always. Alertness and care are needed to avoid false conclusions.

Example 18. Data collected by General Electric once showed that a component of a major appliance was failing at ever higher rates as the appliance became older. Preparations began for the manufacture of a more reliable component. Then a statistician noted that the rate of failure was roughly constant for the first year of service, turned up sharply at exactly 12 months, was roughly constant at this higher rate during the second year of service, and turned up again at exactly 24 months. No appliances in the sample had been in service more than 29 months.

Alerted by this suspicious regularity, the statistician checked into the source of the data. For the first 12 months, all appliances were sampled because all had a one-year warranty. Data for the second 12 months referred only to appliances whose owners had bought a service contract for the second year. Data beyond 24 months were collected only for appliances on a renewed service contract. Because a service contract provides free service, appliances covered by contracts are serviced more often. And owners of troublesome appliances are more likely to buy and renew a service contract. So the higher failure rates in the second and third years were not representative of the entire population of appliances. GE did not have to develop a new component. (I hope they paid part of the savings to their statistician.)[7]

Is the arithmetic faulty?

Conclusions that are wrong or just incomprehensible are often the result of plain old-fashioned blunders. Rates and percentages are the most common causes of crooked arithmetic. Sometimes the matter can be straightened out by some numerical detective work. Here is an example.

Example 19. The BLS report on employment and unemployment for August 1977 noted that the unemployment rate was 6.1% for whites and 14.5% for blacks. The *New York Times* (September 3, 1977) included the following paragraph in its article on this report:

> *The bureau also reported that the ratio of black to white jobless rates "continued its recent updrift to the unusually high level of 2.4 to 1 in August," meaning that 2.4 black workers were without jobs for every unemployed white worker.*

Now 14.5% is 2.4 times as great as 6.1%, so the BLS is correct in stating that the ratio of black to white jobless rates was 2.4 to 1. But the *Times'* interpretation is completely wrong. Because blacks make up only a small part of the labor force, there are fewer jobless blacks than whites even though the percent of blacks who are unemployed is higher than the percent of whites who are without jobs. The *Times* confused percent unemployed with actual counts of the number of unemployed workers.

Calculating the percent increase or decrease in some variable is a common source of arithmetic errors. The percent change in a quantity is found by

$$\text{percent change} = \frac{\text{amount of change}}{\text{starting value}} \times 100$$

Example 20. Last year the price of gold rose from $300 an ounce to $450 an ounce. This was an increase of 50%, because

$$\frac{\text{increase}}{\text{starting value}} = \frac{\$150}{\$300} = 0.5 = 50\%$$

This year, the gold price drops by 50%. What is an ounce now worth? Well, the amount of the decrease is 50% of $450, or $225. So the price of an ounce of gold is now $450 less $225, or $225.

An increase of 50%, followed by a decrease of 50%, does *not* bring us back to the original value. To press the point further, an increase of 100% means that the quantity in question has doubled, since the amount of the increase is 100% of the original value. But a decrease of 100% means that the quantity is now zero: it has lost 100% of its original value, and 100% is all there is.

I concede that percents are a bit mysterious, but if you are going to read or write about statistical subjects you will have to get such things straight. Most subjects these days are statistical subjects.

Data that enlighten

The aim of statistics is to provide insight by means of numbers. To achieve this aim, we must first collect numbers that are *valid* in the sense of being both correct and relevant to the issue at hand. Since we most often have data on only some people or things from a larger population, we distinguish between *internal validity* and *external validity*.

> *Internal validity* **refers to the particular people or things that we have measured.** *External validity* **concerns generalizability of our conclusions to a wider population.**

Valid measurement is part of internal validity. Scores on an employment test that is unrelated to the job to be done are of no value in judging those who took the test, let alone for broader conclusions. Confounding with extraneous variables can also destroy internal validity, as can arithmetic mistakes or incomplete information.

Convenience samples are a common threat to external validity. Even though we have trustworthy information about the sample (internal validity), the data say little about any wider population. Unrealistic experimental treatments or unrepresentative experimental subjects can also make correct numbers useless for broad conclusions.

A naked number, without source or context, can easily mislead. But we need not be misled. The many examples of misleading numbers in this section should reinforce rather than weaken our resolution to master the statistical ideas that produce clear, valid, and accurate data.

Section 4 exercises

3.27. The following quotation appears in a book review in *Science*, Volume 189 (1975), p. 373:

> . . . *a set of 20 studies with 57 percent reporting significant results, of which 42 percent agree on one conclusion while the remaining 15 percent favor another conclusion, often the opposite one.*

Do the numbers given in this quotation make sense? Can you decide how many of the 20 studies agreed on "one conclusion," how many favored another conclusion, and how many did not report significant results?

3.28. The excerpt and graph on the following page are from a United Press International dispatch that appeared in the Middletown, N.Y., *Times Herald Record* of June 7, 1975.

    (a) Is the source of the data given, and is that source trustworthy?

    (b) Because of suspicious regularity, I do not believe that 21.8% of manufacturing workers were unemployed, as the graph claims. Explain why I'm suspicious and how you think this apparent error came about.

3.29. The late English psychologist Cyril Burt is famous for his studies of the IQ scores of identical twins who were raised apart. The high correlation between the IQs of separated twins in Burt's studies pointed to heredity as a major factor in IQ. (Correlation is a measure of the connection or association between two variables. We will become better acquainted with it in Chapter 5.) Burt wrote several accounts of his work, each reporting on more pairs of twins. Here are his reported correlations as he published them.

Unemployment hit a 34-year high of 9.2 percent in May, the government reported Friday, but a labor analyst insisted there are new signs the recession may be over.

The Bureau of Labor Statistics said the number of unemployed rose 362,000 in May to a total of 8,538,000 – 9.2 percent of the labor force compared to 8.9 percent in April.

It was the first time the rate exceeded 9 percent since the final Depression year of 1941 when the 12-month average was 9.9 percent.

. . .

The report showed unemployment in construction trades hit an all-time high of 21.8 percent in May.

Reprinted with permission of U.P.I.

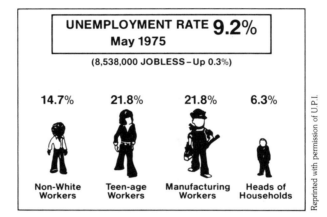

UNEMPLOYMENT RATE 9.2%
May 1975

(8,538,000 JOBLESS – Up 0.3%)

| 14.7% | 21.8% | 21.8% | 6.3% |
| Non-White Workers | Teen-age Workers | Manufacturing Workers | Heads of Households |

Reprinted with permission of U.P.I.

Date of publication	Twins reared apart	Twins reared together
1955	0.771 (21 pairs)	0.944 (83 pairs)
1958	0.771 ("over 30" pairs)	0.944 (no count)
1966	0.771 (53 pairs)	0.944 (95 pairs)

What is suspicious here? (Further investigation made it almost certain that Burt fabricated his data.)

3.30. A series of measurements of weights in grams is recorded as

$$11.25 \quad 13.75 \quad 12.00 \quad 13.25 \quad 10.75$$
$$12.50 \quad 12.25 \quad 11.00 \quad 13.25 \quad 11.75$$

Because two decimal places are given, we might conclude that the measurements are precise to two decimal places, that is, are rounded to the nearest 0.01 gram. A closer look at the data suggests that they are much less precise than this. How precise do you think these data are? Why?

3.31. The *New York Times* (February 22, 1989) quoted a sociologist as saying that for every 233 unmarried women in their 40s in the U.S., there are only 100 unmarried men in their 40s. These numbers point to an unpleasant social situation for women of that age. Are the numbers plausible?

3.32. Here are some citations containing numbers that may not be plausible. In each case, identify the data that seem implausible. Try to decide if the numbers are actually right. (Not all of them are wrong!)

(a) From the *New York Times*, May 8, 1976, p. 25: "Altogether, in some 30 associations and groups of independents, there are almost 80 million Baptists in the nation. They are outnumbered only by the Roman Catholics." (The total population of the U.S. in 1976 was about 215 million.)

(b) From *Audubon*, May 1976, p. 28, writing of Alaska: "Throughout the state each year, hunters harvest upward of 10,000 moose, skin them out, pack them out, and dress them out—five, six million pounds of meat in their freezers . . . ." (For comparison, beef cattle yield an average of around 300 pounds of beef each.)

(c) The *Statistical Abstract of the United States*, 1973 edition, p. 83, reports that on December 31, 1971, there were 82,294 narcotics addicts in the united States. (Later editions of the *Statistical Abstract* contain no such data.)

3.33. An advertisement for the pain reliever Tylenol was headlined: "Why Doctors Recommend Tylenol More Than All Leading Aspirin Brands Combined." A counteradvertisement by the makers of Bayer Aspirin, headlined "Makers of Tylenol, Shame On You!" accused Tylenol of misleading by giving the truth but not the whole truth. You be the detective again: How is Tylenol's claim misleading even if true?

3.34. An article on adding organic matter to soil in *Organic Gardening*, March 1983, said, "Since a 6-inch layer of mineral soil in a 100-square-foot plot weighs about 45,000 pounds, adding 230 pounds of compost will give you an instant 5% organic matter." What percent of 45,000 is 230? Does the error lie in the arithmetic, or is that 45,000 pounds too heavy?

3.35. The newspaper article on marijuana reprinted in Review Exercise 2.70 contains numbers that appear to be inconsistent. Find the inconsistency.

3.36. Data on accidents in recreational boating in the *Statistical Abstract* show that the number of deaths remained steady between 1970 and 1980 (1418 deaths in 1970 and 1360 in 1980). But the number of injuries reported jumped from 780 in 1970 to 2650 in 1980. Why do you think the injury count rose when deaths did not? Why are there so few injuries in these government data relative to the number of deaths? Which count (deaths or injuries) is probably more accurate?

3.37. On "Black Monday" October 19, 1987, the Dow Jones Industrial Average of common stock prices fell from 2244 to 1736. What percent decrease was this?

3.38. In 1982, the Census Bureau gave a simple test of literacy in English to a random sample of 3400 people. The *New York Times* (April 21, 1986) printed some of the questions under the headline "113% of Adults in U.S. Failed This Test." Why is the percent in the headline clearly wrong?

3.39. An advertisement claims that a new toothpaste "cuts cavities by 200 percent." Explain why this claim is numerical nonsense.

3.40. The question-and-answer column of a campus newspaper was asked what percent of the student body was "Greek." The answer given was that "the figures for the fall semester are approximately 13 percent for the girls and 15–18 percent for the guys, which produces a 'Greek' figure of approximately 28–31 percent of the undergraduates at Purdue" (*Purdue Exponent*, September 21, 1977). Discuss the campus newspaper's arithmetic.

3.41. A newspaper story on housing costs (*Lafayette Journal and Courier*, September 21, 1977) noted that in 1975 the median price of a new house was $39,300 and the median family income was $13,991. (Half of all families earn less than the median income and half earn more. We will become better acquainted with the median in Chapter 4.) The writer then claimed that "the ratio of housing prices to income is much lower today than it was in 1900 (2.8 percent in 1975 vs. 9.8 percent in 1900)."

I do wish that I could buy a new house for 2.8% of my income. Where did that 2.8 come from? What is the correct expression for $39,300 as a percent of $13,991?

3.42. Below is a table from *Smoking and Health Now*, a report of the British Royal College of Physicians. It shows the number and percent of deaths among men age 35 and over from the chief diseases related to smoking. One of the entries in the table is

	Lung cancer	Chronic bronchitis	Coronary heart disease	All causes
Number	26,973	24,976	85,892	312,537
Percent	8.6%	8.0%	2.75%	100%

incorrect, and an erratum slip was inserted to correct it. Which entry is wrong, and what is the correct value?

3.43. The January 1982 issue of *Playboy* magazine contained a 133-question survey on sexual habits. A total of 80,324 usable responses (65,396 men and 14,928 women) were mailed in. The results showed that women claimed to be sexually active earlier than men: Of the respondents under age 21, 58% of the women and 38% of the men claimed intercourse before age 16. Discuss the internal and external validity of these results.

3.44. Find in a newspaper or magazine an example of one of the following; explain in detail the statistical shortcomings of your example.

Meaningless numbers or comparisons

Lack of internal consistency

Spurious precision

Implausible numbers

Omission of essential information

Faulty arithmetic

NOTES

1. Reported in the *Chronicle of Higher Education*, September 6, 1989.
2. A more extensive discussion, in the context of the New York scholarship case, is Constance Holden, "Court Ruling Rekindles Controversy Over SATs," *Science*, Volume 243 (1989), pp. 885–887.
3. These data appear in Harry H. Ku, "Statistical Concepts in Metrology," in Harry H. Ku (ed.), *Precision Measurement and Calibration* (Washington, D.C., National Bureau of Standards Special Publication 300, 1969), p. 319.
4. Quoted from Barbara Yuncker, "The Strange Case of the Painted Mice," *Saturday Review/World*, November 30, 1974, p. 53.
5. Quoted from Eliot Marshall, "San Diego's Tough Stand on Research Fraud," *Science*, Volume 234 (1986), pp. 534–535.
6. *Consumer Reports*, October 1971, pp. 584–586.
7. This example appears in Wayne B. Nelson, "Data Analysis with Simple Plots," General Electric Technical Information Series, April 1975.

Review exercises

3.45. Why does the Current Population Survey ask a lengthy series of Questions to determine whether a person is employed or unemployed, rather than simply asking, "Are you employed right now?"

3.46. Scientists who study human growth use different measures of the size of a person. Weight, height, and weight divided by height are three common measures of size. If you are interested in studying the short-term effects of digestive illnesses on the growth of children, which of these three variables would you use? Why?

3.47. You are writing an article for a consumer magazine based on a survey of the magazine's readers that asked about the reliability of their household appliances. Of 13,376 readers who reported owning Brand A dishwashers, 2942 required a service call during the past year. Only 192 service calls were reported by the 480 readers who owned Brand B dishwashers. Describe an appropriate variable to measure the reliability of a make of dishwasher, and compute the values of this variable for Brand A and for Brand B.

3.48. A sociologist measures many variables on each of a sample of urban households. In what type of scale is each of the following variables measured?

(a) Annual household income in dollars.

(b) The race of the householder.

(c) The educational level of the householder (elementary school, 1 to 3 years of high school, completed high school, and so on).

3.49. Here are data on the population of several states and the numbers of prisoners on death row awaiting execution (as of July 1989) in those states.

State	Population (thousands)	Death row prisoners
California	28,168	247
Florida	12,377	294
Illinois	11,544	120
New Jersey	7,720	25
Nevada	1,060	45
Pennsylvania	12,027	115
Texas	16,780	283

California, Florida, and Texas lead the nation in the number of death-row prisoners. Because these are all large states, they could be expected to have many such prisoners. Find the rate of death-row prisoners per million population for the states in the table. Because population is given in thousands, you can find the rate per million as

$$\text{rate per million} = \frac{\text{prisoners}}{\text{population in thousands}} \times 1000$$

Which state has the highest number of prisoners relative to its population? Are any of California, Florida, and Texas still high by this measure?

3.50. An article in a midwestern newspaper about flight delays at major airports said,

> *According to a Gannett News Service study of U.S. airlines' performance during the past five months, Chicago's O'Hare Field scheduled 114,370 flights. Nearly 10 percent, 1,136, were canceled.*

Check the newspaper's arithmetic. What percent of the 114,370 flights were actually canceled? (From the *Lafayette Journal and Courier*, October 23, 1988.)

3.51. Westchester County is a prosperous suburban area outside New York City. *Fine Gardening* magazine (September/October 1989, p. 76) claimed that the county is home to 800,000 deer. Is this number plausible?

3.52. Here's a numerical puzzle. A government report contained a table showing what percent of households fell in each of several classes when the number of people per room in the household was calculated. For example, four people living in a five-room residence have $4/5 = 0.80$ people per room. The classes used in the table were

· · ·

$0.40\leq$	people per room	<0.45
$0.45\leq$	people per room	<0.50
$0.50\leq$	people per room	<0.55

· · ·

There were many households in the first and third of these classes, but none in the second. Can you explain why? (*Hint*: All the households had 10 or fewer rooms.)

"*Tonight, we're going to let the statistics speak for themselves.*"

Part II

Organizing data

Data, like words, speak clearly only when they are organized. Also like words, data speak more effectively when well organized than when poorly organized. Again like words, data can obscure a subject by their quantity, requiring a brief summary to highlight essential facts. The second of statistics' three domains is the organizing, summarizing, and presenting of data.

Data are produced in many forms: completed questionnaires, laboratory notebooks, electronically stored readings from recording instruments. A completed set of data is usually stored as a table, either in printed form or electronically in a computer. Our task is to digest such sets of raw data, to organize and summarize them for human use. A few general principles can help us understand data. *The first step in data analysis is to display the data in a graph.* The human eye and mind can see more in a graph than any computer, so we begin by taking advantage of our ability to see and understand. When looking at a graph, avoid getting lost in the details. *Look first for an overall pattern in the data and then for any striking exceptions to that pattern.* Numerical calculations can help us describe specific aspects of data, such as the average value of a variable. Another principle of data analysis is therefore to *move from graphs to well-chosen numerical descriptions.* What we see in the graph will help us choose numerical summaries.

General principles become clear only when put into practice. The following three chapters apply our principles in detail, first in Chapter 4 to data on a single variable. Chapter 5 then looks at relationships among several variables. Chapter 6 considers data that are collected regularly over time and focuses on understanding some of the economic and social statistics that attract such attention in the news media.

Summarizing and presenting a large body of facts offers to ignorance or malice ample opportunity for distortion. This is no less (but also no more) the case when the facts to be summarized are numbers rather than words. We shall therefore take due note of the traps that the ignorant fall into and the tools of duplicity that the malicious use. Those who picture statistics as primarily a piece of the liar's art concentrate on the part of statistics that deals with summarizing and presenting data. Misleading summaries and selective presentations go back to that after-the-apple conversation between Adam, Eve, and God. Don't blame statistics. Do remember the saying "Figures don't lie, but liars figure," and beware.

Chapter 4

Describing distributions

W̶hat mental picture does the word *statistics* call to mind in most
people? Very likely a picture of tables crowded with numbers. And
close behind tables come graphs, zigging up or zagging down. Now,
tables do lack sex appeal, and you may have exhausted your interest in
graphs when you mastered pie charts in grade school. Yet tables remain
the first step in organizing data, and data are most vividly presented in
graphs.

There is an art to presenting complex data clearly. This chapter begins
with some principles of data display in Section 1, with emphasis on
common graphs. The remaining sections of the chapter provide tools
and ideas for displaying and understanding data on a single measured
variable. The *distribution* of a variable describes what values the variable
takes and how often each value occurs. It is no accident that we met our
first table and an accompanying graph in discussing the sampling distribution of statistics computed from an SRS in Chapter 1. Distributions of
variables are central to statistics. Section 2 presents graphical tools for
describing distributions, and Sections 3 and 4 supplement graphs with
numerical descriptions of specific aspects of a distribution. In Section 5,
we meet a particularly important class of distributions, the normal
distributions.

1. Displaying data

We use data, like words, to communicate facts and to support conclusions. Like words, data must be well-organized if they are to communicate clearly. All but the smallest sets of data must be summarized, boiled

down into simpler form, if they are to be at all clear. Data are summarized in tables or, with greater impact, in graphs. There is a place for tables of the original, unsummarized data, but that place is usually in the archives or in the memory of a computer. The tables we refer to when seeking information are almost always summary tables that organize and greatly condense the original data. I invite you to look at the *Statistical Abstract of the United States*, an annual volume packed with every variety of numerical information. Has the number of private elementary and secondary schools grown over time? What about minority enrollments in these schools? How many college degrees were given in each of the past several years, and how were these degrees divided among fields of study and by the age, race, and sex of the students? All this and more can be found in the education section of the statistical abstract. All the tables that present this information are summary tables; we don't want to see information on every college degree individually, only the counts in categories of interest to us.

The *Statistical Abstract* offers a source of examples of how to organize complex tables, though we will have to confront only rather simple tables. When you present a table, be certain that it is clearly *labeled* so that readers can see at once what the subject of your data is. The main heading should describe the general subject of the data and give the date if the data will change over time. Labels within the table should identify the variables and state the *units* in which they are measured. You should state the *source* of the data, usually at the foot of the table. Table 4-1 is an

Table 4-1. Farms by size (1987)

Size of farm (acres)	Number of farms (thousands)	Percent of farms
Under 10	183	8.8
10–49	412	19.8
50–99	311	14.9
100–179	334	16.0
180–259	192	9.2
260–499	286	13.7
500–999	200	9.6
1000–1999	102	4.9
2000 and over	67	3.2

SOURCE: 1987 Census of Agriculture.

example of clearly presented summary data on the size of American farms.

Frequency tables

One of our first acts in organizing a set of data is usually to count how often each value occurs. After completing a sample survey of 1537 students that included the question "Do you agree or disagree that users of small amounts of cocaine should be imprisoned?" we are eager to tabulate the answers and learn that 928 agreed, 543 disagreed, and 66 had no opinion. (Check that $928 + 543 + 66 = 1537$, so all the answers are accounted for. That's a check for internal consistency.) Because rates or proportions are often more useful than totals, we go on to compute that

$$\frac{928}{1537} = 0.60, \text{ or } 60\%, \text{ agreed,}$$

$$\frac{543}{1537} = 0.35, \text{ or } 35\%, \text{ disagreed,}$$

and $\qquad \dfrac{66}{1537} = 0.04, \text{ or } 4\%, \text{ had no opinion.}$

Again you should check for internal consistency. There should be 100% in all. We got $60\% + 35\% + 4\% = 99\%$. What happened? The arithmetic is right, but when we rounded the fractions off to two decimal places, a little precision was lost, and the results do not quite add up to 100%. Such *round-off errors* will be with us from now on as we do more arithmetic.

Totals and percentages of this kind occur so often that they deserve formal names:

The *frequency* of any value of a variable is the number of times that value occurs in the data; that is, a frequency is a count.

The *relative frequency* of any value is the proportion or fraction or percent of all observations that have that value.

In the sample survey, the frequency of students in the sample who agreed that cocaine users should be imprisoned was 928. The relative frequency was 0.60. Relative frequencies are usually expressed in this decimal form, but we can just as correctly say that the relative frequency was 60%. (Remember that 1% is 1/100 or 0.01. A number in decimal

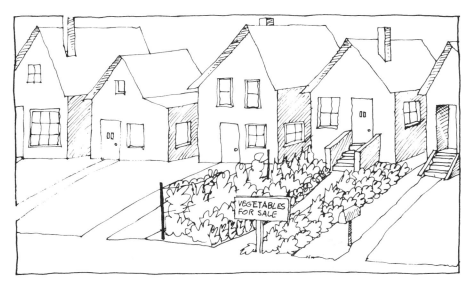

"8.8% of American farms are under 10 acres."

form can be changed to a percent by moving the decimal point two places to the right. So 0.60 is 60%.)

Frequencies and relative frequencies are a common way of summarizing data when a nominal scale is used, as was the case for our question about cocaine users. Even when an interval/ratio scale is used and the variable has numerous possible values, we often summarize data by giving frequencies or relative frequencies for groups of values. Such a summary is conveniently presented in a *frequency table*. Table 4-1 is an example. From this table we can learn much about the size of American farms. For example, the most common size category in 1987 was 10–49 acres. There were 412,000 farms in that size range. (Did you read the heading carefully enough to see that the number of farms is given in *thousands?*) This was 19.8% of all U.S. farms. The number column gives frequencies, while the percent column gives relative frequencies; it is good practice to follow the *Statistical Abstract* in *not* using the technical terms so that the table is easier for untrained persons to read. Finally, it is important that the source is given. Remembering how essential it is to know the definitions of variables, we should ask what the definition of a farm is. Without knowing what makes a piece of land a farm, we don't know what is being counted in Table 4-1. The source cited gives the exact definition: A farm is any place from which $1000 or more of agricultural products are normally sold in a year.

Graphing data

The purpose of a graph is to provide a visual summary of data. Graphs are the most effective way to communicate with data. A good graph frequently reveals facts about the data that would be difficult or impossible to detect from a table. What is more, the immediate visual impression of a graph is much stronger than the impression made by data in numerical form—so strong, in fact, that we must guard against false impressions.

Like tables, graphs should be clearly labeled to reveal the variables plotted, their units, the source of the data, and so on. There are some additional general principles that good graphs should obey.[1] First, *make the data stand out*. The actual data, not the labels, scale markings, or grids, should draw the viewer's attention first. Figure 4-1 demonstrates

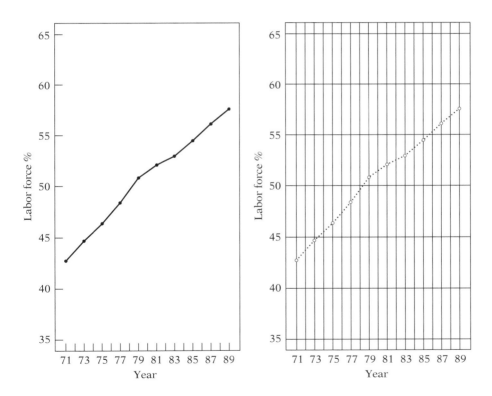

Figure 4-1. Unnecessary clutter makes the graph on the right harder to read than the identical graph on the left. The graphs display the percent of women at least 16 years old who are in the labor force. [Data from the 1989 *Statistical Abstract of the United States.*]

how a combination of weak plotting of the data and unneeded grid lines on the right obscure the trend that stands out clearly on the left. The figure also illustrates a related guideline for making good graphs, *avoid clutter*. A background grid is only needed if your audience must read numbers from the graph rather than simply see the overall picture. Otherwise, leave out the grid lines. A good graph uses no more ink than is needed to present the data clearly. Always ask if your graph is more complicated than is necessary.

Graphs are powerful tools for presenting data because they immediately convey an overall visual impression. We see the data at once without having to untangle them bit by bit. To draw effective graphs you must *pay attention to how visual perception works*. A look at some common types of graphs will apply this principle in specific cases.

Pie charts and dotcharts

Pie charts show how a whole is divided into parts. The left-hand graph in Figure 4-2 is a pie chart from the *Statistical Abstract* that shows the breakdown of state and local government spending. To make a pie chart, first draw a circle. Wedges within the circle represent the parts, with the

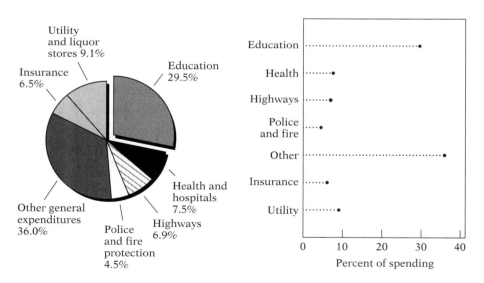

Figure 4-2. A dot chart (right) and a pie chart (left) of the percent of 1986 state and local government expenditure by sector. The pie chart is reproduced from the 1989 *Statistical Abstract of the United States.*

angle spanned by each wedge proportional to the size of that part. For example, education makes up 29.5% of state and local government spending. Because there are 360 degrees in a circle, the education wedge spans an angle of

$$0.295 \times 360 = 106 \text{ degrees.}$$

Pie charts force us to see that the parts do make a whole. But because we don't see angles as clearly as we see lengths, a pie chart is not a good way to compare the sizes of the various parts of the whole.

The right-hand graph in Figure 4-2 is a *dotchart* of the same data. The dotchart makes it clear that other general expenditures are larger than education spending; it's hard to see this from the wedges in the pie chart. The dotchart is also easier to draw than the pie chart and expends less ink to present the same data. The comparison in Figure 4-2 points up a common conflict of aims in graphing. The pie chart seems more impressive to many people than the simpler dotchart, especially when the wedges are in bright colors. Pie charts are a staple of business presentations, where flash is valued. Statisticians prefer the clarity of dotcharts.

Line graphs

Line graphs show the behavior of a variable over time. Time is marked on the horizontal axis, and the variable being plotted is marked on the vertical axis. Figure 4-3 is an example. The variable here is the average retail price of fresh oranges, collected monthly from a probability sample of 2300 food outlets as part of the government's effort to measure consumer prices. It would be difficult to see patterns in the long table of monthly prices. The line graph makes the pattern clear. We can see at once that:

- The price of oranges is highest in late summer and early fall, and lowest early in the year. This is *seasonal variation* caused by the fact that each year's orange crop is harvested early in the year.

- Prices were generally increasing in the 1983-to-1988 period, once the seasonal variation is ignored. This is a long-term *trend*.

Just because graphs speak so strongly, they can mislead the unwary. The intelligent reader of a line graph looks closely at the *scales* marked off on both axes. Figure 4-4 displays the same data as does Figure 4-1. The data show that the percent of adult women who are in the labor force has steadily increased over this 18-year period. The two line graphs

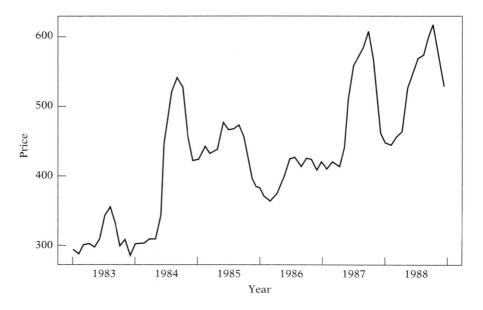

Figure 4-3. Average retail price of fresh oranges, 1983 to 1988. Price is given as a percent of the 1967 price. [From Bureau of Labor Statistics, *CPI Detailed Reports* for 1983 to 1988.]

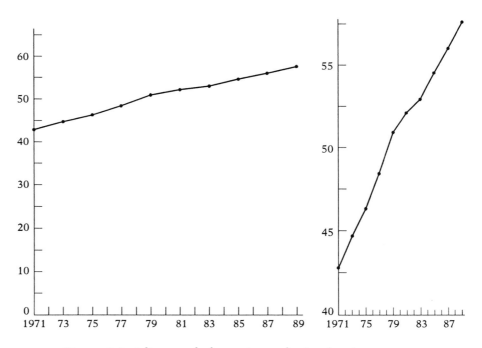

Figure 4-4. A line graph drawn to emphasize the change over time. The data are the same as in Figure 4-1.

give quite different visual impressions of the rate of this increase. You can transform the left-hand graph into the right-hand graph by stretching the vertical axis, compressing the horizontal axis, and cutting off the vertical axis at a value above zero. Now you know the trick of giving an exaggerated impression with a line graph. Because there is no one "right" scale for a line graph, perfectly honest sources can give somewhat different impressions by their choices of scale. Moral: Look carefully at the scales.

Once in a while you may encounter a more barbarous line graph. Figure 4-5 is the least civilized I have seen. Note first that time is on the vertical axis. So when the graph goes straight up (1940 to 1946), this means that taxes were not increasing at all. And when the graph is quite flat (1965 to 1972), this means that taxes were thundering upward. The graph gives an impression exactly the reverse of the truth. That's why time always belongs on the horizontal axis. Second, the time scale does not have equal units; equal lengths on the vertical axis represent first two years, then four years, then six years, then two years again, then whatever the interval between 1962 and 1964–1965 is; and so on. The graph is stretched and squeezed haphazardly by the changing time scale, so the rise in taxes in different time periods cannot be compared by looking at the steepness of the graph. These barbarisms were perpetrated by the

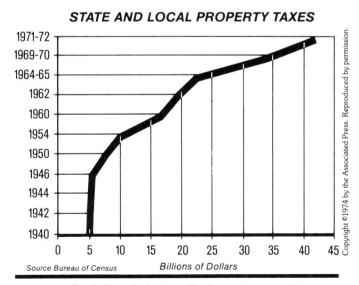

Chart indicates rise in state and local property taxes over 12 years

Figure 4-5.

Associated Press (not by the Census Bureau, which was only the source of the data). The newspaper then provided the caption below the chart, which no doubt refers to the "12 years" from 1940 to 1972.

Bar graphs

Bar graphs compare the values of several variables. Often the values compared are frequencies or relative frequencies of outcomes of a nominal variable. For example, Figure 4-6 is a bar graph that displays the number of degrees earned by men and women at each of three levels. As usual, the graph makes the overall pattern of the data instantly clear. We can see how the number of advanced degrees compares with the count of bachelor's degrees. We notice that women earn slightly more bachelor's and master's degrees than men do, but fewer doctorates.

A dot chart can replace a bar chart. The dot chart uses less ink to convey the same information, because the width of the bars in a bar chart doesn't add any information. Most people find bar charts more attractive and (unlike pie charts) they present data clearly. So I recommend bar charts rather than dot charts even though this violates the principle that graphs should be as uncluttered as possible.

The bars in a bar chart may be vertical (as in Figure 4-6) or horizontal. They may touch each other (as some do in Figure 4-6) or be separated. But all bars must have the same width, for our eyes respond to the *area* of the bars. When the bars have the same width and a height that varies

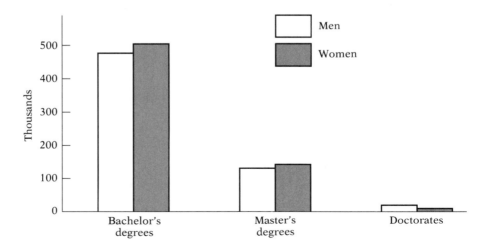

Figure 4-6. Earned degrees, 1987. [Data from 1990 *Statistical Abstract of the United States.*]

with the variable being graphed, then the area (height times width) also varies with the variable and our eyes receive the correct impression. The most common abuse of bar graphs is to replace the bars by pictures and to change both the width and the height of the picture in proportion to the variable graphed. Figure 4-7, an advertisement placed by *Time* magazine in the commercial section of the *New York Times*, illustrates this.

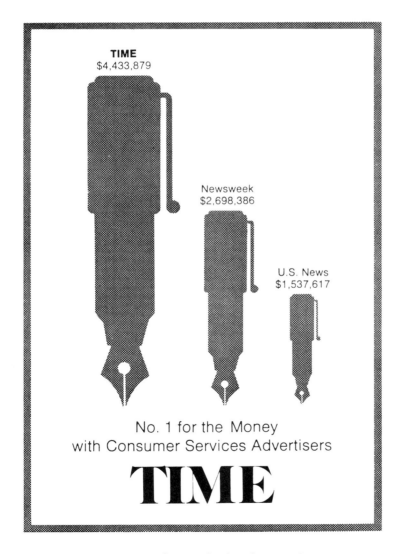

Figure 4-7. An attractive but misleading bar graph. [Copyright © 1971 by Time, Inc. Reproduced by permission.]

Figure 4-8. An attractive and accurate bar graph. [Copyright © 1972 by Time, Inc. Reproduced by permission.]

Time enjoys a lead of less than two to one over *Newsweek* in dollar value of consumer-services advertising (the dollar amounts appear over the pictures). But because *both* the height and the width of the pen representing *Time* are almost double those of *Newsweek's* pen, the area of *Time's* pen is almost four times that of *Newsweek's*. Our eyes receive the misleading impression that *Time* has a four-to-one, rather than a two-to-one, edge.

In case you are wondering if graphic attractiveness must always be sacrificed to accuracy, consider Figure 4-8. This is another *Time* ad making the same point as Figure 4-7, and it is at least as attractive as Figure 4-7. Unlike Figure 4-7, the graph in Figure 4-8 is accurate. Congratulations to the designer who managed to combine accuracy with graphic impact.

Section 1 exercises

4.1. Answer the following questions using the data in Table 4-1.
  (a) What percent of all American farms in 1987 were at least 500 acres in size?
  (b) How many farms were there in 1987? About how precise is this figure (nearest 1, nearest 100, or nearest 1000)?
  (c) Why do the percents in the last column have sum 100.1% rather than 100%?

4.2. Make a dot chart to display the percent of American farms in each of the size categories in Table 4-1.

4.3. An article in the *New York Times* (December 31, 1989) discussed homicide in the 1980s. During the years 1980 to 1987, 46.8% of all murders were committed with handguns, 16.7% with other firearms, 20.4% with knives, 6.4% with a part of the body (usually the hands or feet), and 5.4% with blunt objects. Make a dot chart to display these data. Do you need an "other methods" category?

4.4. In 1988, doctorate-granting universities spent over $11 billion on research. The federal government provided 60.6% of this money, 8.4% came from state and local governments, 6.4% from industry, and 17.7% from the universities themselves. About what percent came from other sources, such as private foundations? You are preparing a talk in which you want to show these data. Make a clearly labeled pie chart of the data. For comparison, make a dot chart as well.

4.5. Here are data on the rate of deaths from cancer (deaths per 100,000 population) in the United States over the 50-year period 1935 to 1985. The data are from the *Statistical Abstract* and *Historical Statistics of the United States, Colonial Times to 1970*. (The *Historical Statistics* volumes supplement the *Statistical Abstract* by providing data from earlier years on many of the same subjects.)

Year	1935	1940	1945	1950	1955	1960	1965	1970	1975	1980	1985
Deaths	108.2	120.3	134.0	139.8	146.5	149.2	153.5	162.8	169.7	183.9	193.3

(a) Draw a line graph of these data designed to emphasize the rise in the cancer death rates. (Imagine you are trying to persuade Congress to appropriate more money to fight cancer.)

(b) Draw another line graph of the same data designed to show only a moderate increase in the death rate.

(c) Describe in words the overall pattern of these data.

4.6. Use Figure 4-5 to make an approximate table of the total amount of state and local property taxes in the years 1940, 1942, 1944, 1946, 1950, 1954, 1960, 1962, 1964, 1969, and 1971. Then draw a line graph from your table to see what a correct version of Figure 4-5 would look like.

4.7. Figure 4-9 shows a graph that appeared in the Lexington, Kentucky, *Herald-Leader* of October 5, 1975. Discuss the correctness of this graph.

4.8. Plotting two line graphs on the same axes is a useful way of comparing two time series. Here (again from the *Statistical Abstract*) are the death rates per 100,000 population for women age 45 to 54 from the two types of cancer that kill the most women, breast cancer and respiratory/intrathoracic (mainly lung) cancer. Plot the two together.

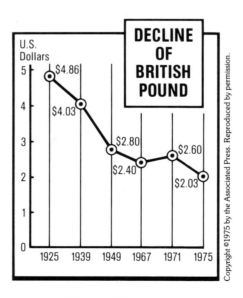

Figure 4-9.

Year	1940	1950	1960	1970	1975	1980	1985
Breast	47.5	46.9	51.4	52.6	50.4	48.9	46.7
Lung	6.2	6.7	10.1	22.2	28.0	34.8	35.9

Describe the long-term trends in cancer death rates for women as revealed by your graph. Now extrapolate your plots (this is risky) to estimate when lung cancer will pass breast cancer as the leading killer of women.

4.9. Here are data on the percent of females among people earning doctorates in 1987 in several fields of study (from data compiled by the Department of Education). Present these data in a well-labeled bar graph. For comparison, make a dot chart of the same data.

All fields	35.2%
Life sciences	35.0%
Education	54.9%
Engineering	6.9%
Physical sciences	17.3%
Psychology	53.3%

4.10. In Exercise 3.28 in Section 4 of Chapter 3, there is a newspaper graph of unemployment rates for several groups of workers. Comment on the correctness of the graph as a presentation of the numbers given. (We already saw that some of those numbers are probably incorrect.)

4.11. Figure 4-10 is a full-page advertisement for *Fortune* magazine. It contains eight separate graphic presentations of data. Comment briefly on the correctness of each one.

4.12. The table below shows the trend of U.S. imports of crude oil (in millions of barrels) from 1970 to 1988. The data are taken from the *Statistical Abstract*. Present them in a clear, well-labeled graph. Then describe the most important changes in oil imports since 1970.

Year	1970	1972	1974	1976	1978	1980	1982	1984	1986	1988
Oil imports	483	811	1269	1935	2320	1926	1263	1254	1525	1869

4.13. In 1988, 83.4% of the students at American colleges were white, 11.7% were black, 1.6% were Asian, 0.7% were American Indian, and 2.5% were of other races. (Because Hispanics may be of any race, the 6.4% of students who were Hispanic are included among these racial categories.)

    (a) Why don't the percents belonging to different races add to 100.0%

    (b) Present these data in a clear graph of your choice.

We asked America's top businessmen about business magazines. This is what they said.

Which one contains the best writing?
Fortune 56% Forbes 23% Business Week 19%

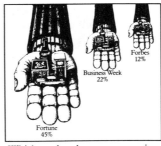

Which one has the most persuasive advertising?
Fortune 45% Business Week 22% Forbes 12%

Which one is easiest to read?
Business Week 45% Forbes 41% Fortune 11%

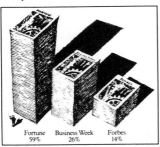

Which one best keeps its readers up to date on business events?
Business Week 82% Forbes 15% Fortune 2%

Which one carries the most interesting advertising?
Fortune 59% Business Week 20% Forbes 10%

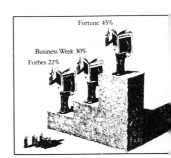

Which one is least accurate?
Fortune 8% Business Week 22% Forbes 37%

In which one would you like to see a major story on your company?
Fortune 59% Business Week 26% Forbes 14%

Which one is the most authoritative?
Fortune 45% Business Week 30% Forbes 22%

Erdos and Morgan recently asked officers of the top one thousand companies—chairmen, presidents, vice presidents, treasurers, secretaries and controllers—for their opinions of Business Week, Forbes and Fortune. 999 executives responded.

You can see the results for yourself. In nearly every instance, Fortune was the winner. Not just by a hair—but overwhelmingly.

Most authoritative? Best writing? Where would they most like to see their company story? Of course they named Fortune. You'd expect them to.

But why did they see the advertising in Fortune as more persuasive and more interesting—when the same advertising often runs in all three magazines?

Obviously, the Fortune climate makes something happen to advertising that doesn't happen anyplace else. It's a valuable edge.

Business leaders get more involved with Fortune, so they get more involved with the advertising. They respond to Fortune, so they respond to the advertising. The survey proves it.

The conclusion is clear and simple: dollar for dollar, your advertising investment gets more impact in Fortune.

You get more than mere advertising exposure in Fortune. You get real communication with the people who can *act* on your business or consumer message. Isn't that what advertising is all about?

FORTUNE
Nobody takes you to the top like Fortune.

Figure 4-10.

4.14. The 1989 *Encyclopedia Britannica Book of the Year* estimates that
the number of adherents to the larger religious groups in the world
are as follows.

Group	Number (millions)
Christians	1670
Moslems	881
Nonreligious or atheists	1116
Hindus	663
Buddhists	312
Chinese folk religionists	172

Present these data in a graph.

2.   Displaying distributions

The sampling distribution of a statistic describes the frequency or relative frequency with which the statistic takes each possible value in
repeated sampling. The idea of a distribution of values is not confined to
sample statistics, but is useful whenever our data are values of a variable
measured in an interval/ratio scale. The distribution of a variable can be
displayed by a table of frequencies of each value or, if these are numerous, of groups of possible values of the variable. Table 4-1 shows the
distribution of sizes of American farms. There are several effective ways
of displaying distributions graphically.

Histograms

Frequency or relative frequency distributions are most commonly displayed by *histograms*. Table 4-2 presents the percent of residents aged 65
years and over in each of the 50 states.

Example 1. To make a histogram of this distribution, proceed as
follows.

1. *Divide the range of the data into classes of equal width.* In this
    case our classes are

$$3.0 < \text{percent over } 65 \leq 4.0$$

$$4.0 < \text{percent over } 65 \leq 5.0$$

$$\cdot$$
$$\cdot$$
$$\cdot$$

$$17.0 < \text{percent over } 65 \leq 18.0$$

Table 4-2. Resident population 65 years old and over, by state (1987)

Alabama	12.4	Kentucky	12.3	N. Dakota	13.3
Alaska	3.6	Louisiana	10.8	Ohio	12.5
Arizona	12.7	Maine	13.4	Oklahoma	12.8
Arkansas	14.6	Maryland	10.7	Oregon	13.7
California	10.6	Massachusetts	13.7	Pennsylvania	14.8
Colorado	9.2	Michigan	11.5	Rhode Island	14.7
Connecticut	13.4	Minnesota	12.6	S. Carolina	10.7
Delaware	11.6	Mississippi	12.1	S. Dakota	14.0
Florida	17.8	Missouri	13.8	Tennessee	12.4
Georgia	10.0	Montana	12.5	Texas	9.7
Hawaii	10.1	Nebraska	13.8	Utah	8.2
Idaho	11.5	Nevada	10.6	Vermont	11.9
Illinois	12.1	New Hampshire	11.5	Virginia	10.6
Indiana	12.1	New Jersey	13.0	Washington	11.8
Iowa	14.8	New Mexico	10.0	W. Virginia	13.9
Kansas	13.6	New York	13.0	Wisconsin	13.2
		N. Carolina	11.8	Wyoming	8.9

SOURCE: *Statistical Abstract of the United States*, 1989.

We have specified the classes precisely so that each observation falls in exactly one class. A state with 4.0% of its residents aged 65 or older would fall in the first class, while 4.1% falls in the second.

2. *Count the number of observations in each class.* These counts are the frequencies of the classes.

3. *Draw the histogram.* The data scale is horizontal and the frequency scale vertical. Each bar represents a class. The base of the bar covers the class, and the bar height is the class frequency. The graph is drawn with no horizontal space between the bars (unless a class is empty, so that its bar has 0 height). Figure 4-11 is our histogram.

The histogram of Figure 4-11 shows at once that most states have between 8% and 15% of their residents over 65, most commonly about 12%. Two states stand out from the crowd: Florida with 17.8% of its residents over age 65 and Alaska with 3.6%. It is much easier to see both the general pattern and these deviations in the histogram than in the original table.

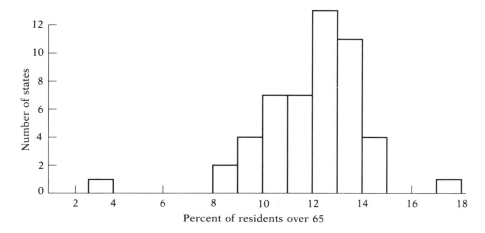

Figure 4-11. Histogram of the distribution of the percent of residents at least 65 years of age in the 50 states. Data from Table 4-2.

Histograms look like bar graphs, but they differ from bar graphs in several respects. First, the bars in a histogram should be vertical, and the base scale is marked off in equal units; there is no base scale in a bar graph (such as Figure 4-6). Second, the widths of the bars in a histogram have meaning: The base of each bar covers a class of values of the variable, the class whose frequency is the height of that bar; the widths of the bars in a bar graph have no meaning. Third, the bars in a histogram touch each other (unless some class has frequency zero), because their bases must cover the entire range of observed values of the variable, with no gaps. Even when the possible values of a variable have gaps between them, we extend the bases of the bars to meet halfway between two adjacent possible values. For example, in a histogram of the number of home runs hit by major-league baseball players, the bars representing 20 to 24 and 25 to 29 would meet at 24.5.

Just as with bar graphs, our eyes respond to the area of the bars in a histogram. If the heights are frequencies, the widths of all bars must be equal in order to avoid false impressions. So in dividing a set of data into classes, you should usually choose classes of equal width if you wish to draw a histogram. There is no single correct choice of class width. The goal is to display the distribution effectively. Avoid either cramming all the observations into a few big classes or sprinkling one or two observations into each of many tiny classes. We did not make a histogram of the farm data in Table 4-1 because the classes are of unequal width and the top class (2000 acres or more) has no upper endpoint. Making an accurate graph of the distribution in such a case is a more complicated affair than I want to deal with.

Stemplots

For small data sets, say less than 100 observations, a different graphical portrayal of the distribution is quicker to make and retains more information than a histogram. This is the *stemplot*. I will illustrate the making of a stemplot by asking how long American presidents live. Table 4-3 lists the ages at death of the presidents. To make a stemplot, we use the first digit of the age as the *stem*. Write the stems vertically, with a vertical line to their right. Then proceed through Table 4-3 writing the second digit of each age as a *leaf* to the right of the proper stem. The first entry made, for George Washington's death at age 67, is a 7 following the stem 6. The plot at step 2 of Figure 4-12 results. The final step is to arrange the leaves following each stem in increasing order from left to right. The final stemplot appears as step 3 in Figure 4-12.

As a display of the shape of a distribution, a stemplot looks like a histogram turned on its side. The chief advantage of a stemplot is that it retains the actual values of the observations. We can see, as we could not from a histogram, that the earliest death of a president occurred at age 46 and that two are tied for the latest death at age 90. The stemplot is also faster to draw, as long as we are willing to use the one or more uppermost digits as stems. This amounts to an automatic choice of classes and

Table 4-3. Age at death of U.S. presidents

Washington	67	Fillmore	74	Roosevelt	60
Adams	90	Pierce	64	Taft	72
Jefferson	83	Buchanan	77	Wilson	67
Madison	85	Lincoln	56	Harding	57
Monroe	73	Johnson	66	Coolidge	60
Adams	80	Grant	63	Hoover	90
Jackson	78	Hayes	70	Roosevelt	63
Van Buren	79	Garfield	49	Truman	88
Harrison	68	Arthur	56	Eisenhower	78
Tyler	71	Cleveland	71	Kennedy	46
Polk	53	Harrison	67	Johnson	64
Taylor	65	McKinley	58		

4		4	9 6
5		5	3 6 6 8 7
6		6	7 8 5 4 6 3 7 0 7 0 3 4
7		7	3 8 9 1 4 7 0 1 2 8
8		8	3 5 0 8
9		9	0 0

4	6 9
5	3 6 6 7 8
6	0 0 3 3 4 4 5 6 7 7 7 8
7	0 1 1 2 3 4 7 8 8 9
8	0 3 5 8
9	0 0

Step 1. The Stems   Step 2. The leaves   Step 3. Order the leaves

Figure 4-12. Making a stemplot of the ages at death of U.S. presidents.

can give a poor picture of the distribution. It is possible to split these stems to approach the flexibility of the classes in a histogram; see Exercise 4.21 for an example. Stemplots do not work well with large data sets, since the natural stems then have too many leaves.

Distributions

Both histograms and stemplots display the overall shape of a distribution of values as well as marked deviations from the overall pattern. One aspect of the shape of a distribution is its symmetry or skewness. A distribution is *symmetric* if its two sides are approximately mirror images of each other about a center line. The distribution of ages at death in Figure 4-12 is quite symmetric. In describing data, we don't insist on perfect symmetry but are content to give an approximate description of the distribution's shape.

Figure 4-13 is a histogram of the distribution of lengths of words used in articles in *Popular Science* magazine. In this case the height of each bar is not the count of words of each length but the proportion of words having that length. This is a *relative frequency histogram*. The shape of a histogram is the same whether it is frequencies or relative frequencies that we plot; this choice only changes the vertical scale. The distribution of word lengths is *skewed to the right*. Short words are common, but the distribution has a long tail at the right because there are a few words of 13, 14, or 15 letters.

In addition to symmetry or skewness, you might ask whether the distribution has a single *peak* (as in both Figures 4-12 and 4-13) or two or more peaks. Also note where the *center* of the distribution is and *how spread out* the values are about this center. The next section will show

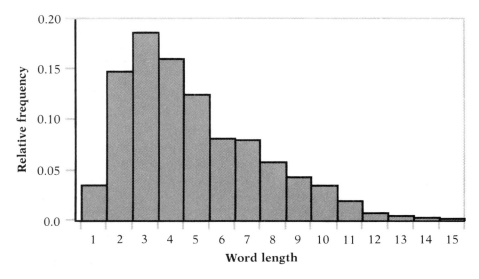

Figure 4-13. The distribution of the lengths of words appearing in articles in *Popular Science*.

how to use numbers to describe the center and spread. Symmetry or skewness, single or multiple peaks, and center and spread are all features of a distribution's overall shape. In examining a stemplot or histogram, you should also look for deviations from the overall shape. Look in particular for *outliers*, individual observations that fall outside the pattern. Outliers point to special cases that are often worth comment. The two outliers in Figure 4-11 point to the unusual concentration of older people in Florida and the unusually young population of Alaska. Errors in writing the data or entering them into a computer often produce outliers, and can be corrected when the outliers are spotted.

Relative-frequency histograms are also used to portray the sampling distributions of statistics in a more general way than we have done thus far. Figure 1-2 in Section 3 of Chapter 1, our first histogram, showed the distribution of the sample proportion $\hat{p}$ in 200 SRSs drawn independently from the same population. It is a frequency histogram of the sampling distribution of $\hat{p}$. Its shape is characteristic of the sampling method and its spread allows us to evaluate the precision of the statistic as an estimate of the corresponding population parameter. But in a strict sense, the sampling distribution of $\hat{p}$ is not this specific observed distribution; it is the ideal distribution that is approached as more and more samples are taken. This distribution can be found by long experiments or, much more quickly, by mathematics. It must be presented in terms of relative frequencies because it does not refer to any specific number of trials.

Figure 4-14 redraws the results of the 200 SRSs of Chapter 1 as relative frequencies (the dashed bars) and also presents the theoretical sampling distribution (solid bars) for comparison. The sampling distribution is similar to the distribution observed in 200 trials, but it differs in several respects. In particular, the sampling distribution is more symmetric than the observed distribution. Some of the deviation between the observed and theoretical distributions is caused by the chance outcome of the particular 200 trials made. But some of the deviation occurs also because the theoretical sampling distribution describes truly random sampling, while the scoop gives only approximately random samples of beads. If we use a table of random digits to select the samples, the observed distribution of values of $\hat{p}$ will always approach the sampling distribution as more and more samples are drawn.

The characteristic shape of the sampling distribution appears even more clearly in Figure 4-15. This is a *frequency polygon*, drawn by placing a dot in the center of the top of each bar of the histogram, then drawing straight lines between the dots. As a final approximation, used to convey quickly the shape of a distribution, we can replace the frequency polygon by a smooth *frequency curve*. It is a remarkable fact that the sampling distribution of $\hat{p}$ from random sampling can always be approximated by a smooth curve of a certain kind. These are the *normal curves*, which we will study further in Section 5. Figure 4-16 is the

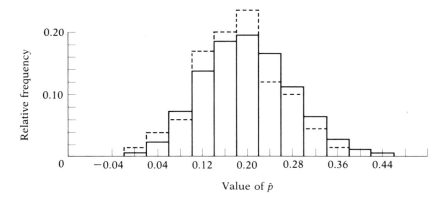

Figure 4-14. Sampling distribution for a sample proportion, observed (broken bars) and theoretical (solid bars).

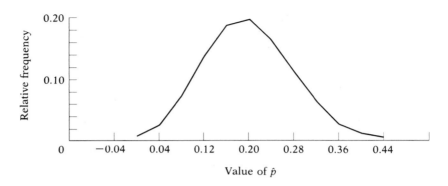

Figure 4-15. A frequency polygon.

normal curve of the sampling distribution of $\hat{p}$ that was described earlier by a histogram in Figure 4-14 and a polygon in Figure 4-15. A frequency curve provides a mathematical description of the distribution. It does not depend on either a particular number of trials or a particular grouping of observations into classes. The symmetry of the sampling distribution, for example, is portrayed clearly. With frequency curves, we have completed our discussion of graphical methods of presenting distributions.

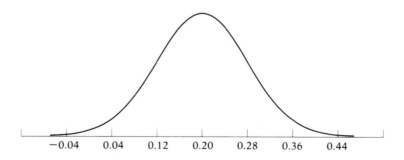

Figure 4-16. A normal curve.

Section 2 exercises

4.15. Here are data on the percentages of words of 1 to 12 letters used by Shakespeare in his plays. (From C. B. Williams, *Style and Vocabulary: Numerical Studies* [London: Griffin, 1970.])

Length	1	2	3	4	5	6	7	8	9	10	11	12
Percent	4.8	17.6	22.5	23.8	12.4	7.1	5.3	3.1	1.8	0.9	0.3	0.2

(a) Make a histogram of this distribution. Then describe its overall shape.

(b) How does Shakespeare's word-length distribution compare with the similar distribution in Figure 4-13 for *Popular Science*? Look in particular at short words (2, 3, and 4 letters) and very long words (more than 10 letters).

4.16. The distribution of the ages of a nation's residents has a strong influence on economic and social conditions. The following table shows the age distribution of U. S. residents in 1950 and 2075, in millions of persons. The 1950 data come from that year's census, while the 2075 data are projections made by the Census Bureau.

Age group	1950	2075
Under 10 years	29.3	34.9
10 to 19 years	21.8	35.7
20 to 29 years	24.0	36.8
30 to 39 years	22.8	38.1
40 to 49 years	19.3	37.8
50 to 59 years	15.5	37.5
60 to 69 years	11.0	34.5
70 to 79 years	5.5	27.2
80 to 89 years	1.6	18.8
90 to 99 years	0.1	7.7
100 to 109 years	—	1.7
TOTAL	151.1	310.6

(a) Make a histogram of the 1950 age distribution, and describe the main features of the distribution. Look in particular at the number of children relative to the rest of the population.

(b) Make a histogram of the projected age distribution for the year 2075. What are the most important changes in the U. S. age distribution in the 125-year period between 1950 and 2075?

4.17. A study of bacterial contamination in milk counted the number of coliform organisms (fecal bacteria) per milliliter in 100 specimens of milk purchased in east-coast groceries. The U.S. Public Health Service recommends no more than 10 coliform bacteria per milliliter. Here are the data:

```
5  8  6  7  8  3  2  4  7  8  6  4  4  8  8  8  6 10  6  5
6  6  6  6  4  3  7  7  5  7  4  5  6  7  4  4  4  3  5  7
7  5  8  3  9  7  3  4  6  6  8  7  4  8  5  7  9  4  4  7
8  8  7  5  4 10  7  6  6  7  8  6  6  6  0  4  5 10  4  5
7  9  8  9  5  6  3  6  3  7  1  6  9  6  8  5  2  8  5  3
```

(a) Make a table of the frequencies of each of the values 0 to 10 in this set of 100 observations.

(b) Draw a frequency histogram of these data. (Graph paper makes this easier.)

(c) On the histogram you have drawn, put a second vertical scale, labeled in relative frequency rather than frequency. Because there are 100 observations, relative frequency = frequency/100. Your histogram is now a relative frequency histogram.

(d) Make a frequency polygon similar to Figure 4-15 by connecting the centers of the tops of the bars in your histogram.

(e) Describe the shape of the distribution. Are the data symmetric? If so, approximately where is the center about which they are symmetric? Are the data strongly skewed? If so, in which direction? Are there any outliers?

4.18. Make a stemplot of the data in Table 4-2. Take the whole numbers as stems and the tenths as leaves. Do you think the stemplot gives as clear a picture of the data as the histogram in Figure 4-11?

4.19. *Consumer Reports* magazine (June 1986, p. 367) presented the following data on the number of calories in a hot dog for each of 17 brands of meat hot dogs.

173	191	182	190	172	147	146	139	175
136	179	153	107	195	135	140	138	

Make a stemplot of the distribution of calories in meat hot dogs and briefly describe the shape of the distribution. Most brands of meat hot dogs contain a mixture of beef and pork, with up to 15% poultry allowed by government regulations. The only brand with a different makeup was *Eat Slim Veal Hot Dogs.* Which point on your stemplot do you think represents this brand?

4.20. A marketing consultant observes 50 consecutive shoppers at a grocery store and records how much each shopper spends in the store. Here are the data (in dollars), arranged in increasing order for convenience.

2.32	6.61	6.90	8.04	9.45
10.26	11.34	11.63	12.66	12.95
13.67	13.72	14.35	14.52	14.55
15.01	15.33	16.55	17.15	18.22
18.30	18.71	19.54	19.55	20.58
20.89	20.91	21.13	23.85	26.04
27.07	28.76	29.15	30.54	31.99
32.82	33.26	33.80	34.76	36.22
37.52	39.28	40.80	43.97	45.58
52.36	61.57	63.85	64.30	69.49

Make a histogram of these data. Then describe the shape of the distribution. Are there any clear outliers?

4.21. To make a stemplot of the data in Exercise 4.20, first *truncate* the data by discarding the cents, leaving only the dollar amounts spent. (You could also *round* the data to the nearest dollar, but truncating is faster.) Make a stemplot with tens of dollars as the stems and dollars as the leaves.

You can make a stemplot with twice as many stems by *splitting the stems*. Write two 0s, two 1s, and so on, as stems at the left of your vertical bar. The first of each pair of stems gets leaves 0 through 4, and the second gets leaves 5 through 9. Spreading out the stemplot in this manner is helpful if many observations fall on only a few stems. Make a stemplot of the truncated grocery spending data with split stems. Which of your two stemplots do you prefer?

4.22. Climatologists interested in flooding gather statistics on the daily rainfall in various cities. The following data set gives the *maximum* daily rainfall (in inches) for each of the years 1941 to 1970 in South Bend, Indiana. (Successive years follow each other across the rows in the table.)

1.88	2.23	2.58	2.07	2.94	2.29	3.14	2.15	1.95	2.51
2.86	1.48	1.12	2.76	3.10	2.05	2.23	1.70	1.57	2.81
1.24	3.29	1.87	1.50	2.99	3.48	2.12	4.69	2.29	2.12

Make a stemplot for these data; truncate and split stems as described in Exercise 4.21 if necessary. Describe the general shape of the distribution and any prominent deviations from the overall pattern.

4.23. Agronomists have developed varieties of corn that have increased amounts of the essential amino acid lysine. In a test of the protein quality of this corn, an experimental group of 20 one-day-old male chicks was fed a corn-soybean ration containing high-lysine corn. A control group of another 20 chicks was fed the same diet except that normal corn was used. The weight gains (in grams) after 21 days are recorded below.

Control group				Experimental group			
380	321	366	356	361	447	401	375
283	349	402	462	434	403	393	426
356	410	329	399	406	318	467	407
350	384	316	272	427	420	477	392
345	455	360	431	430	339	410	326

(a) This experiment was designed using the principles of Chapter 2. Briefly discuss the proper design of this experiment.

(b) To compare the two distributions of weight gains, make a *back-to-back stemplot:* Draw a single column of stems with vertical lines on both sides of it. Now add leaves to these stems on both sides, the experimental group on the right and the control group on the left. Then, order the leaves so that they increase as you move away from the stem.

(c) What does your plot show about the effect of high-lysine corn on weight gain?

4.24. Although the stemplot of Exercise 4.23 is effective for comparing the two weight-gain distributions, the natural stems divide the data into too many classes to show the shape of the distributions clearly.

(a) Make separate frequency tables for the weight gains in the experimental group and the control group. Use the classes 270–299, 300–329, 330–359, and so on.

(b) Draw separate histograms for the two groups. For easy comparison, draw them one above the other, with the scales on the horizontal axes aligned. What does this plot show about the effect of high-lysine corn on weight gain?

(c) Are the distributions of weight gains symmetric or skewed?

4.25. Here are the number of home runs that Babe Ruth hit in each of his 15 years with the New York Yankees, 1920 to 1934.

54   59   35   41   46   25   47   60   54   46   49   46   41   34   22

Ruth's record of 60 home runs in a season was broken by another Yankee, Roger Maris, who hit 61 home runs in 1961. Here are Maris' home-run totals for his 10 years in the American League.

13   23   26   16   33   61   28   39   14   8

Make a back-to-back stemplot to compare Ruth and Maris (see Exercise 4.23 for instructions on back-to-back stemplots). Describe the most important features of the comparison.

4.26. Draw a frequency curve for a distribution that is skewed to the left.

4.27. The entries in the table of random digits have the property that each value 0, 1, 2, 3, 4, 5, 6, 7, 8, and 9 occurs equally often in the long run.
  - **(a)** Make a frequency table and draw a frequency histogram for the entries in the first three rows of Table A (120 digits in all).
  - **(b)** Is this distribution approximately symmetric?
  - **(c)** Draw a frequency curve representing the distribution of values in a very large number of observations from a table of random digits.

3.   Measuring center or average

Tables organize data, and graphs present a vivid overall picture. But more specific aspects of data such as their average and their variability are most succinctly summarized by a few well-chosen numbers. We often read or hear such statements as "The median income of households covered by Medicaid was $9700, less than one-third the median for all households." The median is one of the numerical measures called *descriptive statistics* that summarize specific features of a data set. Descriptive statistics are as important in your statistical vocabulary as random samples and line graphs. We have already met counts (frequencies) and rates or proportions (relative frequencies). Now we will lengthen the list of descriptive statistics we can claim as acquaintances, if not as friends. Our task in this section is to describe by a number a fundamental aspect of the overall pattern of any distribution, its center or average value.

Almost any presentation of data uses averages—average gas mileage, average income, average score on the exam, average absolutely refractory period of the tibial nerves of rats fed DDT. (I am quoting that last one from a report on the effects of persistent pesticides—don't blame me for it.) Everyone has heard that statistics features the mean, the median, and the mode. Those are in fact the three "averages" we will study. But do remember that you absorbed a good deal of statistics before meeting this famous trio. Our subject has other (and more interesting) parts, but to be statistically literate we must add the mean, the median, and the mode to our vocabulary. Here are brief definitions, followed in turn by a discussion of each measure of center.

> The *mean* of a set of *n* observations is the arithmetic *average*; it is the sum of the observations divided by the number of observations, *n*.
>
> The *median* is the *typical value*; it is the midpoint of the observations when they are arranged in increasing order.
>
> The *mode* is the *most frequent value*; it is any value having the highest frequency among the observations.

We have defined these measures for any set of *n* observations. The mean, the median, and the mode are statistics if these observations are a sample; they are parameters if the observations form an entire population.

Calculating measures of center

The *mean* is the usual arithmetic average. To find the mean, add the observations and divide by the number of them. The mean of the 10 observations

$$4, 6, 10, 3, 7, 6, 6, 8, 5, ,9$$

is

$$\frac{4 + 6 + 10 + 3 + 7 + 6 + 6 + 8 + 5 + 9}{10} = \frac{64}{10} = 6.4.$$

When someone says "average," it is usually the mean that is intended. Now that we know the proper vocabulary, be sure to say "mean" and not "average." Means are so common that a compact notation is useful.

Remember the idea from algebra of letting letters stand for numbers so recipes applicable to any set of numbers can be given? We will do this. A set of n observations is denoted by

$$x_1, x_2, \ldots, x_n$$

and the mean of this set of observations is

$$\bar{x} = \frac{x_1 + x_2 + \cdots + x_n}{n}$$

It is read "x bar" and is such common notation that in reading literature in psychology, sociology, medicine, and other fields, you can assume that an $\bar{x}$ or $\bar{y}$ or $\bar{z}$ used without explanation is a mean.

The *median* is the midpoint of the observations when they are arranged in increasing order of magnitude. To find the median of the numbers

$$8, 4, 9, 1, 3$$

arrange them in increasing order as

$$1, 3, 4, 8, 9$$

The median is 4, because it is the midpoint; two observations fall below 4 and two fall above it. Whenever the number of observations is odd, the middle observation is the median. If the number of observations is even, there is no one middle observation. But there is a middle pair, and we take the median to be the mean of this middle pair, the point halfway between them. So the median of

$$8, 4, 1, 9, 1, 5$$

is found by arranging these numbers in increasing order:

$$1, 1, 4, 5, 8, 9$$

and taking the mean of the middle pair:

$$\text{median} = \frac{4 + 5}{2} = 4.5$$

Note that three observations fall below 4.5 and three fall above.

Some words to the wise about finding medians are needed. First, never fail to arrange the observations in increasing order; the middle value in the haphazard order in which the numbers come to you is not the median. Second, never fail to write down all the observations, even if some have the same value. The median of

$$4, 5, 5, 6, 6, 6, 8, 9$$

is 6 because both of the middle pair of numbers have the value 6. Third, a recipe can save you from trying to locate the middle one of 471 observations or the middle pair of 232 observations by counting in from both ends. Here it is.

If there are n observations in all, find the number $(n + 1)/2$. Arrange the observations from smallest to largest, and count $(n + 1)/2$ observations up from the bottom. This gives the *location* of the median in the list of observations.

Returning to the examples, the set of data 1, 3, 4, 8, 9 has $n = 5$ observations. Because $(n + 1)/2 = (5 + 1)/2 = 3$, the median is the third number in the list, which is 4. The data set 4, 5, 5, 6, 6, 6, 8, 9 has $n = 8$ observations. So $(n + 1)/2 = (8 + 1)/2 = 4.5$. This means that the median has location "four and a half," or midway between the fourth and fifth numbers in the list. (Note that 4.5 is not the median but merely its location in the list.) The median of a set of 232 numbers has location $(232 + 1)/2 = 116.5$, or midway between the 116th and 117th numbers when arranged from smallest to largest. The median of 471 observations is the 236th in order of magnitude, because

$$\frac{471 + 1}{2} = \frac{472}{2} = 236$$

The *mode* is any value occurring most frequently in the set of observations. It is convenient to arrange observations in increasing order as an aid to seeing how often each value occurs. The mode of

$$4, 5, 5, 6, 6, 6, 8, 9$$

is 6 because this value occurs three times and no other value appears more than twice. What about

$$1, 4, 5, 5, 5, 6, 8, 9, 9, 9, 12?$$

Both 5 and 9 are modes, because both are "most frequent" in this data set. Such a data set is called *bimodal*. When no value occurs more than once, we could say that all are modes; that's not very helpful, so we prefer to say that such a set of data has no mode.

Using measures of center

Having made the acquaintance of mean, median, and mode, we now ought to inquire how they do their job of describing a set of observations. Certainly arithmetic average, midpoint, and most frequent value are quite different notions with different uses. Here is a survey of their use.

The *mode* is little used, because it records only the most frequent value, and this may be far from the center of the distribution of values. We have also seen that there may be several modes, or none. The chief advantage of the mode is that of our three measures, it alone makes sense for variables measured in a nominal scale. It is nonsense to speak of the median sex or mean race of United States ambassadors, but the most frequent (modal) sex is male and the modal race is white.

The *median* uses only order information in the data. (How many observations are above a point? How many are below?) The median therefore makes sense for ordinal variables as well as for interval/ratio variables. The median does not employ the actual numerical value of the observations; it is not affected by how far above or below it the observations fall but only requires that equal numbers of observations fall above and below. The *mean* alone of our measures of center does use the actual numerical values of the observations. Thus the mean utilizes more of the information in the data than either the mode or the median. For this reason, the mean is the most common measure of center. In a strict sense, the mean makes sense only for interval/ratio data, because it requires adding the observations. But in practice, means are frequently computed for ordinal variables as well.

The median is sometimes a better description of the center of a distribution of values than is the mean exactly because the median does *not* respond to actual numerical values. An example will show how this insensitivity can be a virtue.

"Should we scare the opposition by announcing our mean height or lull them by announcing our median height?"

Example 2. A professional basketball team has four players earning $200,000 per year, five who earn $500,000 and a superstar who earns $3,100,000. The mean salary for the team's 10 players is therefore

$$\frac{(4)(200,000) + (5)(500,000) + 3,100,000}{10} = \frac{6,400,000}{10} = \$640,000.$$

You see that the superstar's salary has pulled the mean well above the amount paid to any other player. The mean is quite sensitive to a few extreme observations. The median, though, is $500,000 and does not change even if the superstar earns $10 million; his salary is just one number falling above the midpoint. Lest you think that extreme differences between the median and the mean are confined to artificial examples, consider the claim of the Insurance Information Institute that the average award in successful product liability lawsuits passed $1 million in 1984. That's the mean. The median award was $271,000, a less shocking figure. Because monetary data like salaries and court awards often have a few extremely high observations, descriptions of these distributions usually use the median: "half earned more than this, and half earned less." Medians are prominent in the income data in the *Statistical Abstract*, for example. This is not to say that the median always should be used for data containing extreme observations. Insurance companies are interested in the total amount they must pay out for product liability lawsuits. The median award is uninformative, while the mean tells us what we want to know, because it is computed from the sum of the awards. Always ask which of most frequent value, midpoint, or arithmetic average will best represent the data for your intended use.

The different response of the mean and the median to extreme values can be pictured by using the frequency curve of a distribution. The mode is the value where the curve peaks. The median is the value such that half the area under the curve lies to the left and half to the right of it. These facts are true because the frequency curve is a smooth version of the frequency histogram. So it is highest at the most frequent value, or mode. And because areas in a histogram represent frequencies, equal areas lie to the left and to the right of the median. Figure 4-17 illustrates these facts.

It is not so clear where to place the mean on a frequency curve or histogram. In fact, if we think of the curve as cut out of solid material, the mean is the point where the shape would balance. Figure 4-18 illustrates this.

If a frequency curve is symmetric, the balance point and the equal-areas point are the same. The mean and median are the same for symmetric distributions. This is one reason that mean scores are often reported for standardized tests even though the scores have an ordinal scale. The distribution of such scores is often roughly symmetric, so the mean and median fall close together. If one tail of a frequency curve is stretched to correspond to a few extreme observations on one side, the frequency curve becomes skewed in that direction. A little weight far out in one direction moves the balance point quite a bit in that direction. So the mean moves farther toward the long tail than does the

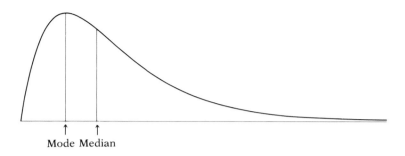

Mode Median

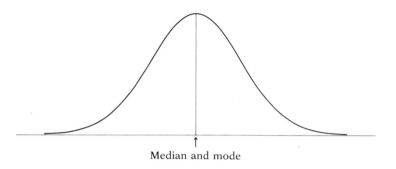

Median and mode

Figure 4-17. The mode and median of a frequency distribution. The mode is the point at which the frequency curve attains its highest value. The median is the point that divides the area under the curve into two equal parts to the left and to the right of it. The median is the center point of any symmetric frequency curve. This normal curve is highest at the center, so the center point is also the mode.

median. We saw a numerical example earlier, and Figure 4-19 gives a pictorial view.

Now you are an expert on mean, median, and mode. I hope they have lived up to your expectations. Be warned, however, that averages of all kinds can play tricks if you are not alert. Some examples of these tricks appear in the exercises. Arithmetic is never a substitute for understanding.

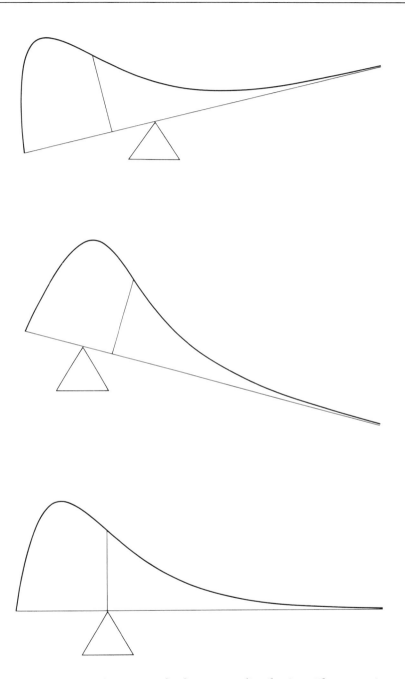

Figure 4-18. The mean of a frequency distribution. The mean is the center of gravity of the frequency curve, the point about which the curve would balance on a pivot placed beneath it.

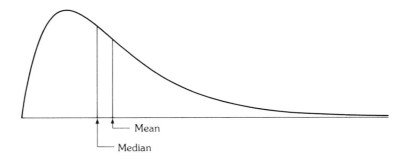

Figure 4-19. The mean and median of a skewed distribution. The mean is located farther toward the long tail of a skewed frequency curve than is the median.

Section 3 exercises

4.28. Compute the mean, the median, and the mode for each of the following sets of numbers:
   (a) 4, 15, 2, 8, 4, 6, 10
   (b) 4, 2, 2, 6, 4, 4, 15, 8, 2, 17, 10, 4, 2, 6
   (c) 6, −3, 0, −11, 7, 120, −3

4.29. Find the mean and median number of home runs hit in a season by Babe Ruth, from the data given in Exercise 4.25. Then do the same for Roger Maris.

4.30. Last year a small accounting firm paid each of its five clerks $22,000, two junior accountants $50,000 each, and the firm's owner $270,000. What is the mean salary paid at this firm? How many of the employees earn less than the mean? What is the median salary? What is the mode of the salaries?

4.31. From the data in Exercise 4.22, find the mean and median of the maximum one-day rainfall amounts. Explain in terms of the shape of the distribution why the mean is larger than the median.

4.32. Return to the sample of 100 counts of coliform bacteria in milk given in Exercise 4.17. You drew a frequency histogram for the data as part of that exercise.
   (a) Compute the mean and the median of the data.
   (b) Explain in terms of the shape of the frequency distribution why these measures of center fall as they do (close together or apart).

4.33. Here is a sample of 100 reaction times of a subject to a stimulus, in milliseconds:

```
10 14 11 15  7  7 20 10 14  9  8  6 12 12 10 14 11 13  9 12
13 11 12 10  8  9 14 18 12 10 10 11  7 17 12  9  9 11  7 10
14 12 12 10  9  7 11  9 18  6 12 12 10  8 14 15 12 11  9  9
11  8 11 10 13  8 11 11 13 20  6 13 13  8  9 16 15 11 10 11
20  8 17 12 19 14 17 12 18 16 15 16 10 20 11 19 20 13 11 20
```

(a) Compute the mean and the median of these data.
(b) Find the frequency of each outcome 6, 7, 8, . . ., 20, and draw a frequency histogram for these data.
(c) Explain in terms of the shape of the frequency distribution why these measures of center fall as they do (close together or apart).

4.34. Identify which measure of center (mean, median, or mode) is the appropriate "average" in each of the following situations:
(a) Someone declares, "The average American is a white female."
(b) Middletown is considering imposing an income tax on citizens. The city government wants to know the average income of citizens so that it can estimate the total tax base.
(c) In an attempt to study the standard of living of typical families in Middletown, a sociologist estimates the average family income in that city.

4.35. As part of its twenty-fifth reunion celebration, the Class of '70 of Central New Jersey University mails a questionnaire to its members. One of the questions asks the respondent to give his or her total income last year. Of the 820 members of the class of '70, the university alumni office has addresses for 583. Of these, 421 return the questionnaire. The reunion committee computes the mean income given in the responses and announces, "The members of the class of '70 have enjoyed resounding success. The average income of class members is $120,000!"

This result exaggerates the income of the members of the Class of '70 for (at least) three reasons. What are these reasons?

4.36. According to the Department of Commerce, the mean and median prices of new houses sold in the United States in 1989 were $129,900 and $159,000. Which of these numbers is the mean, and which is the median? Explain your answer.

4.37. The mean age of 5 persons in a room is 30 years. A 36-year-old person walks in. What is now the mean age of the persons in the room? Suppose that the median age is 30 years and a 36-year-old person enters. Can you find the new median age from this information?

4.38. You wish to measure the average speed of vehicles on the interstate highway on which you are driving, so you adjust your speed until the number of vehicles passing you equals the number you are passing. Have you found the mean speed, the median speed, or the modal speed of vehicles on the highway?

4.39. Figure 4-20 presents three frequency curves, each with several points marked on them. At which of these points on each curve do the mean, the median, and the mode fall? (More than one measure of center may fall at one point.)

4.40. Make up a list of numbers of which only 10% are above the average (that is, above the mean). What percent of the numbers in your list fall above the median?

4.41. Which of the mean, median, and mode of a list of numbers must always appear as one of the numbers in the list?

4.42. In computing the median income of any group, federal agencies omit all members of the group who had zero income. Give an

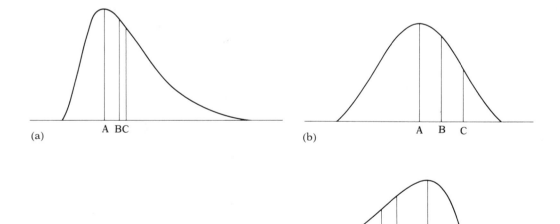

(a)   A BC

(b)   A   B   C

(c)   A B   C

Figure 4-20.

example to show how the median income of a group (as reported by the federal government) can go *down* when the group becomes better off economically.

4.43. The mean age of members of the class of '40 at their fiftieth reunion was 71.9 years. At their fifty-first reunion the next year, the mean age was 71.5 years. How can the mean age decrease when all the class members are a year older?

4.44. You drive 5 miles at 30 miles per hour, then 5 more miles at 50 miles per hour. Have you driven at an average speed of 40 miles per hour? (Your average speed is total miles driven divided by the time you took to drive it.)

4.45. The following paragraph contains three major statistical blunders. Describe them in one sentence each.

> *In response to protests over the firing of coach Rockne, the university administration released the results of a questionnaire mailed to all 90,000 living alumni. Of the 8000 responses, 65% favored firing the coach, showing that the alumni want Rockne to go. And these alumni are supporters of the athletic program — the mean contribution of those responding was over $200 last year, including James Barkaddidy's handsome gift of $1 million for a new scoreboard. What is more, 25% of the faculty want Rockne fired, so that when faculty and alumni are considered together, an overwhelming 90% want a new coach.*

4. Measuring spread or variability

Useful as the measures of center are, they are usually incomplete and often misleading without some accompanying indication of how spread out or dispersed the data are. The median family income of $30,853 in 1987 masks the fact that 11.7% of all families received less than $10,000 and that the top 5% earned over $86,000. The distribution of family incomes is skewed to the right and very spread out. Knowing the median alone gives us an inadequate description of the distribution of incomes for American families. It lumps the Rockefellers in with the welfare cases. Similarly, the mean potency of a drug may be exactly what the doctor ordered, but if the potency of the lot varies too much, many doses will be ineffective owing to low potency or perhaps dangerous owing to high potency. The simplest adequate summary of a univariate distribution usually requires both a measure of center and a measure of variability.

Quartiles and the five-number summary

When the median is used to measure center, both the variability and general shape of a distribution can be described by giving several percentiles.

> The *cth percentile* of a set of numbers is a value such that *c* percent of the numbers fall below it and the rest fall above.

You have met percentiles if after taking a standardized test such as the SAT you received a report of the result in the form "Raw score 590, percentile 81." You scored 590, but more informative is the fact that 81% of those taking the exam had scores lower than yours.

The median is the 50th percentile. Some other percentiles are important enough to have individual names.

> The *lower quartile* is the 25th percentile.
>
> The *upper quartile* is the 75th percentile.

A convenient way of indicating the spread of a data set is to give the quartiles along with the median. Often we also give the extremes (the smallest and largest individual observations). Median, quartiles, and extremes offer a reasonably complete *five-number summary* of a data set. An example will illustrate the details.

Example 3. Table 4-4 shows the Nielsen ratings of all 56 regularly scheduled national prime-time television shows for a week in the summer of 1989. Each rating point represents 1% of all households having a television set. The ratings are already arranged in numerical order, so the median is midway between the 28th and 29th in that order. The median is

$$\frac{10.6 + 10.4}{2} = \frac{21.0}{2} = 10.5.$$

To find the lower quartile, compute the median of all observations falling below the location of the overall median. There are 28 ratings below 10.5. By our usual rule for the median of 28 observations, the median of these ratings is halfway between the 14th and 15th ratings from the bottom of the list. So the lower quartile is

Table 4-4. Nielsen television ratings, July 31–August 6, 1989

Program	Network	Rating	Program	Network	Rating
Roseanne	ABC	18.8	13 East	NBC	10.2
Wonder Years	ABC	16.7	Prime Time Live	ABC	10.0
Who's the Boss	ABC	15.5	MacGyver	ABC	9.5
Cheers	NBC	15.4	Kate & Allie	CBS	9.2
Coach	ABC	15.4	48 Hours	CBS	9.2
Golden Girls	NBC	14.9	Gregory	CBS	9.0
Murder, She Wrote	CBS	14.6	Harrison		
			Amen	NBC	8.9
Empty Nest	NBC	14.5	Hooperman	ABC	8.5
Designing Women	CBS	14.5	My Two Dads	NBC	8.4
			Knight & Daye	NBC	8.3
Dear John	NBC	14.4	Robert	ABC	8.3
Cosby Show	NBC	13.9	Guillaume		
Murphy Brown	CBS	13.8	China Beach	ABC	8.2
Hunter	NBC	13.6	Mission:	ABC	7.4
Different World	NBC	13.5	Impossible		
Newhart	CBS	13.4	Family Ties	NBC	7.2
60 Minutes	CBS	13.0	Highway to	NBC	7.1
Unsolved Mysteries	NBC	12.2	Heaven		
			Wise Guy	CBS	7.1
Full House	ABC	12.0	West 57th	CBS	6.9
Yesterday	NBC	12.0	Disney	NBC	6.7
Head of the Class	ABC	11.9	A Man Called	ABC	6.5
Growing Pains	ABC	11.8	Hawk		
Hogan Family	NBC	11.6	Paradise	CBS	5.9
Mr. Belvedere	ABC	11.5	Curse of the	CBS	5.5
L. A. Law	NBC	11.5	Corn People		
Hot Prospects	CBS	11.3	Tour of Duty	CBS	5.5
Thirtysomething	ABC	11.3	Incredible	ABC	5.4
Alf	NBC	11.2	Sunday		
Perfect Strangers	ABC	10.6	Protect and Surf	ABC	5.1
			Beauty and the	CBS	5.0
Jake and the Fatman	CBS	10.4	Beast		
Just the Ten of Us	ABC	10.3	Chain Letter	ABC	4.5

SOURCE: A. C. Nielsen Co. Published by permission

$$\frac{7.4 + 8.2}{2} = \frac{15.6}{2} = 7.8$$

Similarly, *the upper quartile is the median of all observations falling above the location of the overall median.* Check that this is

$$\frac{13.5 + 13.4}{2} = \frac{26.9}{2} = 13.45$$

The extreme ratings are 4.5 and 18.8. The five-number summary of the distribution of ratings is therefore 4.5, 7.8, 10.5, 13.45, 18.8.

When several observations are tied, be sure to pay attention to the *location* of the overall median, as given by the $(n + 1)/2$ rule, when computing quartiles. For example, the median of the 12 observations

$$2 \ 4 \ 7 \ 11 \ 11 \ 11 \mid 11 \ 14 \ 16 \ 16 \ 24 \ 29$$

is located as marked, between the sixth and seventh observations from the bottom of the ordered list. The value of the median is 11. The lower quartile is the median of the six observations falling below the location of the overall median, not of the three observations with values less than 11. The lower quartile of these data is 9, and the upper quartile is 16.

The extremes indicate the overall spread of the data but are sensitive to outliers. The quartiles and the median (which is the middle quartile) divide the data into quarters. The lower and upper quartiles show the spread of the middle half of the data and are a good description of the variability of a distribution. The spacing of the quartiles about the median also gives an indication of the symmetry or skewness of the distribution. In a symmetric distribution, the lower and upper quartiles are equally distant from the median. But the upper quartile will be farther above the median than the lower quartile is below it in most distributions that are skewed to the right. Figure 4-21 illustrates these facts.

The extremes of a large set of data may be very far from the median and not descriptive of any but outlying observations. In such cases we can replace the extremes by the lower and upper *deciles* in the five-number summary. The lower decile is the 10th percentile, and the upper decile is the 90th percentile. We will not give precise rules for computing the deciles and other percentiles but will be satisfied with approximate answers. For example, to find the upper decile of the 56 Nielsen ratings in Table 4-4, note that 90% of 56 is

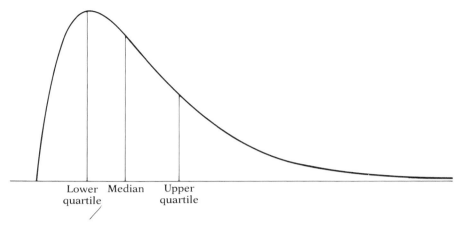

Lower   Median   Upper
quartile   quartile

Figure 4-21.

$$(0.90)(56) = 50.4.$$

This rounds off to 50, so we will take the 50th rating (in increasing order) as the upper decile. That rating is the 14.6 earned by "Murder, She Wrote."

Boxplots

The five-number summary of a distribution lends itself to a graphical portrayal that is particularly helpful in comparing two or more distributions. This is the *boxplot*. Figure 4-22 gives boxplots for the distributions of Nielsen ratings for the three networks. Each central box has its ends at the quartiles, and the median of the distribution is marked by the line within the box. The "whiskers" at either end extend to the extremes. For example, the 21 ABC programs rated have the five-number summary 4.5, 7.8, 10.3, 11.95, 18.8. These numbers determine the ABC boxplot in Figure 4-22.

Figure 4-22 allows immediate comparison of the popularity of the three networks' shows. Overall, NBC won the ratings battle that week, as comparison of the boxes shows. But the upper whiskers remind us that ABC had the highest-rated individual program. Figure 4-22 is another example of the effectiveness of graphs in conveying information.

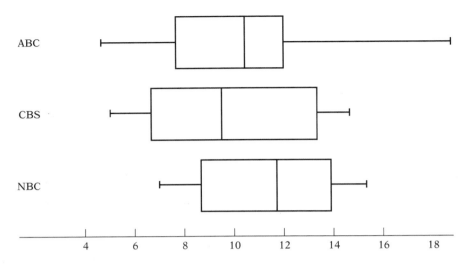

Figure 4-22. Boxplots of the Nielsen TV ratings for prime-time programs of the three major networks for the week of July 31 to August 6, 1989.

Standard deviation

Though the five-number summary is the most generally useful numerical description of a distribution, it is not the most common. That distinction belongs to the combination of the mean with the *standard deviation*. The mean, like the median, is a measure of center; the standard deviation, like the quartiles and extremes in the five-number summary, is a measure of spread. The standard deviation and its close relative, the *variance*, measure spread about the mean as center. They should be used only in the company of the mean.

> **The *variance* is the mean of the squares of the deviations of the observations from their mean.**
>
> **The *standard deviation* is the positive square root of the variance.**

These definitions require an example to see what they say, and then some commentary to see why they say it. Here is the example.

Example 4. Find the variance and standard deviation of the 5 observations

$$6, 7, 5, 3, 4$$

(a) First compute the mean.

$$\bar{x} = \frac{6 + 7 + 5 + 3 + 4}{5} = \frac{25}{5} = 5$$

(b) The deviation of any observation x from the mean is the difference $x - \bar{x}$, which may be either positive or negative. Use this arrangement to compute the variance:

Observation x	Deviation $x - \bar{x}$	Squared deviation $(x - \bar{x})^2$
6	$6 - 5 = 1$	$(1)^2 = 1$
7	$7 - 5 = 2$	$(2)^2 = 4$
5	$5 - 5 = 0$	$(0)^2 = 0$
3	$3 - 5 = -2$	$(-2)^2 = 4$
4	$4 - 5 = -1$	$(-)^2 = 1$
	sum $= 0$	sum $= 10$

The sum of the squares of the deviations from the mean is 10. The variance is therefore

$$\text{variance} = \frac{\text{sum of squared deviations}}{\text{number of observations}} = \frac{10}{5} = 2$$

(c) The standard deviation is the square root of the variance.

$$\sqrt{2} = 1.4$$

A calculator with $\sqrt{}$ key makes square roots easy. In fact, many calculators will compute both the mean and the standard deviation from keyed-in data. If you plan to do many statistical calculations, you should obtain such a calculator and spare yourself the arithmetic.

In the language of algebra, the recipe for the standard deviation is as follows:

Observations $\qquad$ $x_1, x_2, \ldots, x_n$

Mean $\qquad$ $\bar{x} = \dfrac{x_1 + x_2 + \cdots + x_n}{n}$

Variance $\qquad$ $s^2 = \dfrac{(x_1 - \bar{x})^2 + (x_2 - \bar{x})^2 + \cdots + (x_n - \bar{x})^2}{n}$

Standard deviation $\quad s = \sqrt{s^2}$

If you have difficulty following the algebraic recipe, look back at the example for guidance.* Some comments on the standard deviation will help you to interpret it.

1. The standard deviation s makes sense as a measure of spread or variability. It is a kind of average deviation of the observations from their mean. If the observations are spread out, they will tend to be far from the mean, both above and below. Some deviations will be large positive numbers, and some will be large negative numbers. But the squared deviations will all be large and positive, so both s^2 and s will be large when the data are spread out. And s^2 and s will be small when all the data are close together. (The deviations from the mean will always be both positive and negative and will always have sum zero. Look back at the example to check that the sum of the "Deviation" column is zero. This helps you check your arithmetic. Remember that the square of a negative number is positive, so squared deviations are never negative.)

2. The standard deviation s is always zero or positive. It can be zero only if all the squared deviations are zero, which means that every observation x_i has the same value as $\bar{x}$, and so all observations x_i are the same. *Standard deviation zero means no spread at all; otherwise the standard deviation is positive and increases as the spread of the data increases.*

3. When the observations x_i are measured in some units (seconds, centimeters, grams), their variance s^2 is measured in the square of those units because it is an average squared deviation. The standard deviation s has the same units as the original observations. This is one reason why s is used more often than s^2. (Another reason appears in Section 5.)

---

*There is a complication that you should be aware of if you study more statistics or if you use a calculator that gives you the standard deviation directly from keyed-in data. When the observations form an SRS of size n, s^2 is usually found by dividing the sum of the squared deviations by $n - 1$ rather than n. There are arguments in favor of both n and $n - 1$. Since the difference is quite small when n is of moderate size, we use n for simplicity.

4. Like the mean, the variance and standard deviation are heavily influenced by outliers. If, for example, the largest value in Example 4 were 17 instead of 7, the mean would increase to $\bar{x} = 7$, and the variance would increase to $s^2 = 26$. Even the standard deviation s increases from 1.4 to 5.1.

5. The standard deviation is often not a useful measure of the spread of a skewed distribution. Because the two sides of a strongly skewed distribution have different spreads, no single number describes the spread well. The five-number summary, with its two quartiles and two extremes, does a better job. In most cases, $\bar{x}$ and s should be reserved for reasonably symmetric distributions. If outliers or strong skewness make $\bar{x}$ and s seem inappropriate, use the five-number summary.

6. To allow more meaningful comparisons of the variability of different distributions, the *coefficient of variation* is often used. This is just the standard deviation expressed as a percent of the mean: $CV = s/\bar{x}$ converted to percent. An appendix in the *Statistical Abstract* gives CVs for many government sample surveys. For example, the CV is 2% for the annual average size of the labor force from the Current Population Survey and 5.5% for the count of robberies from the National Crime Survey. CVs are preferable to standard deviations for comparing the accuracy of these surveys because the surveys are measuring variables that differ in both kind and size. A standard deviation of 500,000 would be small when measuring the labor force (over 125 million persons) but large when counting robberies (about 900,000). Comparison of variability requires some care.

Section 4 exercises

4.46. Calculate the five-number summary of the Nielsen ratings for the 19 NBC programs in Table 4-4. Verify your answers by comparison with the boxplot in Figure 4-22.

4.47. Calculate five-number summaries for each of the following sets of data, and make a boxplot of each distribution. What does the boxplot show about the shape of the distribution?
    (a) The coliform bacteria counts in Exercise 4.17.
    (b) The reaction times in Exercise 4.33.

4.48. Compute five-number summaries for the weight gains of the experimental and control groups in the chicken-nutrition experiment of Exercise 4.23. By how much does the weight gain of a typical chick fed high-lysine corn exceed that of a typical chick fed normal corn?

Draw boxplots for the two groups with a common scale, as in Figure 4-22. Describe the conclusions that can be drawn from your plot. (Are the distributions symmetric? Skewed? Do they differ in variability?)

4.49. Here are five-number summaries of the distributions of weekly earnings (in pounds sterling) of full-time male and female workers in Britain as of 1982. (The data are from the annual New Earnings Survey, which reports similar information for many occupational and age groups. U.S. income data are collected by the Current Population Survey, but the United States does not yet use the convenient five-number summary.)

	Lower decile	Lower quartile	Median	Upper quartile	Upper decile
Men	89.7	109.9	139.1	180.5	233.8
Women	60.2	71.7	90.0	116.5	152.0

(a) Why does the New Earnings Survey report deciles rather than extremes in its summaries?

(b) Make boxplots of these two distributions on a common scale. Use whiskers that extend to the deciles. Write a brief description of the differences between the distributions of earnings of men and women workers in Britain.

(c) The median weekly earnings of full-time American workers were $370 for men and $240 for women in 1982. Based on these limited data, were male/female differences in income comparable in the United States and Britain in the early 1980s? Explain your answer.

4.50. Calculate the mean, the variance, and the standard deviation of each of the following sets of numbers:

(a) 4, 0, 1, 4, 3, 6

(b) 5, 3, 1, 3, 4, 2

Which of the sets is more spread out? Draw a histogram of set (a) and one of set (b) to see how the set with the larger variance is more spread out.

4.51. Suppose that we add 2 to each of the numbers in the first set in Exercise 4.50. That gives us the set

6, 2, 3, 6, 5, 8.

(a) Find the mean and the standard deviation of this set of numbers.

(b) Compare your answers with those for set (a) in Exercise 4.50. How did adding 2 to each number change the mean? How did it change the standard deviation?

(c) Can you guess, without doing the arithmetic, what will happen to the mean and standard deviation of set (a) in Exercise 4.50 if we add 10 to each number in that set?

This exercise should help you see that the standard deviation (or variance) measures only spread about the mean and ignores changes in where the data are centered.

4.52. This is a variance contest. You must give a list of six numbers chosen from the whole numbers 0, 1, 2, 3, 4, 5, 6, 7, 8, and 9, with repeats allowed.

(a) Give a list of six numbers with the largest variance such a list can possibly have.

(b) Give a list of six numbers with the smallest variance such a list can possibly have.

(c) Does either part (a) or part (b) have more than one correct answer?

4.53. Scores on the Stanford-Binet IQ test are approximately normally distributed with mean 100 and standard deviation 15. What is the variance of scores on this test?

4.54. If two distributions have exactly the same mean and standard deviation, must their frequency curves look exactly alike? If they have the same five-number summary, must their frequency curves be identical? Explain.

4.55. A school system employs teachers at salaries between $18,000 and $40,000. The teachers' union and the school board are negotiating the form of next year's increase in the salary schedule.

(a) If every teacher is given a flat $1000 raise, what will this do to the mean salary? To the median salary? To the extremes and quartiles of the salary distribution?

(b) What would a flat $1000 raise do to the standard deviation of teachers' salaries? (Do Exercise 4.51 if you need help.)

(c) If, instead, each teacher receives a 5% raise, the amount of the raise will vary from $900 to $2000, depending on the present salary. What will this do to the mean salary? To the median salary?

(d) A flat raise would not increase the spread of the salary distribution. What about a 5% raise? Specifically, will a 5% raise increase the distance of the quartiles from the median? Will it increase the standard deviation?

4.56. Exercise 4.22 gives the maximum daily rainfall each year for a 30-year period at South Bend, Indiana. Based on the shape of this distribution, do you prefer the five-number summary or $\bar{x}$ and s as a brief numerical description? Why? Compute the summary you chose.

4.57. Exercise 4.20 presents data on spending by shoppers at a grocery store. Based on the shape of this distribution, do you prefer the five-number summary or $\bar{x}$ and s as a brief numerical description? Why? Compute the summary you chose.

4.58. Another measure of the spread of a set of data is the *range*, which is the difference between the largest and smallest observations.
   (a) Find the range of the reaction times in Exercise 4.33.
   (b) The range is rarely used for any but very small samples. Can you explain why?
   (c) Can you give an example in which the range is the most appropriate measure of spread?

5. The normal distributions

Frequency curves offer a means of quickly describing the shape of a distribution. The normal curves are particularly important in statistics because the sampling distributions of many statistics (including the proportion $\hat{p}$ and the mean $\bar{x}$) are well described by normal curves for moderate and large sample sizes. There is an entire family of normal curves, two of which are shown in Figure 4-23. All normal curves share several characteristics: They are symmetric. They are bell-shaped. Their tails drop off quickly, so that a set of normally distributed data will have few outliers.

Because normal distributions are symmetric, the mean and median lie together at the center of the curve. This is also the peak of the curve, so the mean, median, and mode of a normal distribution are all identical.

The standard deviation of a normal distribution also can be located on the normal curve. Notice that near its center, a normal curve falls ever more steeply as we move away from the center, like this:

But in either tail, the curve falls every less steeply as we move away from the center, like this:

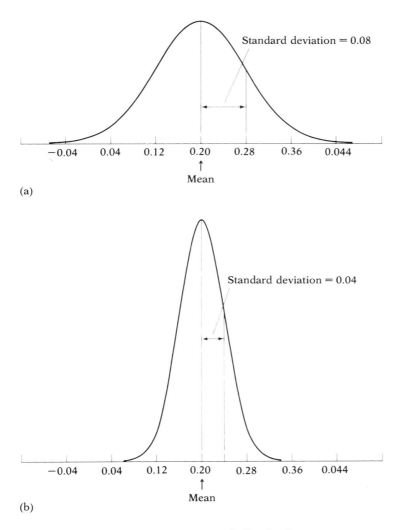

Figure 4-23. Two normal distribution curves.

The points at which the curvature changes from the first type illustrated to the second are located one standard deviation on either side of the mean. With a bit of practice, you can learn to find these points by running a pencil along the curve and feeling where the curvature changes. So both the mean and the standard deviation (but not the variance) of a normal frequency distribution are visible on the normal curve. Because normal curves are common, here is another reason why the standard deviation is often preferred to the variance as a measure of spread.

Figure 4-23 presents two normal curves with the means and standard deviations marked. Study those curves. The shape of a normal curve is completely determined by its mean and standard deviation; there is only one normal curve with mean 0.2 and standard deviation 0.08, for example, and it appears in Figure 4-23(a). The mean fixes the center of the curve, while the standard deviation determines its shape. Changing the mean of a normal distribution does not change its shape, only its location on the axis. Changing the standard deviation does change the shape of a normal curve, as Figure 4-23 illustrates. The distribution with the smaller standard deviation is less spread out and more sharply peaked. But in both cases almost the entire distribution of values described by the curve lies within three standard deviations on either side of the mean. This is true of any normal curve, no matter what mean and standard deviation it has.

The normal curves in Figure 4-23 in fact describe sampling distributions. The first describes the distribution of the sample proportion $\hat{p}$ in SRSs of size 25 from the bead population of Chapter 1 with $p = 0.20$ as the population proportion of colored beads. You can see this by looking again at the progression from histogram to polygon to normal curve in Figures 4-14, 4-15, and 4-16. The bottom curve in Figure 4-23 describes the sampling distribution of $\hat{p}$ in SRSs of size 100 from this same population. In both cases, the mean (and also the median and the mode) of the sampling distribution of $\hat{p}$ falls at 0.20, the true value of the parameter being estimated. So $\hat{p}$ is as likely to fall below p as above it: there is no systematic tendency to overestimate or underestimate the parameter. This is our final version of lack of bias. *A statistic has no bias when the mean of its sampling distribution is equal to the parameter being estimated.* If the population of beads had proportion of dark beads $p = 0.10$ or $p = 0.30$, the center of the curves in Figure 4-23 would move to those values.

The spread or dispersion of a normal curve is completely described by its standard deviation. Figure 4-23(a), for sample size 25, has standard deviation 0.08. Figure 4-23(b), for sample size 100, has standard deviation 0.04. The smaller spread of the second curve about its mean reflects the greater precision of the statistic $\hat{p}$ for the larger sample size. *When the sampling distribution of a statistic is normal, the precision can be described by giving the standard deviation of its sampling distribution.* So when a sampling distribution is normal, the mean and standard deviation together describe both bias and precision. News releases of sample results usually give precision by way of a confidence statement, but accounts in professional publications often report the standard deviation (sometimes called the standard error in this context) of the sample statistic.

In addition to sampling distributions, many types of data have distributions whose shapes are approximately described by a normal curve. Historically, the normal curves were first applied to data by the great mathematician Carl Friedrich Gauss (1777–1855) to describe the small errors made by astronomers or surveyors in repeated careful measurements of the same quantity. (Recall from Chapter 3 that the results of repeated sampling and repeated measurement have similar properties.) You will sometimes see normal distributions labeled Gaussian in honor of Gauss. For much of the nineteenth century normal curves were called error curves because they were first used to describe the distribution of measurement errors. As it gradually became clear that the distributions of some biological and psychological variables were at least roughly normal, the error-curve terminology was dropped. The curves were first called normal by the American logician Charles S. Peirce in 1873. One kind of data that are often normal is physical measurements of many members of a biological population. As an example, Figure 4-24 presents 2000 Hungarian skulls. Finally, it can be proved mathematically that any variable that is the sum of many small independent effects will have a distribution of values that is close to normal. You should again ponder the harmony of creation, that distributions derived from a mathematical study of random sampling should be also useful in describing the real world.

However, even though many sets of data follow a normal distribution, many do not. Most income distributions are skewed to the right and hence are not normal. Nonnormal data, like nonnormal people, meet us quite often and are sometimes more interesting than their normal counterparts.

More detail

To make explicit the connection between the standard deviation and spread or variability, here are some facts about the normal distributions:

1. **In any normal distribution, 68% of the observations fall within one standard deviation of the mean. Half of these (34%) fall within one standard deviation above the mean and the other half within one standard deviation below the mean.**

2. **Another 27% of the observations fall between one and two**

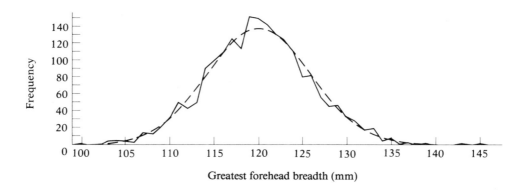

Greatest forehead breadth (mm)

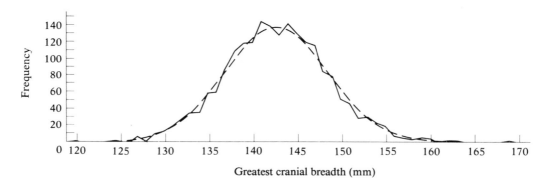

Greatest cranial breadth (mm)

Figure 4-24. 2000 Hungarian skulls. The distribution of each of two measurements on a large sample of male skulls is approximately normal. The solid line is the actual frequency curve, and the broken line is a normal curve that approximates the distribution. [From Karl Pearson, "Craniological notes," *Biometrika*, June 1903, p. 344. Reproduced by permission of the Biometrika Trustees.]

standard deviations away from the mean. So 95% (68% plus 27%) fall within two standard deviations of the mean.

3. In all, **99.7% of the observations fall within three standard deviations of the mean.**

The third fact makes precise the earlier comment that any normal curve is about six standard deviations wide. The second fact states that when a statistic with a normal sampling distribution has no bias, 95% of all samples will give values of the statistic falling within two standard deviations of the true parameter value. The sample proportion from an SRS, for example, has approximately a normal sampling distribution and is an unbiased estimate of the true proportion in the population. Thus, if in a particular case the sample has a standard deviation of 1.5 points, this leads directly to a statement of precision: In repeated sampling, 95% of all samples have sample proportions falling within 3 points (two standard deviations) of the true parameter value. Figure 4-25 illustrates all three facts. Let's call them the *68-95-99.7 rule*. This rule describes the shape of normal curves more exactly than the description ''bell-shaped.'' When using the normal curves to describe the shape of data sets, remember that no set of data is exactly described by a normal curve. The 68-95-99.7 rule will be only approximately true for SAT scores or the forehead breadth of 2000 Hungarian skulls.

The 68-95-99.7 rule shows just how the standard deviation measures the spread or variability for normal distributions. Because any normal distribution satisfies the rule, the point one standard deviation above the mean is always the 84th percentile. (Why? Because 50% of the observations fall below the mean and another 34% within one standard deviation above the mean, as Figure 4-26 shows). Similarly, one standard deviation below the mean is always the 16th percentile, and the point two standard deviations above the mean is the 97.5th percentile. Pictures similar to Figure 4-26 will help you solve problems like the following examples.

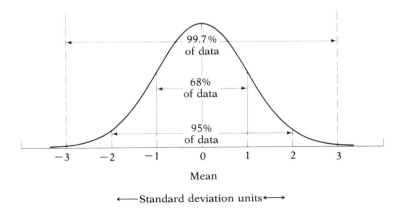

Figure 4-25. The 68-95-99.7 rule for normal distributions.

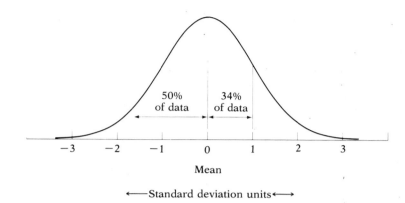

50% of data

34% of data

−3   −2   −1   0   1   2   3

Mean

⟵Standard deviation units⟵⟶

Figure 4-26. The 84th percentile of a normal distribution lies one standard deviation above the mean.

Example 5. The heights of adult American women are approximately normally distributed with mean 64 inches and standard deviation 2.5 inches. What heights contain the central 95% of this population?

Since 95% of all women have heights within two standard deviations of the mean, the central 95% of women's heights lie between 59 inches (that's 64 − 5) and 69 inches (64 + 5).

Example 6. Twenty percent of a large corporation's clerical staff are males. Of the last 100 workers chosen for promotion to administrative positions, 32 were males. How likely is it that 32 or more men would be chosen if the 100 workers promoted were drawn at random?

Here $p = 0.20$ and we want to know how often $\hat{p}$ will exceed 0.32 for an SRS of size 100. The sampling distribution of $\hat{p}$ [Figure 4-23(b)] is approximately normal with mean 0.20 and standard deviation 0.04. So 0.32 is three standard deviations above the mean. Since 99.7% of all SRSs would have $\hat{p}$ closer to the mean, and only half of the remaining 0.3% would fall on the high side, only 0.15% (that's 0.0015) of all possible SRSs would produce 32 or more men. It appears that men are being favored for promotion, either because of discrimination or because the male clerical workers are thought to be more highly qualified.

The standard deviation is the natural unit of measurement for normal distributions. Observations from a normal distribution are often reduced to *standard scores* expressed in standard-deviation units about the mean. The recipe is

$$\text{standard score} = \frac{\text{observation} - \text{mean}}{\text{standard deviation}}$$

A standard score of 1 simply says that the observation in question lies one standard deviation above the mean. An observation with standard score -2 is two standard deviations below the mean. A standard score of 1 is always at the 84th percentile, as Figure 4-26 shows. In fact, every standard score translates into a specific percentile, which is the same no matter what the mean and standard deviation of the original normal distribution are. Table B at the end of this book lists the percentiles corresponding to various standard scores. This table enables us to do calculations in greater detail than the 68-95-99.7 rule.

> **Example 7.** Matt scores 590 on the verbal part of the SAT. Scores on the SAT follow a normal distribution with mean 500 and standard deviation 100.* Matt's standard score is therefore
>
> $$\frac{590 - 500}{100} = \frac{90}{100} = 0.9.$$
>
> Table B shows that this is the 81.59 percentile of the distribution of scores. In plain language, Matt did better than 81% of students who take the test.

Because a standard score translates into the same percentile for all normal distributions, standard scores allow direct comparison of scores from different normal distributions. Standard scores express observations from any distribution as standard deviations away from the mean. Comparisons are less meaningful when the distributions are not normal, however, because in that case equal standard scores need not correspond to the same percentile of the distributions.

---

*This is an oversimplification that is true only for a standardization population used to set up the original scale of SAT scores. The group of students who took the same form of the test as did Matt may have had a mean score higher or lower than 500, depending on their ability.

Section 5 exercises

4.59. Figure 4-27 is a normal frequency curve. What are the mean and the standard deviation of this distribution?

4.60. Figure 4-24 records two frequency distributions for measurements on 2000 Hungarian skulls. Both are approximately normal. Estimate the mean and the standard deviation of each set of data.

4.61. In a study of elite male distance runners, the mean weight was reported to be 139 pounds with a standard deviation of 10.6 pounds. Assuming that the distribution of weights is normal, sketch the frequency curve of the weight distribution with the horizontal axis marked in pounds. (*Hint:* First draw a normal curve, then mark off the units on the horizontal axis. In fact, distributions of weights are usually less close to normal than distributions of heights.)

4.62. Return once more to the coliform counts of Exercise 4.17. You found the mean of these data in Exercise 4.32. The standard deviation can be calculated to be 2.02. What percent of the observations fall within one, two, and three standard deviations of the mean? Does the distribution of coliform counts appear to be approximately normal?

4.63. Explain, by using curves like those in Figures 4-25 and 4-26, why the point one standard deviation below the mean in a normal distribution is always the 16th percentile. Explain why the point two standard deviations above the mean is the 97.5th percentile.

4.64. The heights of women ages 18 to 24 are approximately normally distributed with mean 64.5 inches and standard deviation 2.5 inches.

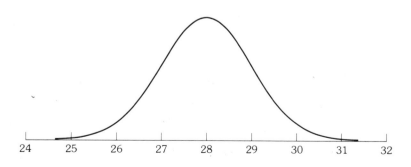

Figure 4-27.

    (a) Draw a frequency curve for this distribution, with the scale on the horizontal axis correctly marked.

    (b) What percent of women in this age group are taller than 62 inches? Taller than 69.5 inches? Shorter than 59.5 inches?

4.65. SAT scores are approximately normal with mean 500 and standard deviation 100. Scores of 800 or higher are reported as 800. (So an SAT score of 800 does not, as you may have thought, imply a perfect performance.) What percent of scores are 800 or higher?

4.66. Scores on the Wechsler Adult Intelligence Scale for the 20 to 34 age group are approximately normally distributed with mean 110 and standard deviation 25. About what percent of people in this age group have scores

    (a) above 110?

    (b) above 160?

    (c) below 85?

4.67. Eleanor scores 700 on the mathematics part of the SAT. Scores on the SAT follow the normal distribution with mean 500 and standard deviation 100. Gerald takes the American College Testing Program test of mathematical ability, which has approximately mean 18 and standard deviation 6. He scores 24. If both tests measure the same kind of ability, who has the higher score?

4.68. Scores on the Wechsler Adult Intelligence Scale for the 60 to 64 age group are approximately normally distributed with mean 90 and standard deviation 25.

    (a) Sarah, who is 30, scores 135 on this test. Use the information of Exercise 4.66 to restate this score as a standard score.

    (b) Sarah's mother, who is 60, also takes the test and scores 120. Express this as a standard score by using the information given in this exercise.

    (c) Who scored higher relative to her age group, Sarah or her mother? Who has the higher absolute level of the variable measured by the test?

4.69. Three landmarks of baseball achievement are Ty Cobb's batting average of .420 in 1911, Ted Williams's .406 in 1941, and George Brett's .390 in 1980. These batting averages cannot be compared directly, because the distribution of major league batting averages has changed over the decades. The distributions are quite symmetric and (except for outliers such as Cobb, Williams, and Brett) reasonably normal. Although the mean batting average has been held roughly constant by rule changes and the balance between batting and pitching, the standard deviation has dropped over time. Here are the facts.

Decade	Mean	Standard Deviation
1910s	.266	.0371
1940s	.267	.0326
1970s	.261	.0317

Compute the standard scores for the batting averages of Cobb, Williams, and Brett to compare how far each stood above his peers. (Data from Stephen Jay Gould, "Entropic homogeneity isn't why no one hits .400 any more," *Discover*, August 1986, pp. 60–66. Gould does not standardize, but gives instead a speculative discussion.)

4.70. The mean height of men 18 to 24 years old is about 69.5 inches, while women that age have a mean height of 64.5 inches. Do you think that the distribution of heights for all Americans 18 to 24 is approximately normal? Explain your answer.

4.71. Example 5 on page 146 reported 11 measurements of the mass standard NB10 conducted by the National Institute of Standards and Technology (NIST). I said there that the NIST is 95% confident that the mean of these measurements is within 0.0000023 of the true mass of NB10. Actually, the NIST reported the standard error of the mean measurement, from which I got this statement of precision by assuming (as is roughly true) that the sample mean has a normal distribution. What value of the standard error did the NIST give?

The following exercises require Table B at the end of this book.

4.72. The Wechsler Intelligence Scale for Children is used (in several languages) in the United States and Europe. Scores in each case are approximately normally distributed with mean 100 and standard deviation 15. When the test was standardized in Japan, the mean was 111. To what percentile of the American-European distribution of scores does the Japanese mean correspond?

4.73. The IQ scores on the Wechsler Adult Intelligence Scale for the 20 to 34 age group are approximately normal with mean 110 and standard deviation 25.
   (a) What percent of persons ages 20 to 34 have IQs below 100? What percent have IQs 100 or above?
   (b) What percent of persons ages 20 to 34 have IQs above 150?

4.74. If only 1% of persons ages 20 to 34 have IQs higher than Eleanor's, what is Eleanor's IQ? (Use the distribution of IQ scores given in the previous exercise and the entry in Table B that comes closest to the point with 1% of the observations above it.)

4.75. Suppose that the proportion of all adult Americans who are afraid to go out at night because of crime is $p = 0.45$. The distribution of the sample proportion $\hat{p}$ in a Gallup Poll probability sample of size 1500 is then approximately normal with mean 0.45 and standard deviation 0.015.

    (a) What percent of all Gallup Poll samples would give a $\hat{p}$ of 0.43 or lower? Of 0.47 or higher?

    (b) What percent of all Gallup Poll samples would give a $\hat{p}$ that misses the true p by two percentage points (± 0.02) or more?

4.76. Consider the opinion poll described in the previous exercise. In approximately what margin of error about $p = 0.45$ can we be 80% confident that $\hat{p}$ will fall? In what margin of error can we be 90% confident that $\hat{p}$ will fall?

NOTES

1. Two excellent books on graphing are William S. Cleveland, *The Elements of Graphing Data* (Monterey, Calif.: Wadsworth, 1985) and Edward R. Tufte, *The Visual Display of Quantitative Information* (Cheshire, Conn.: Graphics Press, 1983). These books elaborate the principles of graphing with abundant examples both good and bad.

Review exercises

4.77. A study of 10,500 students who were high school seniors in 1980 asked what the educational attainment of these students was by 1986. The results for men and women are given below. Present these data clearly in a graph. What are the most important facts about the education achieved by these students in the six years after their senior year of high school?

Sex	No HS diploma	HS only	License	Associate degree	Bachelor's degree	Professional or graduate
Men	1.0%	64.0%	10.5%	5.9%	17.6%	0.9%
Women	0.8%	59.6%	13.3%	7.0%	18.8%	0.6%

4.78. The U.S. Immigration and Naturalization Service reports each year the number of deportable aliens caught by the Border Patrol. Here are the counts for 1970 through 1988. Display these data in a graph. What are the most important facts that the data show?

Year	1970	1975	1980	1982	1984	1986	1988
Count (thousands)	231.1	596.8	759.4	819.9	1138.6	1692.5	971.1

4.79. Colleges announce an "average" Scholastic Aptitude Test score for their entering freshmen. Usually the college would like this "average" to be as high as possible. A *New York Times* article (May 31, 1989) noted that "Private colleges that buy lots of top students with merit scholarships prefer the mean, while open-enrollment public institutions like medians." Use what you know about the behavior of means and medians to explain these preferences.

4.80. Here are the 1989 baseball incomes of the members of the Chicago Cubs, in thousands of dollars. (Data from the *New York Times*, November 1, 1989.)

Player	Income	Player	Income
Sutcliffe	2340	Grace	140
Dawson	2100	Kilgus	130
Sandberg	925	Ramos	130
Dunston	550	Bielecki	123
Webster	550	Lancaster	120
Sanderson	500	Berryhill	115
Wilkerson	500	McClendon	91
Williams	425	Girardi	68
Law	400	Smith	68
Salazar	400	Walton	68
Wynne	350	Wilson	68
Maddux	275	Wrona	68
Assenmacher	226		

(a) Give a graphical display of the distribution of Cubs players' incomes. (Outliers, if any, may fall off your graph.) Then describe in words the major features of the distribution, both overall shape and any outliers.

(b) Is the five-number summary or the mean and standard deviation a better numerical description of this distribution? Calculate the one you think is better.

4.81. Table 4-5 gives the number of home runs hit by the American League leader over a 21-year period. Describe these data with both

Table 4-5. American League home-run leaders, 1970–1990

Year	Player	Home Runs
1970	Frank Howard	44
1971	Bill Melton	33
1972	Dick Allen	37
1973	Reggie Jackson	32
1974	Dick Allen	32
1975	George Scott and Jackson	36
1976	Graig Nettles	32
1977	Jim Rice	39
1978	Jim Rice	46
1979	Gorman Thomas	45
1980	Reggie Jackson	41
1981	Four players	22
1982	Thomas and Jackson	39
1983	Jim Rice	39
1984	Tony Armas	43
1985	Darrell Evans	40
1986	Jesse Barfield	40
1987	Mark McGwire	49
1988	Jose Canseco	42
1989	Fred McGriff	36
1990	Cecil Fielder	51

a graphical display and a numerical summary of your choice. Are there any outliers?

4.82. Government regulations recognize three types of hot dogs: beef, meat, and poultry. Do these types differ in the number of calories they contain? Table 4-6 records the number of calories in one hot dog for several brands of each type, as measured by *Consumer Reports* magazine. Find the five-number summary for each type, and compare the three types in side-by-side boxplots. Then write a brief description of your findings. Are the calorie distributions similar for all three types of hot dogs, or does one type show a different pattern?

4.83. The Acculturation Rating Scale for Mexican Americans (ARSMA) is a psychological test that evaluates the degree to which Mexican Americans are adapted to Mexican/Spanish versus Anglo/English

Table 4-6. Calories in hot dogs, by brand

Beef		Meat		Poultry	
186	152	173	179	129	135
181	111	191	153	132	142
176	141	182	107	102	86
149	153	190	195	106	143
184	190	172	135	94	152
190	157	147	140	102	146
158	131	146	138	87	144
139	149	139		99	
175	135	175		170	
148	132	136		113	

SOURCE: *Consumer Reports*, June 1986, pp. 366–367.

culture. The distribution of ARSMA scores in a population used to develop the test is approximately normal with mean 3.0 and standard deviation 0.8. The range of possible scores is 1.0 to 5.0; higher scores show more Anglo/English acculturation. Between what values do the ARSMA scores of the central 95% of Mexican Americans lie? How high must an ARSMA score be in order to fall in the top 2.5% of the population?

4.84. The ARSMA test is described in the previous exercise. A researcher believes that Mexicans will have an average score near 1.7 and that first generation Mexican Americans will average about 2.1 on the ARSMA scale. What proportion of the original population has scores below 1.7? Between 1.7 and 2.1? (Use Table B.)

Chapter 5

Understanding relationships

A medical study finds that short women are more likely to have heart attacks than women of average height, while tall women have fewer heart attacks. An insurance group reports that heavier cars have fewer deaths per 10,000 vehicles registered than lighter cars. These and many other statistical studies look at the relation between two variables. Once we recognize the importance of relations involving more than one variable, two is often not enough. The study of women's height and heart attacks, for example, also examined many other variables. To conclude that shorter women are at greater risk, the researchers first had to eliminate the effect of other variables such as weight and exercise habits.[1] Our topic in this chapter is relations among variables. One of our main themes is that the relationship between two variables can be strongly influenced by yet other variables. Here is some basic vocabulary.

> Data are *univariate* when only one variable is measured on each unit.
>
> Data are *bivariate* when two variables are measured on each unit.
>
> Data are *multivariate* when more than one variable is measured on each unit.

Notice that bivariate data carry much more information than two separate sets of univariate data. If we measure both the height and the weight of each of a large group of people, we know which height goes with each weight. These bivariate data allow us to study the connection between height and weight. A list of the heights and a separate list of the weights, two sets of univariate data, don't show the connection between the two variables. Bivariate data on height and weight will show that the taller people also tend to be heavier. To give another example, people who smoke more cigarettes per day tend not to live as long as those who smoke fewer. We say that pairs of variables such as height and weight or smoking and life expectancy are *associated*.

> ***Association*** **in bivariate data means that certain values of one variable tend to occur more often with some values of the second variable than with other values of that variable.**

Association between variables is easiest to understand in the physical sciences, where values of one variable are often connected to values of another by a "law." For example, one of the laws of motion states that if you drop a ball from a height, the downward speed of the ball is directly proportional to the time it has been falling. After six seconds it is moving twice as fast as after three seconds. This association can be graphed as a perfect straight-line relation between time and velocity.

Statistics is less concerned with ironclad relationships of the kind expressed by physical laws than with relationships that hold on the average. There is an association between the height and the weight of individuals because, on the average, tall people are heavier than shorter people. There is an association between the sex of workers and their pay, because, on the average, women earn less than do men. Yet many individual women earn more than do many individual men, just as many individual short people are heavier than many individual taller people.

Our first goal is to describe association, keeping in mind its on-the-average character. When both variables are nominal (or their values are grouped into classes), a bivariate frequency table displays the association between them. This case occupies Section 1.

When both variables have an interval/ratio scale, we can display the association in a graph, a scatterplot. Just as in the case of stemplots or histograms, we examine a scatterplot by looking first for an overall pattern and then for striking deviations from the pattern. The simplest overall pattern for a relationship between two variables is a straight line. There is an important descriptive statistic, the correlation coefficient, that describes the strength of straight-line association. Scatterplots and correlation are discussed in Section 2.

Once a strong association is discovered, it is natural to seek an explanation. There is an association between cigarette smoking and early death from several causes, an association so strong that the U.S. surgeon general states that cigarette smoking is "the largest avoidable cause of death and disability in the United States."[2] But does this association really reflect a cause-and-effect relationship, as the surgeon general suggests? There are many possible relationships between variables other than cause and effect that can explain an observed association. The most intense controversies in which statistics is involved center on explaining observed associations: Does smoking cause lung cancer? Does discrimination against women account for the fact that women earn less than men on the average? Section 3 sheds some statistical light on the hard question of causation.

Another way of stating that two variables are associated is that knowledge of one helps to predict the other. Since nonsmokers live (on the average, of course) longer than do smokers, many life insurance firms charge lower rates for nonsmokers. In doing so, the firms use one variable (smoking) to help predict a second (life expectancy). This prediction is useful whether or not the two variables are linked by cause and effect. Prediction, with emphasis on straight-line relationships between two variables, is discussed in Section 4.

1. Cross-classified data

How do women and men compare in the pursuit of academic degrees? More specifically, we want to examine the relationship between the level of a degree and the sex of the recipient. The information we want appears in Table 5-1. This is a frequency table because the entries are counts, in this case counts of earned degrees. It is a *bivariate frequency table* because the degrees are categorized by *two* nominal variables, the level of the degree and the sex of the recipient. For example, women earned about 12,000 doctorate degrees in 1987. (Did you notice that the entries are in thousands?) Because one variable is arranged horizontally and the other vertically, a bivariate frequency table is an example of *cross-classified data*. More elaborate sets of cross-classified data group the observations according to the values of three or more variables.

Table 5-1. Earned degrees, 1987, by level and sex (thousands)

	Bachelor's	Master's	Professional	Doctorates
Male	481	141	47	22
Female	510	148	25	12
Total	991	290	73	34

SOURCE: U.S. Department of Education.

Describing relations

The nature of the relations among variables in cross-classified data is described by calculating and comparing appropriate percents. Let's return to Table 5-1 for an example.

> **Example 1.** A glance at Table 5-1 shows a relationship between degree level and sex: women earn more than half of all bachelor's and master's degrees, but a smaller fraction of professional and doctorate degrees. To put this association in numerical terms, we calculate and compare several percents. Women earn
>
> $$\frac{510}{991} = 0.515 = 51.5\%$$
>
> of all bachelor's degrees. Similarly, women earn 51% of all master's degrees (check this result from the table). But women's share of professional degrees such as law and medicine is 34% and their share of academic doctorates is 35%. Women have achieved parity with men in the lower degrees but still earn markedly fewer than half of the most advanced degrees.

The percents that we calculated in Example 1 are *relative frequencies* in the language of Chapter 4. What is new here is that in a bivariate frequency table each entry has not one but three relative frequencies. We can find the entry as a percent of the table total, as a percent of the total for its row, or as a percent of the total for its column. These percents answer different questions. Compare these questions:

1. What fraction of all doctorate degrees are earned by women?

2. What fraction of all degrees earned by women are doctorates?

3. What fraction of all degrees are doctorates earned by women?

These questions sound alike, but they are not. Recognizing which question we want to ask in a given situation and learning how to answer all of them from the table is part of developing statistical skill. Question 1 says: Look only at the "Doctorate" column in the table; what is the relative frequency of women among the degrees in this column? The answer is 12/34, which is the 35% that we found in the example.

Question 2 says: Look only at the "Female" row in the table, and find the relative frequency of doctorates among the degrees in this row. The table does not give the total number of degrees earned by women. This row total is

$$510 + 148 + 25 + 12 = 695$$

The answer to Question 2 is then

$$\frac{12}{695} = 0.017$$

About 1.7% of all degrees earned by women were doctorates.

Finally, Question 3 asks the relative frequency of doctorates earned by women among all degrees. The table does not give the total number of degrees awarded, so we must compute it by adding the totals for each level. Because,

$$991 + 290 + 73 + 34 = 1388$$

there were 1,388,000 academic degrees awarded in 1987. Of these,

$$\frac{12}{1388} = 0.0086$$

or about 9/10 of 1%, were doctorates earned by women.

Answering these questions was mainly a matter of straight thinking, complicated a bit because the table did not give all of the totals we needed. The column totals (bottom row) are a univariate frequency table of degrees by level. The row totals, which don't appear in our table, are a univariate frequency table of degrees by sex of the recipient. Both of these univariate tables can be obtained by addition from the bivariate table (the eight entries in the body of the table), but the bivariate table *cannot* be obtained from the two univariate tables (see Exercise 5.3). That's a reminder that bivariate data carry more information than two sets of univariate data.

Simpson's paradox

The nature of the observed relation between two variables can change radically when we take into account other variables that lie hidden in the situation. Let's look at a surprising example of this fact.

> **Example 2.** A university offers only two degree programs, one in electrical engineering and one in English. Admission to these programs is competitive, and the women's caucus suspects discrimination against women in the admissions process. The caucus obtains the following data from the university, a two-way classification of all applicants by sex and admission decision:
>
> | | Male | Female |
> |--------|------|--------|
> | Admit | 35 | 20 |
> | Deny | 45 | 40 |
> | Total | 80 | 60 |

The data in Example 2 do show an association between the sex of applicants and their success in obtaining admission. To describe this association more precisely, we compute some percents from the data.

$$\text{Percent of male applicants admitted} = \frac{35}{80} = 44\%$$

$$\text{Percent of female applicants admitted} = \frac{20}{60} = 33\%$$

Aha! Almost half the males but only one-third of the females who applied were admitted.

The university replies that although the observed association is correct, it is not due to discrimination. In its defense, the university produces a *three-way table* that classifies applicants by sex, admission decision, and the program to which they applied. Such a three-way table is conveniently presented as several two-way tables side by side, one for each value of the third variable. In this case there are two two-way tables, one for each program.

Engineering	Male	Female
Admit	30	10
Deny	30	10

English	Male	Female
Admit	5	10
Deny	15	30

Check that these entries do add to the entries in the two-way table. The university has simply broken down that table by department. We now see that engineering admitted exactly half of all applicants, both male and female, while English admitted one-fourth of both males and females. There is *no association* between sex and admission decision in either program.

How can no association in either program produce strong association when the two are combined? Easily: English is hard to get into, and mainly females applied to that program. Electrical engineering, on the other hand, is easy to get into and attracted mainly male applicants. Look at the three-way table, and you will see this clearly. English had 40 female and 20 male applicants, while engineering had 60 male and only 20 female applicants.

The discussion of Example 2 illustrates again how we describe association in a two-way table by computing and comparing well-chosen percents. It also demonstrates how to display cross-classified data with three variables as several two-way tables side by side. But the point of the example is to demonstrate that an observed association between two variables can be misleading when there is another variable that interacts strongly with both variables but was not reported. In fact, Example 2 illustrates *Simpson's paradox*: Classify two groups with respect to the incidence of some attribute, such as admission in the example; if the groups are then separated into several categories, the group with the *higher* incidence overall can have the same or even *lower* incidence within every one of the categories.

Although Example 2 was artificial for the sake of simplicity it is based on a real controversy over admission to graduate programs.[3] The effect of other variables on an observed association is at the heart of most controversies over alleged discrimination in both public debate and court cases.

Example 3. Women who are employed full-time earn (on the average) about 70% as much as men. Does this difference reflect discrimination against women? Careful investigation is needed, because women workers (on the average) differ from male workers in age, years of schooling, labor-force experience, and so on. In particular, many women have spent time outside the labor force for family reasons. Since pay rises with experience, the average pay of men would exceed that of women even in the absence of discrimination. The differing characteristics of male and female workers appear to explain roughly half of the earnings gap. Outright discrimination — different pay for the same work — is illegal and therefore rare. Why then does the remaining earnings gap persist? The major reason is that some jobs (like clerical work) are primarily held by women and others (like maintenance work) are largely male. Jobs that are traditionally male pay more in general than jobs dominated by women.

What to do about this state of affairs is controversial. One approach is to assess the skill, training, and responsibility required by different jobs and to insist that jobs that are similar in these variables carry equal pay. A number of public employers, including the states of Iowa and Minnesota, have adopted this "comparable-worth" approach. The pay of clerical workers is increased to match the pay of "comparable" maintenance jobs. The opposing view holds that it is rarely possible to measure all of the variables that are important in determining pay. Jobs that can be held by part-time workers, for example, will pay less than "comparable" jobs that are restricted to full-time workers, because more people are competing for the available jobs. Part of the comparable-worth debate concerns statistical issues: Do we know the important variables and can we measure them?[4] The larger issue of the male-female earnings gap is a reminder that in complex situations when many variables affect an outcome, you should be slow to jump to a conclusion suggested by a strong association alone.

Section 1 exercises

5.1. The U.S. Office for Civil Rights has among its jobs the investigation of possible sex discrimination in university hiring. The office wants to know if women are hired for the faculty in the same proportion as their availability in the pool of potential employees. If new university faculty in 1987 were recruited almost entirely from new doctorate degree recipients, which of the three questions asked in the text about Table 5-1 is useful to the office? Why?

5.2. In Table 5-1 the "Total" entry for Master's degrees differs slightly from the sum of the male and female entries. Why do you think this occurs?

5.3. Here are the row and column totals for a bivariate frequency table with two rows and two columns.

	50
	50
60 40	100

Find *two different* sets of entries in the body of the table that give these same totals. This shows that the bivariate frequencies cannot be obtained from the univariate frequencies alone.

5.4. Many studies have suggested that there is a link between high blood pressure and cardiovascular disease. A group of white males aged 35 to 64 were classified according to their systolic blood pressure as low (less than 140 millimeters of mercury) or high (140 or higher), and then followed for five years. The following two-way tables gives the results of the study. [From J. Stamler, "The Mass Treatment of Hypertensive Disease: Defining the Problem," *Mild Hypertension: To Treat or Not to Treat* (New York Academy of Sciences, 1978), pp. 333–358.]

Blood Pressure	Died	Survived	Total
Low	21	2655	2676
High	55	3283	3338

(a) Calculate the mortality rate (deaths as a fraction of the total) for each group of men.

(b) Do these data support the idea that there is a link between high blood pressure and death from cardiovascular disease? Explain your answer.

5.5. Here are data on poverty in the United States.

Persons below poverty level, by age and race, 1988 (thousands)

	White	Black	Other
Under 16 years	6931	3978	641
16 to 21 years	2043	1046	239
22 to 44 years	6379	2573	518
45 to 64 years	2816	1045	187
65 years and over	2595	785	102

SOURCE: *Statistical Abstract of the United States*, 1990.

(a) To understand these data, you need an essential definition that is given in the source. What definition?

(b) How many persons below the poverty level were there in 1988?

(c) What percent of persons below the poverty level were 65 or older?

(d) What percent of persons below the poverty level were black?

(e) Of all whites below the poverty level, what percent were 65 or older?

(f) Of all children under 16 below the poverty level, what percent were black?

(g) You want to know what percent of all people 65 and older were below the poverty level. Can you learn the answer from this table?

5.6. Here is a pleasant little bivariate frequency table.

Suicides, by sex and
method, 1986

	Male	Female
Poison	3516	2520
Hanging	3761	845
Firearms	15,518	2635
Other	1431	678

SOURCE: *Statistical Abstract of the United States*, 1990.

(a) How many suicides were reported in 1986?

(b) Give a univariate frequency table of suicides by method. What method was most commonly used, and what percent of all suicides were committed by this method?

(c) What percent of all women who committed suicide used poison?

(d) Describe the chief differences between men and women in their choice of suicide methods, referring to the table to support your statement.

5.7. Here is a table of earned degrees in mathematics for 1987, classified by level and by the sex of the recipient. Write a brief description of the relation between sex and level. What are the most important differences between mathematics degrees and the overall picture given by Table 5-1?

	Bachelor's	Master's	Doctorates
Male	8834	2024	599
Female	7655	1297	126
Total	16,489	3321	725

5.8. Do seat belts and other restraints prevent injuries in automobile accidents? Some evidence is provided by a study of reported crashes of 1967 and later-model cars in North Carolina in 1973 and 1974. There were 26,971 passengers under the age of 15 in these cars. Here are data on their conditions. [Adapted from data of Williams and Zador in *Accident Analysis and Prevention*, Volume 9 (1977), pp. 69–76.]

	Restrained	Unrestrained
Injured	197	3844
Uninjured	1749	21,181

(a) What percent of these 26,971 young passengers were wearing seat belts or were otherwise restrained?

(b) Compute appropriate percents to show the association between wearing restraints and escaping uninjured.

5.9. The study in Exercise 5.8 also looked at where the passengers were seated. Here is a three-way table of passengers by seat location, restraint, and condition.

Front seat			Back seat		
	Restrained	Not		Restrained	Not
Injured	121	2125	Injured	76	1719
Uninjured	981	9679	Uninjured	768	11,502

Unlike Example 2, the third variable here does *not* greatly change the observed association between the first two. Compute appropriate percents to demonstrate this.

5.10. Here is a three-way table classifying all persons employed in the United States in 1987 by sex, race, and age. The data come from the *Statistical Abstract*.

	Male		Female	
	White	Other	White	Other
16 to 19	2999	382	2900	360
20 to 34	22,234	3311	17,943	3140
35 to 64	27,725	3605	21,251	3550
65 and over	1690	160	1047	144

   (a) How many white females ages 16 to 19 were employed in 1987?

   (b) What proportion of all employed persons ages 20 to 34 were females?

   (c) From this three-way table, construct a two-way table of employed persons by sex (horizontal) and race (vertical).

   (d) White and other races differ slightly in male versus female employment. Describe the difference in words, and compute appropriate percents to describe this association between race and sex among employed persons.

5.11. The National Science Foundation publication *Science Indicators 1980* contains the results of a survey showing that young women scientists and engineers earned, on the average, only 77% as much as young male scientists and engineers. But when we look at the results for separate fields of science and engineering, we find that women earned at least 92% as much as men in every field. Explain how this apparent paradox can happen.

5.12. A study of the effect of parents' smoking habits on the smoking habits of students in eight Arizona high schools produced the following counts. [From S.V. Zagona (ed.), *Studies and Issues in Smoking Behavior* (Tucson, Ariz.: University of Arizona Press, 1967), pp. 157–180.]

	Student smokes	Student does not smoke
Both parents smoke	400	1380
One parent smokes	416	1823
Neither parent smokes	188	1168

Describe in words the association between the smoking habits of parents and children, and compute appropriate percents to back up your statements.

5.13. A news report (the *New York Times*, January 12, 1990) said:

> *The results of a government study on death rates in nearly 6000 hospitals were challenged today by researchers who said the Federal analysis failed to account for variations in the severity of patients' illness when they were hospitalized.*
>
> *As a result, they said, some hospitals were treated unfairly in the findings, which named hospitals with higher-than-expected death rates.*

We will examine a simplified example to show how overlooking the severity of the illness can indeed create an incorrect impression. Your community is served by two hospitals, Hospital A and Hospital B. Here are data on the survival of patients after surgery in these two hospitals. All patients undergoing surgery in a recent time period are included; "survived" means that the patient lived at least six weeks following surgery.

Good condition			Poor condition		
	Hosp A	Hosp B		Hosp A	Hosp B
Died	6	8	Died	57	8
Survived	594	492	Survived	1443	192

(a) Combine the data in this three-way table to form a two-way table of outcome (died or survived) by hospital (A or B). Use your two-way table to find the percent of patients who died at each hospital. Which hospital has the higher death rate?

(b) Now return to the three-way table. Find the percent of patients in good condition who died at each hospital. Then find the percent of patients in poor condition who died at each hospital. Which hospital has the higher death rate for patients in good condition? For patients in poor condition?

(c) This is an example of Simpson's paradox: Hospital A has a lower death rate for both classes of patients, but has a higher overall death rate than hospital B. Explain carefully, from the data given, how this can happen.

2. Scatterplots and correlation

When our data are nominal, we describe them using counts and percents. Variables measured in an interval/ratio scale, such as height in inches or age in years, contain more detailed information and allow more elaborate statistical descriptions. Data on a single variable are pictured by a stemplot or histogram and summarized numerically by the five-number summary or by the mean and standard deviation. When we have bivariate data, these univariate descriptions are supplemented by graphs and numerical measures that describe the relation between the variables.

When both of two associated variables are measured in ordinal or interval/ratio scales, the association between the variables can have a direction:

> Two variables are *positively associated* when larger values of one tend to be accompanied by larger values of the other.

> The variables are *negatively associated* when larger values of one tend to be accompanied by smaller values of the other.

In a large group of people there will be a positive association between height and weight and a negative association between packs of cigarettes

"He says we've ruined his positive association between height and weight."

smoked and length of life. It makes no sense to speak of a positive or negative association between sex and earnings, because sex is a nominal variable that has no direction. Both our graphs and our numerical descriptions for relations between ordinal or interval/ratio variables can reveal the direction of an association.

Scatterplots

Scatterplots graph bivariate data when both variables are measured in an interval/ratio or ordinal scale. Units for one variable are marked on the horizontal axis and units for the other variable on the vertical axis. The explanatory variable should always go on the horizontal axis when one of the variables is an explanatory and one a response variable. Each bivariate observation is represented by a point with horizontal coordinate equal to the value of the first variable and vertical coordinate equal to the value of the second.

> **Example 4.** Figure 5-1 is a scatterplot of data from an agricultural experiment. The explanatory variable is planting rate for corn (in thousands of plants per acre) and the response variable is yield (in bushels per acre). The scatterplot shows observations on 13 plots of land. One plot received 12,000 plants per acre and yielded 130.5 bushels per acre. This is indicated by the circled dot on the scatterplot at 12 on the horizontal scale and 130.5 on the vertical scale. The connection between planting rate and yield is clear from the

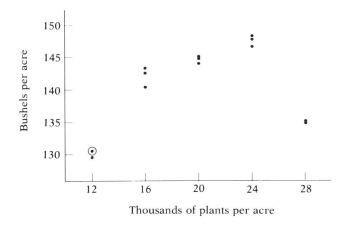

Figure 5-1. Corn planting rate versus yield, a scatterplot.

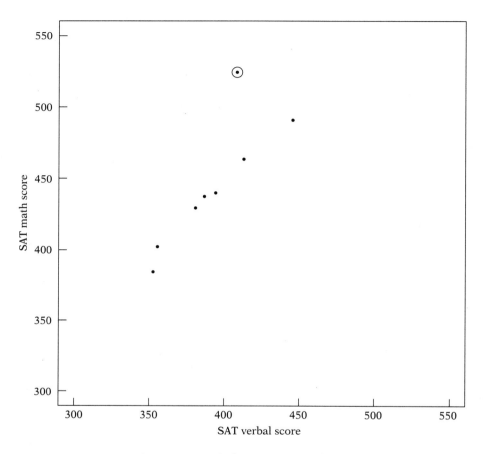

Figure 5-2. Average Scholastic Aptitude Test verbal and mathematics scores for eight racial or ethnic groups.

graph: Yield increases with planting rate until about 24,000 plants per acre are planted, then drops off. This is neither a positive nor a negative association because the relationship changes direction as the planting rate continues to increase.

Scatterplots not only display the relationship between variables but also highlight individual observations that deviate from the overall relationship. These observations appear as *outliers*, data points that stand apart from the rest.

Example 5. Figure 5-2 is a scatterplot of average scores on the math and verbal Scholastic Aptitude Tests for students in eight racial or ethnic groups. In this case there is no clear explanatory variable, so either score could be plotted on the horizontal axis. We suspect that these scores will be positively associated, that is, that verbal and math scores will rise and fall together. The lower left to upper right pattern in the scatterplot confirms this suspicion. The overall pattern of the relationship is approximately a straight line, since seven of the eight points fall in a rough line from lower left to upper right. However, the circled point is an outlier. This point represents the average scores of Asian students. These students score lower on the verbal SAT, or higher on the math SAT, than would be expected from the pattern formed by the other seven groups.

Correlation

Association can be described in many ways. We will give only a single numerical measure of association, the *correlation coefficient*. By a strict interpretation of our scales of measurement, correlation only makes sense when both variables have an interval/ratio scale. Just as with the mean and standard deviation, however, the correlation coefficient is often used with ordinal scales as well. Correlation is almost as common as mean and standard deviation in summaries of data.

The correlation coefficient is the algebraic high point of this book. Be of good courage. In mastering this, you have passed the worst. The notation of algebra is essential for a brief definition of correlation. Here is the definition of the correlation coefficient.

Given n bivariate observations on variables x and y,

$$x_1, x_2, \ldots, x_n$$

$$y_1, y_2, \ldots, y_n$$

compute the correlation coefficient as follows:

(1) Find the mean $\bar{x}$ and standard deviation s_x of the values x_1, $x_2, \ldots, x_n$ of the first variable.

(2) Find the mean $\bar{y}$ and standard deviation s_y of the values y_1, $y_2, \ldots, y_n$ of the second variable.

(3) The *correlation coefficient* r is

$$r = \frac{\frac{1}{n}[(x_1-\bar{x})(y_1-\bar{y}) + (x_2-\bar{x})(y_2-\bar{y}) + \cdots + (x_n-\bar{x})(y_n-\bar{y})]}{s_x s_y}$$

This definition deserves a short explanation, then an example to show what it says, and then a commentary on exactly what r measures. First, the explanation of the definition. The data are bivariate, so x_1 goes with y_1 ("my height and my weight"), x_2 goes with y_2 ("your height and your weight"), and so on. The numerator in r is the mean of the products of the deviations of each x and the corresponding y from their means. The denominator in r is the product of the standard deviation of the x's alone and the standard deviation of the y's alone. Because both numerator and denominator are built up from deviations from the mean, there is a convenient arrangement for putting the definition into practice. The following example illustrates the arithmetic.

Example 6. Suppose that the data are:

x	6	2	2	−2
y	5	5	−3	−3

A scatterplot (Figure 5-3) shows positive association as a trend from lower left (both x and y small) to upper right (both x and y large). The means are

$$\bar{x} = (6 + 2 + 2 - 2)/4 = 8/4 = 2$$
$$\bar{y} = (5 + 5 - 3 - 3)/4 = 4/4 = 1$$

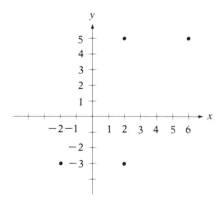

Figure 5-3.

Now make the following table.

(1) x	(2) y	(3) $x - \bar{x}$	(4) $y - \bar{y}$	(5) $(x - \bar{x})(y - \bar{y})$	(6) $(x - \bar{x})^2$	(7) $(y - \bar{y})^2$
6	5	4	4	16	16	16
2	5	0	4	0	0	16
2	−3	0	−4	0	0	16
−2	−3	$\dfrac{-4}{0}$	$\dfrac{-4}{0}$	$\dfrac{16}{32}$	$\dfrac{16}{32}$	$\dfrac{16}{64}$

Columns (3) and (4) are the deviations from the mean for the x's and y's obtained from columns (1) and (2) by subtracting the proper mean. Note that the sum of each set of deviations is 0, as is always the case. Columns (6) and (7) are the squares of these deviations. If we divide the sums of these columns by the number of observations, 4, we see that the variances of x and y are

$$s_x^2 = 32/4 = 8$$
$$s_y^2 = 64/4 = 16$$

and the standard deviations are

$$s_x = \sqrt{8} = 2.8$$
$$s_y = \sqrt{16} = 4$$

Column (5) contains the product of each entry in column (3) and the corresponding entry in column (4). Be careful of the signs here! Both positive and negative products may occur. The sum of the entries in column (5) is the sum in the numerator of r. We need only divide by the number of observations $n = 4$ (*not* $n = 8$, because n is the number of *bivariate* observations). So, finally

$$r = \frac{32/4}{(2.8)\,(4)} = \frac{8}{(2.8)\,(4)} = \frac{2}{2.8} = 0.71$$

The primary purpose of Example 6 is not to show you how to calculate r. You will no doubt use a computer or a special statistical calculator if you must do such calculations for any but the smallest sets of data. Rather, I have shown by example what the recipe for the correlation coefficient r

says. Be sure that you now understand the definition of r. I won't blame
you if the thought of computing the correlation of IQ and grade index for
1200 college freshmen makes you a bit ill. Let a machine do that. But to
interpret the result $r = 0.38$, we must know what r is. Now we know, so
let's relax and interpret r. There are six points to be made.

1. *The correlation coefficient r makes sense as a measure of association; it is positive when the association is positive, and negative when the
association is negative*. To understand how r measures association, look
at the sum of products of deviations in the numerator of r. When x and y
are positively associated, above-average values of x tend to go with
above-average values of y, and below-average values of both variables
also tend to occur together. So the deviations $x - \bar{x}$ and $y - \bar{y}$ are usually
both positive or both negative. In either case, the product $(x - \bar{x})(y - \bar{y})$
is positive, and it is large whenever both x and y are far from their means
in the same direction. So "on the average" for all observations (that's
what the numerator of r does), we get a large positive result when the
association is strong. If, though, the association between x and y is
negative, corresponding values of x and y tend to be on opposite sides of
their means; when one is large, the other is small. The numerator of r is
then negative, and more negative if large x's go with small y's and vice
versa. So the numerator of r does seem to measure association. What
about the denominator of r, which we have been ignoring? That brings
us to the second point.

2. *The correlation coefficient r always has a value between* -1 *and* $+1$.
It turns out that the standard deviations in the denominator standardize r
in this way. That's why they are there. Can you see that the units in both
numerator and denominator of r are the same? If x is height in centimeters and y is weight in grams, both numerator and denominator have the
units centimeters times grams. Therefore r itself has no units; it is a pure
number between -1 and 1. In addition, r does not depend on the choice
of units for x and y. If x were measured in inches (not centimeters) and y
in ounces (not grams), the correlation between height and weight would
be unchanged as long as the same objects were being measured.

3. *The extreme values* $r = -1$ *and* $r = +1$ *indicate perfect straight-line
association*. In particular, $r = -1$ means that all of the points in the
scatterplot fall exactly on a straight line having negative slope (that is,
when x increases, y decreases). Similarly, $r = +1$ means that all the
points fall exactly on a straight line with positive slope (as x increases, y
also increases).

4. *The correlation coefficient r measures how tightly the points on a
scatterplot cluster about a straight line*. That is, r does not measure

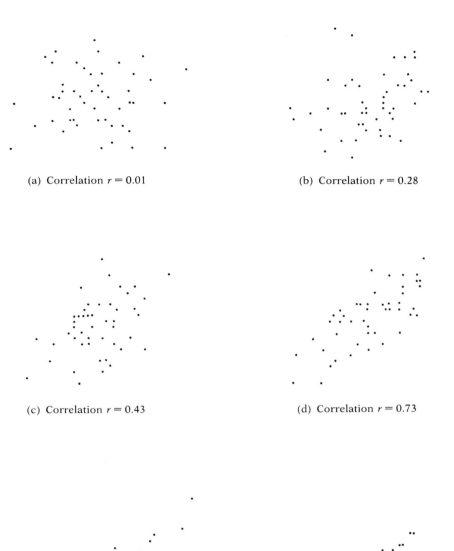

(a) Correlation $r = 0.01$

(b) Correlation $r = 0.28$

(c) Correlation $r = 0.43$

(d) Correlation $r = 0.73$

(e) Correlation $r = 0.91$

(f) Correlation $r = 0.99$

Figure 5-4. Correlation made visible.

association in general but only straight-line association. Correlations near either $+1$ or -1 indicate that the points fall close to a straight line. When $r > 0$, the scatterplot shows a trend from lower left to upper right, and the line about which the points cluster has positive slope. For $r < 0$, the trend is from upper left to lower right, and the slope is negative. The scatterplots in Figure 5-4 illustrate how r measures straight-line association. Study them carefully. (These scatterplots are arranged so that in each case the two variables have the same standard deviation. It is not as easy to guess r from a scatterplot as these examples suggest, because changing the standard deviations can change the appearance of the scatterplots quite a bit.)

5. There is a specific way in which r measures straight-line association: *The square r^2 of the correlation coefficient is the proportion of the variance of one variable that can be explained by straight-line dependence on the other variable.* Notice that r^2 always falls between 0 and 1, so it can be interpreted as a proportion. To understand the meaning of the italicized fact, think first about perfect straight-line association with $r = -1$, such as appears in Figure 5-5(a). The variable y is completely tied to x; when x changes, y moves, so the point (x, y) moves along the line. The eight values of y in Figure 5-5(a) have fairly large variance, but that variance is entirely due to the occurrence of different x values that bring with them different y values. When x is held fixed, there is *no* variation in y, because the data points all fall on the line. Straight-line dependence on x accounts for *all* of the variability in y, and $r^2 = 1$.

The y values in Figure 5-5(b) also show a large variance. Some of this variance can again be explained by the fact that changing x brings with it (on the average) a change in y. To see this, look at the quite different values of y that accompany the two different x values, x_1 and x_2, in Figure 5-5(b). But in this case r is not ± 1, and the association between x and y explains only part of the variability in y. Even when x remains fixed, y still varies. It is possible by mathematics to separate out the part of the variance of y that is explained by straight-line dependence on x. This is the fraction r^2 of the variance of the y values. In Figure 5-5(b), for example, $r = 0.7$ and so $r^2 = 0.49$. The straight-line dependence of the y's on the x's accounts for 49% of the variance of the y's in that scatterplot. Because r^2 has this specific interpretation, it is used almost as much as is r itself. Of course, r^2 only measures the *strength* of the association, not whether it is positive or negative.

6. *Like the mean and the standard deviation, the correlation coefficient is heavily influenced by outliers.* If in Example 6 the value $y = 5$ in the

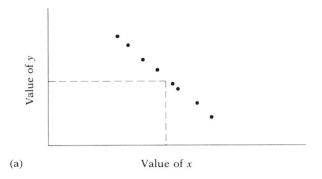

(a)

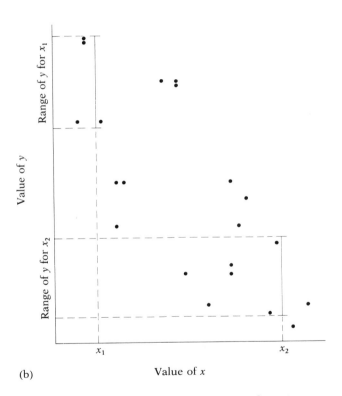

(b)

Figure 5-5. (a) Correlation $r = -1$. For a given x, there is no variation in y, hence y is perfectly predictable from x.
(b) Correlation $r = -0.7$. For a given x, y takes on a range of values, hence y is only approximately predictable from x. Because the range of values of y does change with x (compare x_1 and x_2), x is of some use in predicting y.

first data point were mistakenly entered as $y = 50$, the correlation would increase from $r = 0.71$ to $r = 0.86$. This is a substantial increase, for the proportion of variance explained rises from $r^2 = (0.71)^2 = 0.50$ to $r^2 = (0.86)^2 = 0.74$.

Section 2 exercises

5.14. Figure 5-6 is a scatterplot of the heights of the mothers and the fathers in a sample of parents. Answer the following questions from this graph.

    (a) What is the smallest height of any mother in the sample? How many women had that particular height? What were the heights of the husbands of these women?

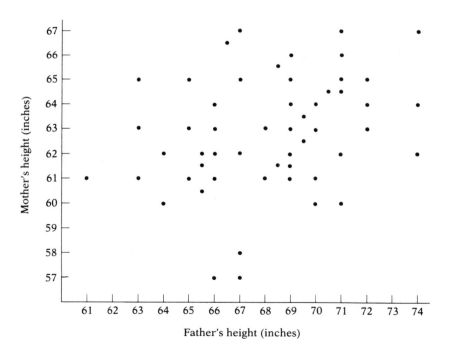

Figure 5-6. Data for an SRS of 53 pairs of parents from a group of 1079 pairs whose heights are recorded in Table XIII of Karl Pearson and Alice Lee, "On the Laws of Inheritance in Man," *Biometrika*, November 1903, p. 408.

(b) What is the greatest height of any father in the sample? How many men had that height, and what were the heights of their wives?

(c) Does the scatterplot show a positive or negative association between the heights of mothers and fathers? Is the association very strong?

5.15. Auto manufacturers are required to test their vehicles for the amount of each of several pollutants in the exhaust. The amount of a pollutant varies even among identical vehicles, so that several vehicles must be tested. Figure 5-7 is a scatterplot of the amounts of two pollutants, carbon monoxide and oxides of nitrogen, for 46 identical vehicles. Both variables are measured in grams of the pollutant per mile driven. [Data from Thomas J. Lorenzen, "Determining Statistical Characteristics of a Vehicle Emissions Audit Procedure," *Technometrics*, Volume 22 (1980), pp. 483–493.]

(a) Describe the nature of the relationship. Is the association positive or negative? Is the relation close to a straight line or clearly curved? Are there any outliers?

(b) A car magazine says, "When an engine is properly built and properly tuned, it emits few pollutants. If the engine is out of tune, it emits more of all important pollutants. You can find

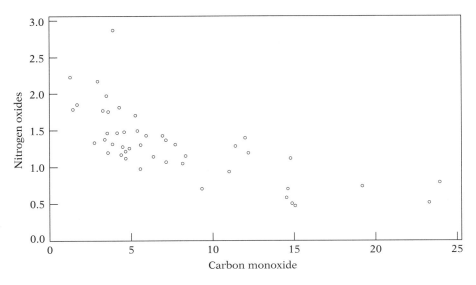

Figure 5-7. Two pollutants in vehicle exhaust.

out how badly a vehicle is polluting the air by measuring any one pollutant. If that value is acceptable, the other emissions will also be OK." Do the data in Figure 5-7 support this claim?

5.16. The Wheat Industry Council surveyed 3368 people, asking them to estimate the number of calories in several common foods. Here is a table of the average estimated calorie level and the correct number of calories.

Food	Estimated calories	Correct calories
8 oz whole milk	196	159
5 oz spaghetti with tomato sauce	394	163
5 oz macaroni with cheese	350	269
One slice wheat bread	117	61
One slice white bread	136	76
2-oz candy bar	364	260
Saltine cracker	74	12
Medium-size apple	107	80
Medium-size potato	160	88
Cream-filled snack cake	419	160

SOURCE: *USA TODAY*, October 20, 1983.

(a) Make a scatterplot with correct calories as the explanatory variable. (Because both variables have the same units, you should use the same scale on both axes of your plot, as in Figures 5-2 and 5-3.)

(b) Is there an approximate straight-line relationship that fits most of the points? Are there any outliers from this pattern?

5.17. Here are the golf scores of 12 members of a college women's golf team in two rounds of tournament play. (A golf score is the number of strokes required to complete the course, so that low scores are better.)

Player	1	2	3	4	5	6	7	8	9	10	11	12
Round 1	89	90	87	95	86	81	102	105	83	88	91	79
Round 2	94	85	89	89	81	76	107	89	87	91	88	80

(a) Make a scatterplot of the data, taking the first-round score as the explanatory variable. (Because both variables have the same scale, you should use the same scale on both axes of your plot, as in Figures 5-2 and 5-3.)

(b) Is there an association between the two scores? If so, is it positive or negative? Explain why you would expect scores in two rounds of a tournament to have such an association.

(c) Circle the outlier on your plot. The outlier may occur because a good golfer had an unusually bad round or because a weaker golfer had an unusually good round. Can you tell from the data given whether the outlier is from a good player or a poor player? Explain your answer.

5.18. For each of the following sets of data,

- Draw a scatterplot.
- Compute the mean, the variance, and the standard deviation of each variable x and y separately.
- Compute the correlation coefficient r, and also r^2.

For your own instruction, compare the values of r^2 and the closeness of the points in the scatterplots to a line.

(a)

x	4	4	−4	−4
y	−4	4	4	−4

(b)

x	4	3	0	−3	−4
y	−4	−2	0	2	4

(c)

x	4	2	−2	−4
y	4	−2	2	−4

5.19. *Archaeopteryx* is an extinct beast having flight feathers like a bird but teeth and a long bony tail like a reptile. Only six fossil specimens are known. Because these specimens differ greatly in size, they have sometimes been classified as different species rather than as individuals from the same species. Statistics can help decide the question. If the specimens belong to the same species and differ in size because they are at different stages of growth, there should be a positive straight-line association between the lengths of a pair of bones from all individuals. Outliers from this relationship would suggest a different species. Here are data on the lengths of the femur (a leg bone) and the humerus (a bone in the forearm) for the five specimens which preserve both bones.

Femur	38	56	59	64	74
Humerus	41	63	70	72	84

(a) Make a scatterplot of these data. Because the femur length is often used as an indicator of body size in birds, take that as the explanatory variable.

(b) Describe the pattern in the scatterplot. Is there a strong straight-line relationship? Are there any outliers?

(c) Calculate the correlation coefficient r between femur length and humerus length. What percent of the variance in humerus length can be explained by straight-line dependence on femur length?

[The data are from Marilyn A. Houck, et al., "Allometric Scaling in the Earliest Fossil Bird, *Archaeopteryx lithographica*," *Science*, Volume 247 (1990), pp. 195–198. The authors conclude from a variety of evidence that all specimens represent the same species.]

5.20. Make a scatterplot of the following data.

x	−5	−3	0	3	5
y	0	4	5	4	0

Show that the correlation coefficient is zero. (You can do this by showing that the *numerator* of r is zero. You need not compute any standard deviations.) The scatterplot shows a strong association between x and y. Explain how it can happen that $r = 0$ in this case.

5.21. Figure 5-8 shows four scatterplots. Which (if any) of these have approximately the same r? Which (if any) have approximately the same r^2?

5.22. Make a scatterplot of the following data.

x	1	2	3	4	10	10
y	1	3	3	5	1	11

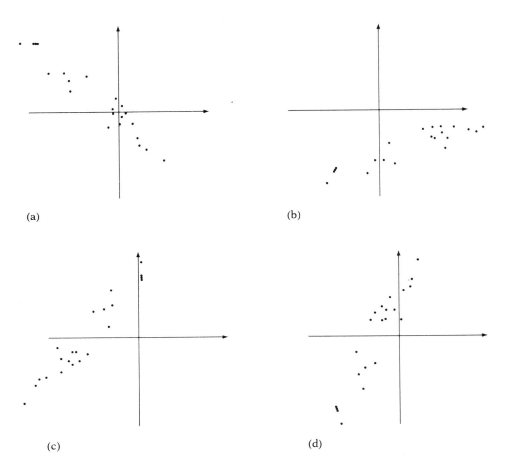

(a)   (b)

(c)   (d)

Figure 5-8. Recognizing r and r^2.

Show that the correlation is about 0.5. What feature of the data is responsible for reducing the correlation to this value despite a strong straight-line association between x and y in most of the observations?

5.23. Figure 5-9 contains five scatterplots. Match each to the r below that best describes it. (Some r's will be left over.)

$$r = -0.9 \qquad\qquad r = 0.3$$
$$r = -0.7 \qquad r = 0 \qquad r = 0.7$$
$$r = -0.3 \qquad\qquad r = 0.9$$

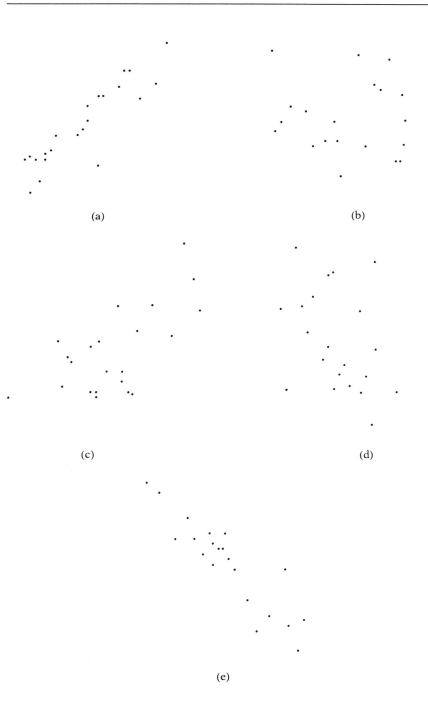

(a)

(b)

(c)

(d)

(e)

Figure 5-9. Match the correlation.

5.24. Your data consist of bivariate observations on the age of the subject (measured in years) and the reaction time of the subject (measured in seconds). In what units are each of the following descriptive statistics measured?
  (a) The mean age of the subjects.
  (b) The variance of the subjects' reaction times.
  (c) The standard deviation of the subjects' reaction times.
  (d) The correlation coefficient between age and reaction time.
  (e) The median age of the subjects.

5.25. If the lengths of the two bones in Exercise 5.19 were measured in inches rather than in centimeters, how would the correlation r change? (Don't do any calculations; 1 inch = 2.54 centimeters.)

5.26. Figure 5-2 displays the relation between average math and verbal Scholastic Aptitude Test scores for seven racial or ethnic groups. The correlation coefficient for these data is $r = 0.84$. What percent of the variation in average math scores among these groups can be explained by the variation in their average verbal scores?

5.27. A psychologist speaking to a meeting of the American Association of University Professors recently said, "The evidence suggests that there is nearly correlation zero between the teaching ability of a faculty member and his or her research productivity." The student newspaper reported this as: "Professor McDaniel said that good teachers tend to be poor researchers and good researchers tend to be poor teachers."
  Explain what (if anything) is wrong with the newspaper's report. If the report is not accurate, write your own plain-language account of what the speaker meant.

5.28. Measurements in large samples show that the correlation
  (a) between father's height and son's adult height is about _____.
  (b) between husband's height and wife's height is about _____
  (c) between a male's height at age 4 and his height at age 18 is about _____.

  The answers (in scrambled order) are
$$r = 0.2 \quad r = 0.5 \quad \text{and} \quad r = 0.8$$

  Match the answers to the statements and explain your choice.

5.29. If women always married men who were two years older than themselves, what would be the correlation between the ages of husband and wife? (*Hint:* Draw a scatterplot for the ages of husbands and wives when the wife is 20, 30, 40 and 50 years old.)

5.30. For each of the following pairs of variables, would you expect a substantial negative correlation, a substantial positive correlation, or a small correlation?

(a) The age of a secondhand car and its price.

(b) The weight of a new car and its overall miles-per-gallon rating.

(c) The height and the weight of a person.

(d) The height of a person and the height of that person's father.

(e) The height and the IQ of a person.

5.31. Each of the following statements contains a blunder. Explain in each case what is wrong.

(a) There is a high correlation between the sex and income of American workers.

(b) Since student ratings of professors' teaching and colleagues' ratings of their research have correlation $r = 1.21$, the better teachers also tend to be the better researchers.

(c) The correlation between pounds of nitrogen fertilizer applied to the field and the bushels per acre of corn harvested was $r = 0.63$ bushels. So applying more fertilizer increases yields.

3. Association and causation

There is a strong association between cigarette smoking and death rate from lung cancer. A study of British doctors found that smokers had 20 times the risk of nonsmokers, and a large study of American men ages 40 to 79 found 11 times higher death rates among smokers (see Figure 5-10). Does this mean that cigarette smoking causes lung cancer?

We are asking whether a specific association is due to changes in one variable (lung cancer) being *caused* by changes in another variable (smoking). Some observed associations are due to cause and effect. But others are not. Here are some examples of different relationships among variables that can explain an observed association.

> **Example 7.** In the Australian state of Victoria, a law compelling motorists to wear seat belts went into effect in December 1970. As time passed, an increasing percentage of motorists complied.
>
> A study found high positive correlation between the percent of motorists wearing belts and the percent of reduction in injuries from the 1970 level (see Figure 5-11). This is a clear instance of *causation*. Seat belts prevent injuries when an accident occurs, so an increase in their use causes a drop in injuries. The top diagram in Figure 5-12 outlines this relationship between two variables.

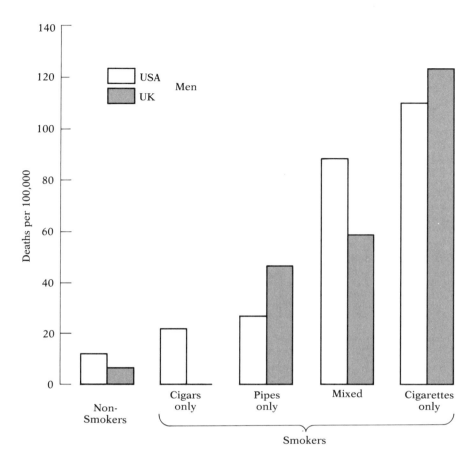

Figure 5-10. Deaths from lung cancer among men with different smoking habits. From two large studies, one of British doctors, one of American men. [From Royal College of Physicians, *Smoking and Health Now* (London: Pitman Medical Publishing, 1971), p. 5.]

Example 8. A moderate correlation exists between the Scholastic Aptitude Test (SAT) scores of high school students and their grade index later as freshmen in college. Surely high SAT scores do not cause high freshman grades. Rather, the same combination of ability and knowledge shows itself in both high SAT scores and high grades. Both of the observed variables are responding to the same unobserved variable, and this *common response* is the reason for the correlation between them. The middle diagram in Figure 5-12 illustrates this type of relationship.

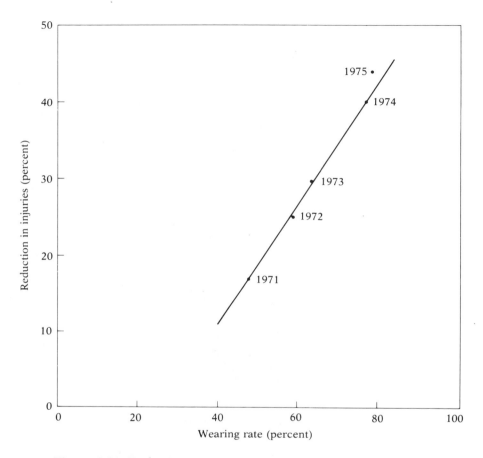

Figure 5-11. Reduction in automotive injuries from 1970 level versus percent of motorists wearing seat belts in Victoria, Australia. [From G. Grime, "A Review of Research on the Protection Afforded to Occupants of Cars by Seat Belts which Provide Upper Torso Restraint," *Accident Analysis and Prevention*, Volume 11 (1979), p. 299.]

Example 9. A study once showed a negative association between the starting salary of persons with a degree in economics and the level of the degree. Persons with a master's degree earned less on the average than those with a bachelor's degree, and Ph.D.s earned less than holders of a master's. So much for the rewards of learning. But wait. Further detective work revealed that there was a positive association between starting salary and degree level among economists who went to work for private industry. There was also a

CAUSATION—Changes in A cause changes in B.

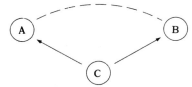

COMMON RESPONSE—Changes in both A and B
are caused by changes in a third variable, C.

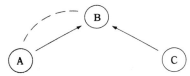

CONFOUNDING—Changes in B are caused both by
changes in A and by changes in third variable C.

Figure 5-12. Some explanations for association. Variables A and B
show a strong association, indicated by the broken line. This
association may result from any of several types of causal
relationships, indicated by solid arrows.

positive association among economists working in government.
And if only economists who took teaching jobs were considered,
there was again a positive association between salary and degree
level. So within every class of job, holders of higher degrees were
better paid.

What happened? Simply another version of Simpson's Paradox.
Teaching salaries were lower than those in government and industry. Few holders of bachelor's degrees in economics chose
teaching, but many holders of advanced degrees did. So average
salaries for advanced-degree holders were lower, even though each
employer paid them more than employees having only a bachelor's
degree. The negative association between salary and level of degree
did not mean that more education depressed salaries; it was due to
the effect of another variable (type of employer) on salaries. This
third variable was *confounded* with the degree held. The bottom
diagram in Figure 5-12 displays this relationship schematically.

An observed association may be due to (in the language of Figure 5-12) either common response or confounding as well as causation. Both common response and confounding involve the influence of another variable (or variables) C on the response variable B. The distinction between these two types of relationship is less important than the common element, the influence of other variables. We reserve "common response" for the situation in which C clearly influences both A and B, while the direct influence of A on B is questionable. Confounding means that A and C both influence B and their effects cannot be distinguished from each other.

Establishing causation

The tobacco industry has argued that the association between smoking and lung cancer may be an instance of common response. In Figure 5-12(b), C might be a genetic factor that predisposes people both to nicotine addiction (A in the diagram) and to lung cancer (B in the diagram). Then smoking and lung cancer would be positively associated even if smoking had no direct effect on health.

How can we evaluate the evidence for causation in disputed cases such as smoking and health? Here are some general words of wisdom. First, *causation is not a simple idea.* Rarely is A "the cause" of B. Smoking is only a *contributory cause* of lung cancer. It is one of several circumstances that make lung cancer more likely. Exposure to asbestos and breathing polluted air may be other contributory causes. Second, as I stressed in Chapter 2, *properly designed experiments are the best means of settling questions of causation.* You should be able to imagine a randomized controlled experiment with human subjects that would determine beyond doubt whether cigarette smoking causes lung cancer. But such an experiment is morally and practically impossible. So the evidence for causation in this case, although very strong, must always fall short of the strongest statistical evidence, which comes from experimentation.

Third, in the absence of an experiment, *to conclude that association is due to causation required evidence and judgment beyond what statistics can provide.* The case for the claim that variable A causes changes in variable B is strengthened if:

- The association between A and B recurs in different circumstances; this reduces the chance that it is due to confounding.

- A plausible explanation is available showing how *A* could cause changes in *B*.
- No equally plausible third factor exists that could cause changes in *A* and *B* together.

The case for the conclusion that smoking causes lung cancer meets these criteria. A strong association has been observed in studies of many groups of people in many places and over long periods of time. The risk of death from lung cancer is related to the number of cigarettes smoked and to the age at which smoking began. The number of men dying of lung cancer has risen in parallel to the prevalence of smoking, with a lag of about 30 years. Lung cancer now kills more men than any other form of cancer. As smoking by women has increased, lung cancer among women (once rare) has risen with the same 30-year lag. Lung cancer is now challenging breast cancer as the leading cause of cancer death among women. Thus unusually good evidence suggest that the association between smoking and lung cancer is not due to confounding or other defects in producing the data.

Second, cigarette smoke contains tars that have been shown by experiment to cause tumors in animals when applied in sufficient quantity. So it is plausible that these same tars, applied in smaller quantities over many years, can cause tumors in human lungs. Finally, such suggested third factors as a genetic predisposition are not plausible. The genetic hypothesis in particular cannot explain the different patterns in the rise of lung cancer in the two sexes or the fact that giving up smoking reduces the risk of cancer.

The evidence that smoking is a cause of lung cancer is about as convincing as nonexperimental evidence can be. And so it is that we read

---

Warning: The Surgeon General Has Determined That Cigarette Smoking Is Dangerous to Your Health.

Is it a coincidence?

We have seen the importance of searching for other variables that may be lurking in the background before drawing conclusions from an observed association. Here is another danger: Casual observation can mislead us by suggesting a cause-and-effect relationship when only the play of chance is present. When we observe an unusual outcome, we tend to seek a cause just because the event was unusual. Yet it may be mere coincidence. After all, if we look around us long enough, we will see something unusual. It is indeed very unlikely that *this particular* unusual event would occur, but it is certain that *some* unusual event would eventually occur, simply by chance.

> **Example 10.** In 1986, Evelyn Marie Adams won the New Jersey state lottery for the second time, adding $1.5 million to her previous $3.9 million jackpot. The *New York Times* (February 14, 1986) claimed that the odds of one person winning the big prize twice were about 1 in 17 trillion. Nonsense, said two Purdue statistics professors in a letter that appeared in the *Times* two weeks later. The chance that Evelyn Marie Adams would win twice in her lifetime is indeed tiny, but it is almost certain that *someone* among the many millions of regular lottery players in the U.S. would win two jackpot prizes. The statisticians estimated even odds of another double winner within seven years. Sure enough, Robert Humphries won his second Pennsylvania lottery jackpot ($6.8 million total) in May 1988.

Example 10 makes clear the distinction between the chance of *some* unusual event and the chance of *this particular* unusual event in a setting where the chances can actually be calculated. The same distinction arises in more important and more complex settings, as the next example illustrates.

> **Example 11.** In 1984, residents of a neighborhood in Randolph, Massachusetts, counted 67 cancer cases in their 250 residences. This cluster of cancer cases seemed unusual, and the residents expressed concern that runoff from a nearby chemical plant was contaminating their water supply and causing cancer. In 1979, two of the eight town wells serving Woburn, Massachusetts, were found to be contaminated with organic chemicals. Alarmed citizens began counting cancer cases. Between 1964 and 1983, 20 cases of childhood leukemia were reported in Woburn. This is an unusual number of cases of this rather rare disease. The residents believed that the well water had caused the leukemia, and proceeded to sue two companies held responsible for the contamination.

Cancer is a common disease, accounting for about 22% of all deaths in the United States. That cancer cases sometimes occur in clusters in the same neighborhood is not surprising; there are bound to be clusters *somewhere* simply by coincidence. But when a cancer cluster occurs in *our* neighborhood, we suspect the worst and look for someone to blame.

Both of the Massachusetts cancer clusters were investigated by statisticians from the Harvard School of Public Health. The investigators tried to obtain complete data on everyone who had lived in the neighborhoods in the periods in question and to estimate their exposure to the suspect drinking water. They also tried to obtain data on other factors that might explain cancer, such as smoking and occupational exposure to toxic substances. Such data may establish a link between exposure to the water and cancer that is better evidence than the cluster alone. The verdict: Chance is the likely explanation of the Randolph cluster, but there is evidence of an association between drinking water from the two Woburn wells and developing childhood leukemia.[5]

Analysis of cancer clusters is another example of the difficulties of trying to study cause and effect when experiments are not possible. But my present point is that most (not all) clusters are just chance happenings, like winning the lottery. You should at least consider coincidence as a possible explanation when data point to something unexpected.

Section 3 exercises

5.32. A study of grade school children ages 6 to 11 years found a high positive correlation between reading ability and shoe size. Explain why common response provides a plausible explanation for this correlation. Present your suggestion in a diagram like one of those in Figure 5-12.

5.33. There is a negative correlation between the number of flu cases reported each week through the year and the amount of ice cream sold that week. It's unlikely that ice cream prevents flu. What is a more plausible explanation for this correlation? Draw a diagram like one of those in Figure 5-12 to illustrate the relations among the variables.

5.34. There is a strong positive correlation between years of schooling completed (call this variable A) and lifetime earnings (variable B) for American men. One possible reason for this association is causation: more education leads to higher-paying jobs. Another explanation is confounding: Men who complete many years of schooling have other characteristics [C in Figure 5-12(c)] that would lead to better jobs even without the education. Suggest several variables that could play the role of *C*.

5.35. A public health survey of 7000 California men found little correlation between alcohol consumption and chance of dying during the 5½ years of the study. In fact, men who did not drink at all during these years had a slightly higher death rate than did light drinkers. This lack of correlation was somewhat surprising. Explain how common response might explain the higher death rate among men who did not drink at all over a short period.

5.36. A study using data from 41 states found a positive correlation between per capita beer consumption and death rates from cancer of the large intestine and rectum. The states with the highest rectal cancer death rates were Rhode Island and New York. The beer consumption in those states was 80 quarts per capita. South Carolina, Alabama, and Arkansas drank only 26 quarts of beer per capita and had rectal-cancer death rates less than one-third of those in Rhode Island and New York. (This study was reported in a Gannett News Service dispatch appearing in the *Lafayette Journal and Courier* of November 20, 1974.)

List some variables possibly influencing state cancer death rates that appear to be confounded with beer consumption. For a clue, look at the high- and low-consumption states given above.

5.37. A study of London double-decker bus drivers and conductors found that drivers had twice the death rate from heart disease as conductors. Because drivers sit while conductors climb up and down stairs all day, it was at first thought that this association reflected the effect of type of job on heart disease. Then a look at bus company records showed that drivers were issued consistently larger-size uniforms when hired than were conductors. This fact suggested an alternative explanation of the observed association between job type and deaths. What is it?

5.38. A study showed that women who work in the production of computer chips have abnormally high numbers of miscarriages. The union claimed that exposure to chemicals used in production caused the miscarriages. Another possible explanation is that these workers spend most of the time standing up. Illustrate these relationships in a diagram like one of those in Figure 5-12.

5.39. The National Halothane Study was a major investigation of the safety of the anesthetics used in surgery. Records of over 850,000 operations performed in 34 major hospitals showed the following death rates for four common anesthetics. (See L. E. Moses and F. Mosteller, "Safety of Anesthetics," in *Statistics: A Guide to the Unknown.*)

Anesthetic	A	B	C	D
Death rate	1.7%	1.7%	3.4%	1.9%

It is possible that the anesthetics cause some deaths, and that anesthetic C is particularly dangerous. Surgeons, for ethical reasons, are unwilling to do an experiment in which similar patients are given different anesthetics. But you have very detailed records for the 850,000 operations to which the given death rates refer. What kinds of information would you seek in these records in order to establish that anesthetic C does or does not cause extra deaths among patients?

5.40. Your friend Julie, during a month of travel in Europe, runs into Bob at a museum in London. Bob was a casual acquaintance from a history course the previous year. "Imagine meeting Bob in a place like that! It must have been fated!" says Julie. Try to convince her that there is nothing extraordinary about such a meeting.

5.41. You are getting to know your new roommate, assigned to you by the college. In the course of a long conversation, you find that both of you have sisters named Deborah. Should you be surprised? Explain your answer.

4. Prediction

A strong relationship between two variables can be used to predict the value of one when the value of the other is known. The basic idea is straightforward: Draw a curve through the points of the scatterplot to represent the relationship and then use this curve for prediction. The simplest curve is a straight line, and this is the case we shall consider. Of course, it would be foolish to draw a straight line when the pattern of the relationship is curved, as in the scatterplot of corn yield versus planting rate in Figure 5-1. The data must show a roughly linear trend like that in Figure 5-11. An example will illustrate how to use a straight-line relation for prediction.

> **Example 12.** Joan is concerned about the amount of energy she uses to heat her home. She keeps a record of the natural gas consumed over a period of nine months. Because the months are not all equally long, she divides each month's consumption by the number of days in the month to get cubic feet of gas used per day.

Demand for heating is strongly influenced by the outside temperature, measured in degree days. (One degree day is accumulated for each degree that a day's average temperature falls below 65°F. For example, a day with an average temperature of 30°F gives 35 degree days.) From local weather records, Joan obtains the number of degree days for each month. She divides this total by the number of days in the month to get the average number of degree days per day during the month. Here are Joan's data, nine measurements on the two variables.

Degree days	15.6	26.8	37.8	36.4	35.5	18.6	15.3	7.9	0
Cubic feet of gas	5.2	6.1	8.7	8.5	8.8	4.9	4.5	2.5	1.1

A scatterplot of these data appears in Figure 5-13(a). Degree days is the explanatory variable x (plotted on the horizontal axis) because outside temperature explains gas consumption. The scatterplot shows a close straight-line relationship. In fact, the correlation is $r = 0.989$, so the points are tightly clustered about a line. To draw a line that describes the relation, lay a transparent straightedge over the points. Since we wish to predict y from x, fiddle the line about until the distances from the line to the points in the *vertical* (y) direction seem smallest. The resulting line is drawn in Figure 5-13(b).

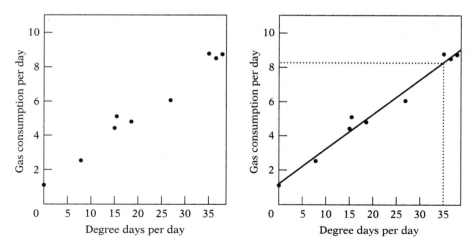

Figure 5-13. Home consumption of natural gas versus outdoor temperature. A fitted line (b) is used for prediction.

"How did I get into this business? Well, I couldn't understand regression and correlation in college, so I settled for this instead."

Now for prediction. How much gas can Joan expect to use in a month that averages 35 degree days per day? First locate $x = 35$ degree days on the horizontal axis. Draw a vertical line up to the fitted line and then a horizontal line over to the gas consumption scale. As Figure 5-13(b) shows, we predict that slightly more than 8 cubic feet per day will be consumed.

Regression equations

When the plot shows a straight-line relationship as strong as that in Figure 5-13, it is easy to fit a line on the graph by using a transparent straightedge. This gives us a line on the graph, but not an *equation* for the line. There is also no guarantee that the line we fit by eye is the "best" line. It is difficult to concentrate on just the vertical distances of the points from the line when fitting a line by eye. There are statistical techniques for finding from the data the equation of the best line (with various meanings of "best"). The most common of these techniques is called *least-squares regression*.

Least-squares regression looks at the vertical deviations of the points in a scatterplot from any straight line. Any line that is a good candidate for describing the data will pass above some points and below others, rather than miss the cloud of points entirely. So some of the deviations will be positive and some will be negative. The squares of the deviations, however, are all positive. The least-squares regression line is the line that makes the sum of the squared deviations as small as possible. Hence the name least-squares. The line in Figure 5-13(b) is in fact the least-squares regression line for predicting y from x. All statistical computer software packages and some calculators will calculate the least-squares line for you, so the equation of a line is often available with little work. You should therefore know how to use a regression equation even if you do not learn the details of how to get the equation from the data.

In writing the equation of a line, x stands as usual for the explanatory variable and y for the response variable. The equation has the form

$$y = a + bx$$

The number b is the *slope* of the line, the amount by which y changes when x increases by 1 unit. The number a is the *intercept*, the value of y when $x = 0$. To use the equation for prediction, just substitute your x value into the equation and calculate the resulting y value.

Example 13. A computer program tells us that the least-squares regression line computed from Joan's data for predicting gas consumption from degree days is

$$y = 1.23 + 0.202x$$

The slope of this line is $b = 0.202$. This means that gas consumption increases by 0.202 cubic feet per day when there is one more degree day per day. The intercept is $a = 1.23$. When there are no degree days (that is, when the average temperature is 65°F or above), gas consumption will be 1.23 cubic feet per day. The slope and intercept are of course estimates based on fitting a line to the data in Example 12. We do not expect every month with no degree days to average exactly 1.23 cubic feet of gas per day. The line represents only the overall pattern of the data.

Joan's predicted gas consumption for a month with $x = 35$ degree days per day is

$$y = 1.23 + (0.202) \ (35)$$
$$= 8.3 \text{ cubic feet per day}$$

This prediction will almost certainly not be exactly correct for the next month that has 35 degree days per day. But because the past data points lie so close to the line, we can be confident that gas consumption in such a month will be quite close to 8.3 cubic feet per day.

Understanding prediction

Computers make prediction easy and automatic, even from very large sets of data. Anything that can be done automatically is often done thoughtlessly. The computer will happily fit a straight line to a curved relationship, for example. Also, the computer cannot decide which is the explanatory variable and which is the response variable. This distinction is important, because the least-squares line for predicting gas consumption from degree days is not the same as the line for predicting degree days from gas consumption. And, as you might expect, the least-squares regression line fitted automatically by the computer can be heavily influenced by a few outliers in the data. Before you begin to calculate, choose your explanatory and response variables and then plot your data to check that there is a straight-line overall pattern and that there are no striking outliers. Exercise 5.49 gives a graphic reminder of the perils of numerical descriptions not accompanied by plots.

Prediction is most secure when the explanatory variable x takes a value within the range of the x-values in our data. Look once more at Figure 5-1. If we had data only on the planting rates between 16,000 and 24,000 plants per acre, we might fit a line and from it predict a high yield at 28,000 plants per acre. Our prediction would be dramatically wrong because a straight line does not continue to describe the relationship for higher planting rates. Using a fitted relation for prediction outside the range of the original data is *extrapolation*. Extrapolation is dangerous.

The usefulness of an observed association for predicting a response y given the value of the explanatory variable x does *not* depend on a cause-and-effect relation between x and y. An employer who uses an aptitude test to screen potential employees does not think that high test scores cause good job performance after the person is hired. Rather, the test attempts to measure abilities that will usually result in good performance. It is a matter of common response. If there is a relationship between the test results and later performance on the job, then, in the language of Chapter 3, the test has predictive validity.

The usefulness of the fitted line for prediction does depend on the strength of the association. Not all straight-line associations are as strong as that between degree days and natural gas consumption. Recall that r^2 measures the strength of a straight-line association. Look back at Figure 5-5. Figure 5-5(a) shows perfect straight-line association ($r^2 = 1$); in this case, y is perfectly predictable from x. Figure 5-5(b) shows an association with $r^2 = 0.49$. When a value of x (say x_1 in the figure) is given in this case, y still varies quite a bit, so only approximate prediction is possible. Recipes such as least-squares regression for fitting a line to the data are most useful in cases like Figure 5-5(b), where we would find it difficult to fit a line by eye. But do remember that the resulting predictions can be far from perfect.

How large an r^2—and hence how precise a prediction—you can expect depends on the variables you are interested in. In many fields outside the physical sciences, high correlations are rare. American law schools, for example, use the Law School Admission Test (LSAT) to help predict performance and to aid in admissions decisions.[6] Yet the correlation between LSAT scores and the first-year grades in law school of admitted applicants is only about $r = 0.36$. This is discouraging, for such a correlation means that only 13% of the variance in law school grades is accounted for. Nonetheless, LSAT scores are a better predictor than are the applicants' undergraduate grades, which have a correlation of only about $r = 0.25$ with later grades in law school.

When no single explanatory variable has a high correlation with a response variable, several explanatory variables together can be used for prediction. This is called *multiple regression*. We can no longer fit a relationship among the variables by eye. A recipe that a computer can follow is essential. Fortunately, the computer will produce a *multiple correlation coefficient* (also called r) between the response variable and all the explanatory variables at once. The square r^2 of the multiple correlation coefficient has the same interpretation as in the one-explanatory-variable case. For example, when both LSAT scores and undergraduate grades are used to predict law school grades, the multiple correlation coefficient is about $r = 0.45$. So straight-line dependence on both explanatory variables together accounts for $(0.45)^2$, or about 20%, of the variability in first-year law school grades.

I hope that our tour of association, causation, and prediction has left you with that slightly winded feeling that follows a good workout. Overconfidence in interpreting association is the root of many a statistical sin. Most users of statistics have learned that correlation does not imply causation. Some go to the other extreme and label all correlations not due to causation as "nonsense correlations." That label is inspired by examples such as the alleged strong positive correlation between teachers' salaries and liquor sales, both of which have steadily increased over time as a result of inflation and general prosperity. But this correlation is perfectly real and not at all nonsensical; what is nonsense is the interpretation that teachers are spending their salaries on booze. With sufficient skill and enough information, most associations can be usefully interpreted, even though (as in Example 9) the obvious interpretation may be seriously misleading.

Regression calculations*

Calculating the least-squares regression line from bivariate data is sufficiently tedious that we usually leave the work to a computer or calculator. If, however, you or your electronic slave have already calculated the basic descriptive measures for bivariate data, the least-squares line is within easy reach.

We are given data on an explanatory variable x and a response variable y. A scatterplot shows a generally straight-line relation without prominent outliers, so we are willing to fit the least-squares line to predict y from x. First calculate the mean $\bar{x}$ and the standard deviation s_x of the x values. (The subscript on s_x serves to remind us that this is the standard deviation of the x's.) Then calculate the mean $\bar{y}$ and the standard deviation s_y of the y's. These four numbers describe the two univariate distributions of the x's and the y's, but say nothing about the relationship between the two variables. To describe the strength and direction of the straight-line association between x and y, calculate the correlation r. From these five numbers we can obtain the equation of the least-squares line. Here are the facts.

The least-squares regression line for predicting y from x is given by $y = a + bx$, where the slope b and the intercept a are

---

*This section is optional reading.

$$b = r\frac{s_y}{s_x}$$

$$a = \bar{y} - b\bar{x}$$

Example 14. Return once more to the gas consumption data in Example 12. You can organize the required calculations as in Example 6 if you do not have a calculator that gives you means, standard deviations, and the correlation from keyed-in data. First we find from the nine x (degree day) values that

$$\bar{x} = 21.544 \quad \text{and} \quad s_x = 12.652$$

Next we turn to the nine y (gas consumption) values and calculate that

$$\bar{y} = 5.589 \quad \text{and} \quad s_y = 2.587$$

The correlation between degree days and gas consumption is

$$r = 0.989$$

From these basic descriptive measures we find that the slope of the least-squares regression line is

$$b = r\frac{s_y}{s_x}$$

$$= 0.989\frac{2.587}{12.652} = 0.2022$$

The intercept of the line is then

$$a = \bar{y} - b\bar{x}$$

$$= 5.589 - (0.2022)(21.544) = 1.23$$

The equation of the least-squares line is therefore

$$y = 1.23 + 0.202x$$

I don't recommend calculating the equation of the least-squares line with a basic calculator. The equations for the slope and intercept do, however, shed more light on the behavior of the line. You can see from the equation for the intercept a that when x is equal to the mean $\bar{x}$, the y predicted by the line is just the mean $\bar{y}$. *The least-squares regression line always passes through the point $(\bar{x}, \bar{y})$.* The equation for the slope b says that *a change of one standard deviation in x corresponds to a change of r standard deviations in y.* When the variables are perfectly correlated ($r = 1$ or $r = -1$), the change in the response y is the same (when measured in standard deviations) as the change in x. Otherwise, since $-1 \le r \le 1$, the change in y is less than the change in x. As the correlation grows less strong, y moves less in response to changes in x.

Section 4 exercises

5.42. Use the regression line in Figure 5-11 to predict the percent reduction from the 1970 injury level that would occur if 75% of motorists wore seat belts.

5.43. Figure 5-14 displays data on the number of slices of pizza consumed by pledges at a fraternity party (the explanatory variable x) and the number of laps around the block the pledges could run immediately afterward (the response variable y). The line on the

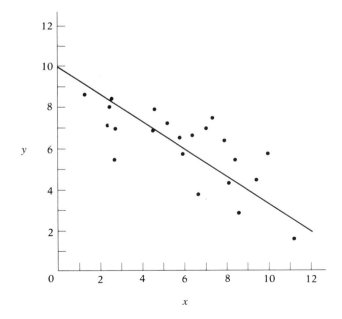

Figure 5-14.

scatterplot is the least-squares regression line computed from these points for predicting y from x.

(a) At the next party, a pledge eats 6 slices of pizza before running. How many laps do you predict he will complete?

(b) Another pledge eats 9 pieces of pizza. Predict how many laps he will complete.

(c) A third pledge shows off by eating 25 pieces of pizza. You should refuse to predict his performance from the scatterplot and regression line. Explain why.

5.44. Exercise 5.16 in Section 2 presents data on estimated versus true calorie content for 10 foods. You wish to predict what calorie count people will guess for a food that has 150 calories.

(a) Make a scatterplot of the data. (Which is the explanatory variable?)

(b) You decide to ignore the two outliers for which the people sampled guessed much too high. The other foods form a straight-line pattern. Use a transparent straightedge to fit a line to this pattern, and draw the line on your scatterplot.

(c) Use your line to predict the calories guessed for a food that has 150 calories.

(d) A computer least-squares regression routine using all 10 points predicts about 253 calories for the guess. Your value should be lower. Why?

5.45. Manatees are large, gentle sea creatures that live along the Florida coast. Many manatees are killed or injured by power boats. Here are data on powerboat registrations (in thousands) and the number of manatees killed by boats in Florida in the years 1977 to 1987.

Year	Boats	Manatees killed
1977	447	13
1978	460	21
1979	481	24
1980	498	16
1981	513	24
1982	512	20
1983	526	15
1984	559	34
1985	585	33
1986	614	33
1987	645	39

(a) Which is the explanatory variable? Make a scatterplot of these data. Describe the overall pattern and any outliers or other important deviations.

(b) Fit a line to the scatterplot by eye and draw the line on your graph. If in a future year powerboat registrations increase to 700,000, predict the number of manatees that will be killed by boats.

5.46. A computer software package tells us that the least-squares regression line for predicting manatees killed from powerboat registrations, calculated from the data in Exercise 5.45, is

$$y = -38.29 + 0.1187x$$

Use this equation to predict the number of manatees killed in a future year in which 700,000 powerboats are registered. Compare this prediction with the graphical prediction you found in Exercise 5.45.

5.47. Data on the average Scholastic Aptitude Test verbal and mathematics scores for high school seniors in each of the 50 states show a strong straight-line association. The least-squares regression line for predicting a state's average math score y from the average verbal score x is

$$y = 27 + 1.03x$$

(a) Is the association between math and verbal scores positive or negative? (That is, do states with high average verbal score tend to have high or low average math score?) How does the regression equation tell you the answer to this question?

(b) The average SAT verbal score in New York was 422. Use the regression line to predict New York's average math score. (New York's actual average math score was 466.)

5.48. Concrete road pavement gains strength over time as it cures. Highway engineers use regression lines to predict the strength after 28 days (when curing is complete) from measurements made after 7 days. Let x be strength after 7 days (in pounds per square inch) and y the strength after 28 days. One set of data gave the least-squares regression line to be

$$y = 1389 + 0.96x$$

A test of some new pavement after 7 days shows that its strength is 3300 pounds per square inch. Predict the strength of this pavement after 28 days.

5.49. Table 5-2 presents three sets of bivariate data prepared by the

Table 5-2. Three data sets with the same correlation and regression line

Data A

x	10	8	13	9	11	14	6	4	12	7	5
y	8.04	6.95	7.58	8.81	8.33	9.96	7.24	4.26	10.84	4.82	5.68

Data B

x	10	8	13	9	11	14	6	4	12	7	5
y	9.14	8.14	8.74	8.77	9.26	8.10	6.13	3.10	9.13	7.26	4.74

Data C

x	8	8	8	8	8	8	8	8	8	8	19
y	6.58	5.76	7.71	8.84	8.47	7.04	5.25	5.56	7.91	6.89	12.50

SOURCE: F.J. Anscombe, "Graphs in Statistical Analysis," *The American Statistician*, Volume 27 (1973), pp. 17–21.

statistician Frank Anscombe to illustrate the dangers of calculating without first plotting the data. *All three sets have the same correlation and the same least-squares regression line* to several decimal places. The regression equation is

$$y = 3.00 + 0.500x$$

(a) Make a scatterplot for each of the three data sets and draw the regression line on each of the plots. [To draw the regression line, substitute two convenient values of x into the equation to find the corresponding values of y. Plot the two (x, y) points on the graph and draw the line through them.]

(b) In which of the three cases would you be willing to use the fitted regression line to predict y given that $x = 14$? Explain your answer in each case.

5.50. "When $r = 0.7$, this means that y can be predicted from x for 70% of the individuals in the sample." Is this statement true or false? Would it be true if $r^2 = 0.7$? Explain your answers.

5.51. Exercise 5.19 in Section 2 gives data on the length of two bones in five specimens of an extinct species that represents the transition from reptiles to birds. Find from these data the equation of the least-squares regression line for predicting humerus length from femur length. A new fossil is found with femur length 50 centimeters, but the humerus bone is missing. Use your equation to predict the length of the missing humerus.

5.52. Exercise 5.45 gives data on the relationship between the number of powerboats registered in Florida (x, in thousands) and the number of manatees killed by boats (y). The basic descriptive statistics for these data are

$$\bar{x} = 530.909 \qquad s_x = 60.288 \qquad r = 0.854$$
$$\bar{y} = 24.727 \qquad s_y = 8.373$$

Use this information to find the equation of the least-squares line for predicting manatees killed from boat registrations. Compare your result with the equation given in Exercise 5.46.

5.53. Exercise 5.17 gives the scores of the 12 members of a college women's golf team in two rounds of tournament play. We are interested in using first-round scores (which are therefore values of the explanatory variable x) to predict second-round scores y. The scatterplot suggests that Player 8 is an outlier.

(a) The correlation between the two scores is $r = 0.687$ when all players are included and $r = 0.842$ when Player 8 is dropped from the calculation. Why are these two values so different?

(b) We decide to leave Player 8 out of our prediction calculations. The basic descriptive statistics for the remaining 11 players are

$$\bar{x} = 88.273 \qquad s_x = 6.166 \qquad r = 0.842$$
$$\bar{y} = 87.909 \qquad s_y = 7.833$$

Use this information to obtain the equation of the least-squares regression line for predicting second-round scores from first-round scores.

(c) In the next tournament, a player shoots 85 on the first round. Predict her second-round score.

5.54. The growth of children follows a straight-line pattern between early childhood and adolescence. Here are data on Sarah's growth between the ages of 36 months and 60 months.

Age (months)	36	48	51	54	57	60
Height (cm)	86	90	91	93	94	95

(a) Calculate the equation of the least-squares regression line for predicting Sarah's height from her age.

(b) Sarah's growth rate is the number of centimeters she grows each month (on the average). According to your regression equation, what is her growth rate?

(c) Make a scatterplot of the data and draw the least-squares line on the plot. [To draw the line, use the equation to find y for two values of x, such as $x = 40$ and $x = 60$. Plot the two (x, y) points on your graph and draw the line through them.]

(d) Use your regression line to estimate Sarah's height at age 42 months.

5.55. Calculate the correlation coefficient and the least-squares regression line for each of the three data sets of Exercise 5.49. Verify that the results are the same in each case. What is the common value of the correlation r? (Don't try this exercise without a computer or calculator that will do the hard work for you.)

NOTES

1. The study, published in the *American Journal of Epidemiology*, is reported in Natalie Angier, "Study Reports that Short Women Have Higher Risk of Heart Attacks," *New York Times*, June 28, 1990.

2. From *The Health Consequences of Smoking: 1983*, Washington, D.C.: U.S. Public Health Service.

3. P. J. Bickel and J. W. O'Connell, "Is There a Sex Bias in Graduate Admissions?" *Science*, Volume 187 (1975), pp. 398–404.

4. For a glimpse of the statistical complexities, see the studies collected in *Pay Equity: Empirical Inquiries* (Washington, D.C.: National Academy Press, 1989).

5. For Woburn, see S. W. Lagakos, B. J. Wessen, and M. Zelen, "An Analysis of Contaminated Well Water and Health Effects in Woburn, Massachusetts," *Journal of the American Statistical Association*, Volume 81 (1986), pp. 583–596 and the following discussion. For Randolph, see R. Day, J. H. Ware, D. Wartenberg, and M. Zelen, "An Investigation of a Reported Cancer Cluster in Randolph, Ma." Harvard School of Public Health technical report, June 27, 1988.

6. The facts about the LSAT, along with much useful information about the alleged cultural bias of such tests, is found in David Kaye, "Searching for the Truth About Testing," *The Yale Law Journal*, Volume 90 (1980), pp. 431–457.

Review exercises

5.56. A study of the salaries of full professors at Upper Wabash Tech showed that the median salary for female professors was considerably less than the median male salary. Further investigation showed that the median salaries for male and female full professors were about the same in every department (English, physics, etc.) of the university. Explain how equal salaries in every department can still result in a higher overall median salary for men.

5.57. The influence of race on imposition of the death penalty for murder has been much studied and contested in the courts. The three-way table below classifies 326 cases in which the defendant was convicted of murder. The three variables are the defendant's race, the victim's race, and whether or not the defendant was sentenced to death. [Data from M. Radelet, "Racial Characteristics and Imposition of the Death Penalty," *American Sociological Review*, Volume 46 (1981), pp. 918–927.]

White defendant			Black defendant		
	Death	Not		Death	Not
White victim	19	132	White victim	11	52
Black victim	0	9	Black victim	6	97

(a) Make from these data a two-way table of defendant's race by death penalty.

(b) Show that Simpson's paradox holds: A higher percent of white defendants are sentenced to death overall, but for both black and white victims a higher percent of black defendants are sentenced to death.

(c) Using the data, explain why the paradox holds in language that a judge could understand.

5.58. Recent studies have shown that earlier reports seriously underestimated the health risks associated with being overweight. The error was due to overlooking some important variables. In particular, smoking tends both to reduce weight and to lead to earlier death. Illustrate Simpson's paradox by a simplified version of this situation. That is, make up a three-way table of overweight (yes or no) by early death (yes or no) by smoker (yes or no) such that:

• Overweight smokers and overweight nonsmokers both tend to die earlier than those not overweight.

• But when smokers and nonsmokers are combined into a two-way table of overweight by early death, persons who are not overweight tend to die earlier.

5.59. To demonstrate the effect of nematodes (microscopic worms) on plant growth, a botanist prepares 16 identical planting pots and then introduces different numbers of nematodes into the pots. A tomato seedling is transplanted into each plot. Here are data on the increase in height of the seedlings (in centimeters) during the 16 days after planting.

Nematodes	Seedling growth			
0	10.8	9.1	13.5	9.2
1000	11.1	11.1	8.2	11.3
5000	5.4	4.6	7.4	5.0
10,000	5.8	5.3	3.2	7.5

(a) Make a scatterplot of the response variable (growth) against the explanatory variable (nematode count).
(b) Briefly describe the conclusions about the effects of nematodes on plant growth that these data suggest.
(c) Do you recommend calculating the correlation coefficient r to describe the strength of the relationship between nematode count and growth? Explain your answer.

5.60. The following table gives data on the lean body mass (kilograms) and resting metabolic rate for 12 women and 7 men who are subjects in a study of obesity. The researchers suspect that lean body mass (that is, the subject's weight leaving out all fat) is an important influence on metabolic rate.

Subject	Sex	Mass	Rate		Subject	Sex	Mass	Rate
1	M	62.0	1792		11	F	40.3	1189
2	M	62.9	1666		12	F	33.1	913
3	F	36.1	995		13	M	51.9	1460
4	F	54.6	1425		14	F	42.4	1124
5	F	48.5	1396		15	F	34.5	1052
6	F	42.0	1418		16	F	51.1	1347
7	M	47.4	1362		17	F	41.2	1204
8	F	50.6	1502		18	M	51.9	1867
9	F	42.0	1256		19	M	46.9	1439
10	M	48.7	1614					

(a) Make a scatterplot of the data for the 12 female subjects. Which is the explanatory variable?
(b) Does metabolic rate increase or decrease as lean body mass increases? What is the overall shape of the relationship? Are there any outliers?

5.61. When observations on two variables fall into several categories, you can display more information in a scatterplot by plotting each category with a different symbol or a different color. Add the data for male subjects to your scatterplot in Exercise 5.60, using a different color or symbol than you used for females. Does the type of relationship you found for females hold for men also? How do the male subjects as a group differ from the female subjects as a group?

5.62. A study of class attendance and grades among freshmen at a state university showed that in general students who attended a higher percent of their classes earned higher grades. Class attendance explained 16% of the variation in grade index among the freshmen studied. What is the numerical value of the correlation between percent of classes attended and grade index?

5.63. Researchers studying acid rain measured the acidity of precipitation in an isolated wilderness area in Colorado for 150 consecutive weeks in the years 1975 to 1978. The acidity of a solution is measured by pH, with lower pH values indicating that the solution is more acid. The acid-rain researchers observed a linear pattern over time. They reported that the least-squares line

$$pH = 5.43 - (0.0053 \times weeks)$$

fit the data well. [See William M. Lewis and Michael C. Grant, "Acid Precipitation in the Western United States," *Science*, Volume 207 (1980), pp. 176–177.]

(a) Draw a graph of this line. Note that the straight-line change in pH over time is downward; this means that acidity is increasing.

(b) According to the fitted line, what was the pH at the beginning of the study (weeks = 1)? At the end (weeks = 150)?

(c) What is the slope of the fitted line? Explain clearly what this slope says about the change in the acidity of the precipitation in this wilderness area.

5.64. The number of people living on American farms has declined steadily during this century. Here are data on the farm population (millions of persons) from 1935 to 1980.

Year	1935	1940	1945	1950	1955	1960	1965	1970	1975	1980
Pop	32.1	30.5	24.4	23.0	19.1	15.6	12.4	9.7	8.9	7.2

(a) Make a scatterplot that shows the change in farm population over time. Is the association positive or negative? Is the pattern roughly a straight line?

(b) The equation of the least-squares regression line for predicting farm population y (in millions) from the year x is

$$y = 1166.93 - 0.5869x$$

According to this fitted line, how much did the farm population decline each year on the average during this period?

(c) Use the regression equation to predict the number of people living on farms in 1990. Explain why the result makes no sense.

Chapter 6

The Consumer Price Index and its neighbors

Concern over the state of the economy rises as economic health declines, but such concern is always in the news. Many of the reports that arouse comment and expressions of official concern are veiled in the language of statistics: "The Consumer Price Index rose 0.6% in August . . ." and "Senator Bean called for increased stimulation of the economy, pointing out that the Composite Index of Leading Indicators had leveled off following six months of growth . . . ." The data that alarm or soothe the politicians are collected by sample surveys. We are therefore well-equipped to understand their basic trustworthiness as well as the occasional comment that last month's rise of 0.1% in the unemployment rate should not be taken seriously since such a change is within the margin of sampling error. In some cases, such as the measurement of unemployment, we have looked at these national economic statistics in more detail. But what is meant by the Consumer Price *Index*, the *index* of leading indicators, the Producer Price *Indexes*? These are *index numbers*, an important type of numerical description of data that we have not yet met. This chapter introduces index numbers and their uses (Section 1); looks in more detail at the most famous index, the Consumer Price Index (Section 2); and then gives a brief overview of national economic indicators and the proposal that official social indicators also be compiled (Section 3). National economic and social statistics almost always take the form of *time series*, that is, data collected regularly over time. Finally the chapter presents some tips on how to interpret (and how to avoid misinterpreting) time series (Section 4).

1. Index numbers

An index number measures the value of a variable relative to its value during a base period. Thus the essential idea of an index number is to give a picture of changes in a variable much like that drawn by saying "The average cost of a hospital stay rose 108% between 1980 and 1987." Such a statement measures change from a base period without giving the actual numerical values of the variable. The recipe for an index number is

$$\text{index number} = \frac{\text{value}}{\text{base value}} \times 100$$

"Yes sir, I know that we have to know where the economy is going. But do we have to publish the statistics so that everyone else does too?"

Here is an example.

> **Example 1.** A pound of oranges that cost 35 cents in 1980 cost 65 cents in 1990. The orange price index in 1990 with 1980 as base is
>
> $$\frac{65}{35} \times 100 = 186$$
>
> The orange price index for the base period, 1980, is
>
> $$\frac{35}{35} \times 100 = 100$$

Example 1 illustrates some important points. The choice of base is a crucial part of any index number and should always be stated. The index number for the base period is always 100, so it is usual to identify the base period as 1980 by writing "1980 = 100." You will notice in press articles concerning the Consumer Price Index the notation "1982–84 = 100." This mysterious equation simply means that the years 1982 to 1984 are the base period for the index.

> **Example 2.** A Mercedes sedan that cost $35,000 in 1980 cost $65,000 in 1990. The corresponding index number (1980 = 100) for 1990 is
>
> $$\frac{65,000}{35,000} \times 100 = 186$$

Comparing Examples 1 and 2 makes it clear that an index number measures only change relative to the base value. The orange price index and the Mercedes price index for 1990 are both 186. That oranges rose from 35 cents to 65 cents and the Mercedes from $35,000 to $65,000 is irrelevant. Both rose to 186% of the base value, so both index numbers are 186.

Index numbers can be interpreted in plain terms as giving the value of a variable as a percent of the base value. An index of 140 means 140% of the base value, or a 40% increase from the base value. An index of 80 means that the current value is 80% of the base, a 20% decrease. Be sure to notice that index numbers can be read as percents only relative to the base value. Do not confuse an increase of so many *points* with an increase of so many *percent* in news accounts of an index.

Example 3. The Consumer Price Index (1982–84 = 100) for the New York metropolitan area increased from 129.6 in May 1989 to 136.6 in May 1990. This was an increase of 7 points, because

$$136.6 - 129.6 = 7$$

To find the *percent* increase in that one-year period, we express the 7-point increase as a percent of the value at the beginning of the period, which was 129.6. Since

$$\frac{7}{129.6} = 0.054$$

the New York area Consumer Price Index rose by 5.4% during that year.

Thus far it may seem that the fancy terminology of index numbers is little more than a plot to disguise simple statements in complicated language. Why say "The Consumer Price Index (1982–84 = 100) stood at 129.2 in May 1990" instead of "consumer prices rose 29.2% between the 1982–84 average and May 1990?" In fact, the term "index number" usually means more than a measure of change relative to a base; it also tells us the kind of quantity whose change we measure. That quantity is a weighted average of several variables, with fixed weights. This idea is best illustrated by an example of a simple price index.

Example 4. A homesteader striving for self-sufficiency purchases only salt, kerosene, and the services of a professional welder. In 1980 the quantities purchased and total costs were as follows (the cost of an item is the price per unit multiplied by the number of units purchased):

Good or service	1980 quantity	1980 price	1980 cost
Salt	100 pounds	$ 0.07/pound	$ 7.00
Kerosene	50 gallons	0.80/gallon	40.00
Welding	10 hours	14.00/hour	140.00
			$187.00

The total cost of this collection of goods and services in 1980 was $187. To find the homesteader price index for 1990, we compute the 1990 cost of this same collection of goods and services. Here is the calculation.

Good or service	1980 quantity	1990 price	1990 cost
Salt	100 pounds	$ 0.10/pound	$ 10.00
Kerosene	50 gallons	0.60/gallon	30.00
Welding	10 hours	25.00/hour	250.00
			$290.00

The same goods and services that cost $187 in 1980 cost $290 in 1990. So the homesteader price index (1980 = 100) for 1990 is

$$\frac{290}{187} \times 100 = 155$$

When prices of a collection of goods and services are being measured, the kind of index number illustrated in Example 4 is called a *fixed market basket price index*. The Consumer Price Index and the government's Producer Price Indexes for wholesale prices are fixed market basket price indexes. The variable whose changes such an index number describes is the total cost of a "market basket" of goods and services. The makeup of the market basket is determined in the base year. In Example 4, the basket contained 100 pounds of salt, 50 gallons of kerosene, and 10 hours of a welder's services. By 1990 the homesteader may have changed his purchases, but no matter. The price index measures the relative change in the cost of the *same* market basket from 1980 to 1990.

The basic idea of a fixed market basket price index is that the weight given to each component (salt, kerosene, and welding) remains fixed over time. The index numbers published by the federal government and featured in our economic news are not all price indexes, but they nonetheless refer to weighted averages of a number of components with weights that remain fixed over time. Thus the term *index number* carries two ideas.

- It is a measure of the change of a variable relative to a base value.

- The variable is an average of many quantities, with the weight given to each quantity remaining fixed over time.

Section 1 exercises

6.1. The average price of unleaded regular gasoline has fluctuated as follows:

> 1979   90¢ per gallon
>
> 1983   124¢ per gallon
>
> 1988   95¢ per gallon

Give the unleaded gasoline price index numbers (1979 = 100) for 1979, 1983, and 1988.

6.2. Sulfur dioxide emitted into the air is a pollutant that contributes to acid rain. The Environmental Protection Agency reports the amounts of several air pollutants emitted each year. In 1970, 28.3 million metric tons of sulfur dioxide were emitted. The figures for 1980 and 1987 were 23.4 and 20.4 million metric tons. Calculate an index number for sulfur dioxide emissions in each of these years, using 1980 as the base year.

6.3. The Department of Agriculture reports that the average retail price of a 12-ounce can of frozen orange juice has changed as follows:

> 1980 $0.88
>
> 1985 $1.32
>
> 1987 $1.11
>
> 1988 $1.33

Give the orange juice price index number (1987 = 100) for all four years.

6.4. In Exercise 6.3,

(a) How many points did the orange juice price index rise from 1980 to 1988? What percent increase was this?

(b) How many points did the orange juice price index rise from 1987 to 1988? What percent increase was this?

6.5. The Consumer Price Index (1982–84 = 100) stood at 123.7 in May 1989 and at 129.2 in May 1990.

(a) How many points did the index gain between the 1982-to-1984 base period and May 1990? What percent increase was this?

(b) How many points did the index gain between May 1989 and May 1990? What percent increase was this?

6.6. The Bureau of Labor Statistics publishes separate price indexes for major metropolitan areas in addition to the national index. The values of two of these indexes (1982–84 = 100) in May 1990 were 118.3 for Houston and 136.3 for Boston.

Can you conclude that prices were higher in Boston than in Houston? Why or why not?

6.7. A bicycle racer must purchase bicycles, helmets, riding shorts, and riding shoes to equip himself. His 1985 purchases and their prices in both 1985 and 1990 are given below.

Commodity	1985 quantity	1985 price	1990 price
Bicycle	2	$350 each	$400 each
Helmet	1	$35 each	$40 each
Shorts	3	$12 each	$15 each
Shoes (pair)	2	$55/pair	$80/pair

Use these data to compute a fixed market basket price index (1985 = 100) for 1990.

6.8. A guru purchases for his sustenance only olive oil, loincloths, and copies of the *Atharva Veda* from which to select mantras for his disciples. Here are the quantities and prices of his purchases in 1980 and 1990.

Commodity	1980 quantity	1980 price	1990 quantity	1990 price
Olive oil	20 pints	$2.50/pint	18 pints	$3.80/pint
Loincloth	2	$2.75 each	3	$2.80 each
Atharva Veda	1	$10.95	1	$12.95

From these data, find the Guru Price Index (1980 = 100) for 1990.

2. The Consumer Price Index

The Consumer Price Index (forever after abbreviated CPI), published monthly by the Bureau of Labor Statistics (BLS), is a fixed market basket index measuring changes in the prices of consumer goods and services. The CPI is the most important of all index numbers. Not only does the CPI make headlines as the most popular measure of inflation (the declining buying power of the dollar), but it directly affects the income of over 60 million Americans. This is so because many sources of income are "indexed" that is, they are tied to the CPI and automatically increase when the CPI increases. (The last year in which the CPI decreased was 1955, so we can ignore that pleasant but unlikely possibility.) Since the United Auto Workers first won a cost-of-living escalator clause in 1948, indexing has become a common feature in union contracts. Social Security payments are indexed, as are federal civil service and military pension payments. Even food stamp allowances are indexed. When dependents are included, the CPI influences the income of half the American population. This state of affairs may well be economically foolish, for every rise in the CPI generates more inflation by driving up government payments and labor costs. But it certainly focuses attention on the Consumer Price Index.[1]

Understanding the CPI

What does the CPI measure? It is a fixed market basket index covering the cost of a collection of 364 goods and services. The CPI is computed just as in Example 4, though the arithmetic is a bit longer with 364 items. There are actually two CPIs. The older index is officially called the Consumer Price Index for Urban Wage Earners and Clerical Workers, abbreviated as CPI-W. The items in the CPI-W market basket, and their weights in the index, are chosen to represent the spending habits of families of city wage earners and clerical workers. If you are a farmer, teacher, medical doctor, or unemployed, the CPI-W does not describe the change in prices of the goods and services typically consumed by people like you. In fact, only about 32% of the population is covered by the CPI-W. Beginning in 1978, the BLS added a broader index, the Consumer Price Index for All Urban Consumers, or CPI-U. The CPI-U uses items and weights that represent the average purchases of all non-farm households living in metropolitan areas. About 80% of the population live in such households. (Metropolitan areas as officially defined include most suburban regions and a good many rural regions in addition to cities.) Teachers, medical doctors, and the unemployed are now included, but not farmers. We will discuss only the CPI-U, which is the index usually mentioned in news reports.

How is the market basket arrived at? By an exercise in probability sampling. The Bureau of the Census, under a contract with the BLS, conducts a Consumer Expenditure Survey (CES) in which a sample of households keeps diaries of all expenses and another sample is interviewed about major purchases. The Consumer Expenditure Survey is a multistage probability sample like those you met in Section 6 of Chapter 1. It provides detailed information on how Americans spend their money. Because the CPI is supposed to be a fixed market basket index, the market basket changes only infrequently. Since 1987, the market basket has been based on the results of the CES for the years 1982 to 1984.

Household expenditures are classified into 69 classes, such as "women's apparel" and "television and sound equipment." These classes are further divided into 207 strata; then a probability sample is taken to choose one or more individual items to represent each stratum. The "food and beverage" class, for example, is represented by 73 individual items. The 364 items in the market basket are therefore a multistage stratified sample of all goods and services available. The weight given to each item in computing the index depends on the share of spending its stratum received in the Consumer Expenditure Survey.

How are the data for the monthly CPI collected? By another exercise in probability sampling. The CES also asks *where* the sample households bought each item, so a sample of retail outlets can be chosen that gives proper weight to drug stores, department stores, discount stores, and so forth. Prices are collected each month in 91 urban areas. These 91 areas are a probability sample of all the nation's metropolitan areas — a stratified sample that contains all of the largest metropolitan areas and a sample of the others. (Recall from Chapter 1 that stratified samples automatically including all the largest units are common in economic sampling.) Prices are obtained from a probability sample of about 25,000 retail and service outlets, distributed over the 91 metropolitan areas. In all, the BLS records the price of about 95,000 items each month. In addition, information about shelter costs is obtained from 40,000 renters and 20,000 owner-occupied housing units. It's a big job.

The result of all this sampling is the monthly CPI, both the two national CPIs and separate CPIs for large metropolitan areas. Probability sampling produces results with little bias and quite high precision for the national CPI. Because the published CPI is rounded off to one decimal point (such as 137.5), a change of 0.1 may be due to rounding off. But a change of 0.2 in the CPI almost certainly reflects a real change in the price of the market basket.

Does the CPI have shortcomings? Indeed it does.[2] The CPI is computed with great statistical expertise, but we have seen again and again that statistics is not a cure-all. The CPI is a fixed market basket price index, and all such indices have some problems with validity of measurement (recall Chapter 3). It is not really possible to keep the market basket fixed year after year. After all, a 1992 car is not the same as a 1982 car, so the "new car" item in the market basket changes each year. The BLS then must decide the issue of *changes in quality*. How much of the yearly rise in passenger car price is due to the better quality of the product, and how much is a genuine price hike? In 1990, for example, manufacturers' list prices for passenger cars rose an average of $804.91. The Bureau decided that $216.40 of this increase paid for higher quality, primarily new passive restraint systems. So only the remaining $588.51 counted as a price increase in computing the CPI.

A related difficulty with a fixed market basket is *changing buying habits*. Men's 100% cotton shirts almost disappeared in the 1970s, replaced by blends of fabrics that required no ironing, and then became popular once more in the 1980s. Spending in the "television and sound equipment" category has shifted rapidly as video recorders and cameras, compact disc players, and digital audio tape players have become popular. Major revisions in the market basket are infrequent, but the BLS must replace outdated items on a regular basis.

Adjustment for quality and replacement of outdated items are carefully done and generally can be ignored in interpreting the CPI. More important is the question of the *costs of owning a home*. The CPI long treated housing as just another consumer expense. Mortgage costs, property taxes, and other expenses of purchasing and financing a home made up about 26% of the entire CPI. But people buy homes as an investment as well as to live in. During inflationary times the prospect of a fat profit on resale makes buyers willing to pay higher prices and interest rates in order to own a home. A better measure of the cost of shelter stripped of its investment aspects is the cost of renting houses or condominiums that are similar to those that are owned. Beginning in 1983, the BLS switched to this "rental equivalence" method. Homeowners' costs dropped from 26% to 15% of the CPI. Until this was done, the CPI consistently overstated the rise in prices during the inflation that began in the late 1960s. (When the market-basket weights were revised in 1987, homeowner's costs rose to almost 20% of the index because the CES showed that more consumers owned homes and were spending a higher fraction of their income on them.)

Finally, a fixed market basket index *does not measure changes in the cost of living*. Our buying habits are directly affected by changing prices. We can keep our cost of living down by switching to substitutes when one item rises in price. If beef prices skyrocket, we eat less beef and more chicken. Or even more beans. The CPI measures price changes, not changes in the cost of living. It probably slightly overstates rises in the cost of living by ignoring our ability to change our buying habits.

Neither does the CPI directly measure the income needed to maintain our standard of living—for one thing, it does not take taxes into account.

Using the CPI

We can use the CPI to adjust dollar amounts for the effects of inflation over time. A dollar in 1990 is not the same as a dollar in 1970, for the decrepit 1990 dollar could buy much less than the more robust 1970 dollar. So an income earned in 1970 will look misleadingly small now unless we restate it in present-day dollars. To compare dollar amounts from different years in "real" terms, we must compare how much the money will buy. We do this by giving all amounts in dollars of the same year. This is called restating the amounts in "constant dollars" because the dollars all have the same buying power when they are dollars of the same year. Here is the recipe for converting dollars of one year into dollars of another year.

"Now this here's a genuine 1960 dollar. They don't make 'em like that anymore."

To convert an amount in current dollars at time *A* to the amount with the same buying power at time *B*,

$$\textbf{dollars at time } B = \textbf{dollars at time } A \times \frac{\textbf{CPI at time } B}{\textbf{CPI at time } A}$$

Example 5. The average weekly earnings of workers in private industry in the United States (excluding agricultural workers) were $120 in 1970 and $322 in 1988. Although the number of dollars a typical worker earned was almost 3 times as high in 1988 as in 1970, these dollars would buy much less. To compare workers' earnings in real terms, let's convert the 1970 earnings into 1988 dollars. Table 6-1 shows that the average CPI (1982–84 = 100) was 38.8 in 1970 and 118.3 in 1988. So the 1988 value of the 1970 average earnings is

$$\$120 \times \frac{118.3}{38.8} = \$366$$

This is higher than the actual 1988 earnings, $322. Workers earned more (in terms of what their money would buy) in 1970 than in 1988.

An advantage of our recipe is that it is not affected by changes in the base year of the CPI. As long as the same base is used for time *A* and time

Table 6-1. Annual average CPI-U, 1982–84 = 100

Year	CPI	Year	CPI	Year	CPI
1915	10.1	1970	38.8	1981	90.9
1920	20.0	1971	40.5	1982	96.5
1925	17.5	1972	41.8	1983	99.6
1930	16.7	1973	44.4	1984	103.9
1935	13.7	1974	49.3	1985	107.6
1940	14.0	1975	53.8	1986	109.6
1945	18.0	1976	56.9	1987	113.6
1950	24.1	1977	60.6	1988	118.3
1955	26.8	1978	65.2	1989	124.0
1960	29.6	1979	72.6		
1965	31.5	1980	82.4		

B, the recipe always gives the same answer. That's handy, because the base for the CPI was changed in 1935, 1953, 1961, 1972, and most recently in 1988. A large CPI reminds us that our money is losing its value, so politicians prefer to have the base changed now and then to keep the index below the clouds. From 1972 to 1987, for example, the base year for the CPI was 1967. By 1987, the CPI (1967 = 100) had passed 340. When the present 1982–84 base was adopted in January 1988, the CPI dropped to about 115. These changes don't affect comparisons using our recipe.

Don't go away unhappy at the steep climb of the CPI in recent years. The German wholesale price index (1914 = 100) reached 234 when World War I ended in November 1918, hit 3490 by December 1921, and finally 126,160,000,000,000 in December 1923. Now *that's* inflation.

Section 2 exercises

6.9. An article in the *New York Times* of January 17, 1989, reported that the earnings of workers without a college education were falling ever farther behind those of college graduates. The article stated that the real earnings of male college graduates between 25 and 34 years of age increased by 9% between 1979 and 1987. The real earnings of male high school graduates in the same age group fell 9% during this period, while the real earnings of high school dropouts fell 15%.

Explain carefully what is meant by "real earnings" in this news report.

6.10. Example 5 shows that the buying power of individual workers has stagnated since 1970. What about family income? Family income could increase even in the face of falling individual income if more family members went to work. Let's look only at families headed by married couples, to avoid the effects of the growing number of single-parent families. From the *Statistical Abstract* we learn that the 1970 median income for these families was $10,516. In 1988, their median income was $36,389. Restate the 1970 family income in 1988 dollars. How did real family income change between 1970 and 1988?

6.11. The "muscle cars" of the 1960s became popular with collectors in the late 1980s, perhaps because the collectors were teens in the 1960s. A 1965 Pontiac GTO hardtop sold new for $3100. The car was worth $17,000 in 1989. Restate the 1965 price in 1989 dollars to see how much value the GTO gained in real terms.

6.12. The median income of 30-year-old men who graduated from high school but not college was $9512 in 1973 and $17,614 in 1986. For

30-year-old male college graduates, the corresponding median incomes were $11,005 in 1973 and $26,347 in 1986.

(a) Restate the 1973 incomes in 1986 dollars to allow comparisons in dollars of the same purchasing power.

(b) Did the median real income of 30-year-old male college graduates increase sharply, decrease sharply, or remain about the same in the years between 1973 and 1986? What about the median income of high school graduates?

(c) Did the gap in real income between high school and college graduates increase sharply, decrease sharply, or remain about the same between 1973 and 1986?

6.13. Much has been made of the inability of middle-income families to buy houses at the high prices of recent years. Here are the median prices of new single-family homes sold in each of three years.

1980	$64,600
1985	$84,300
1989	$119,400

Express the 1980 and 1985 prices in 1989 dollars. What happened to real housing costs in the decade of the 1980s? (But do notice that there is no adjustment for changes in quality in these prices. The typical 1989 house was larger and more luxurious than the typical 1980 house.)

6.14. Tuition for Indiana residents at Purdue University has increased as follows:

Year	1973	1975	1977	1979	1981	1983	1985	1987	1989
Tuition	$700	$750	$820	$933	$1158	$1432	$1629	$1816	$2032

Use the annual average CPIs given in Table 6.1 to restate the tuition in constant 1973 dollars. Make two line graphs on the same axes, one showing current dollar tuition for these years and the other showing constant dollar tuition.

6.15. Few prices have had more influence on the world's economic condition than the benchmark price of crude oil set by the Organization of Petroleum Exporting Countries (OPEC). Here are OPEC benchmark prices (dollars per barrel) at the year end.

Year	1971	1975	1979	1983	1987
Oil price	$2.18	$10.46	$14.54	$29.00	18.00

Use the annual average CPIs given in Table 6.1 to restate OPEC's price in constant 1971 dollars. Make two line graphs on the same axes, one showing current dollar oil prices and the other showing constant dollar prices.

6.16. The Ford Model T sold for $950 in 1910 and for $290 in 1925. Obtain the current value of the CPI, and use it to restate these costs in current dollars.

6.17. What would the CPI be in 1985 if the base period were 1950? [You can find this by changing $100 in 1950 dollars into 1985 dollars. The resulting number of 1985 dollars is the 1985 price index (1950 = 100).] What would the current CPI be if the base period were 1950?

6.18. Here are some examples of how the CPI might overestimate price increases. Comment on each example. In particular, which can be corrected by the ongoing Consumer Expenditure Survey because the overestimates are due to out-of-date information? Which require the BLS to adjust prices to take improved quality into account?

   (a) Exterior house-paint prices in the CPI are up, but many buyers have switched from oil paints to latex water-base paints, a different product.

   (b) Exterior house-paint prices are up, but new paints cover better so that less paint is needed to paint a house once. New paints also last longer, so we need to buy them less often.

   (c) Exterior house-paint prices are up, but buyers have switched from small hardware stores (higher prices) to discount stores (lower prices). So the price actually paid is not up as much as the hardware store price.

   (d) Exterior house-paint prices are up, but new paints are much more convenient; they are easier to use and to clean up. This convenience has no direct money value, but it means we get greater satisfaction and would be willing to pay more than for less-convenient paints.

6.19. In addition to the two national CPIs, the BLS publishes separate CPIs for 27 large metropolitan areas. These local CPIs are considerably less precise (that is, they have considerably more sampling variation) than the national CPIs. Explain why this is so.

6.20. Now that there is more than one CPI, political pressure may lead to the creation of a whole family of price indexes, each with a market basket tailored to a specific group. In particular, some members of Congress want to set up an index for the aged, to be used in place of the general CPI to adjust Social Security payments. Briefly discuss the pros and cons of this proposal, and express your opinion.

3.   National economic and social statistics

Among the more important data in most societies are *national economic indicators*, measures of the state of the economy made at regular intervals over time. We have met two of these economic indicators, the CPI and the unemployment rate, in some detail. Some others, such as common stock price indexes, are regularly in the news. Many others — measures of production, incomes, inventories, construction, farm crops, and much else — escape the public eye but are watched closely by government policy makers and private forecasters.

The rise of national economic indicators to a major role in the statistical enterprise can be traced to the increasing intervention of governments in the economy following the great depression of the 1930s and to the advance of economic theory, which suggests ever more variables that can be usefully measured. While economic data are primarily produced by governments, they are widely used for business planning as well. Practices such as the indexing of government and private payments to the CPI give economic statistics a direct role in the lives of many citizens. And the financial markets, ever more sensitive to hints of changing conditions, lurch up or down as new economic data are released. Economic indicators have gradually moved from the business pages to the front page.*

In most nations a central statistical agency, like the aptly named Statistics Canada, is responsible for producing and publishing government economic and social data. The United States government lacks a central statistics office. There are instead eight major statistical agencies, of which the Bureau of the Census in the Department of Commerce and the Bureau of Labor Statistics in the Department of Labor are the most important. These agencies have a good record for statistical professionalism and (equally important) for the independence that allows them to report facts that politicians may find unpleasant.

Yet there is agreement among users of government economic and social data that the quality of the U.S. national statistical system has slipped. Budget problems and a feeling that the government should be

---

*More detail on economic indicators can be found in Geoffrey H. Moore, "The Development and Analysis of Economic Indicators," in *Statistics: A Guide to the Unknown.*

less involved in the economy damaged statistical agencies during the 1980s, when their budgets and staffs were reduced. Lack of coordination due to the absence of a single statistical office compounded the problem. Some of the resulting weaknesses were clearly visible by the end of the decade. The sample size of the Current Population Survey was reduced; the CPI did not have a good measure of soaring medical costs; the 1990 Census attracted considerable criticism, including some of the first allegations of political interference lodged against a federal statistical agency. Lack of resources meant that the statistical agencies could not keep up the research needed better to measure rapidly changing economic and social conditions. Given the importance of reliable data in an "information economy," squeezing the sources of these data seems shortsighted.

Leading economic indicators

The best-publicized economic indicators are those which, like price indexes and the unemployment rate, measure the present state of the economy. These are called *coincident indicators*, because their movements coincide with those of the economy as a whole. Less known to the public, but critical to business and government policy makers, are the *leading indicators*. These are economic statistics whose movements tend to lead (occur before) movements of the overall economy; thus they can help to forecast future economic conditions.

There are 11 official leading economic indicators, chosen for their past success in changing direction ahead of turning points in overall economic activity. Most of the leading indicators are measures of *demand* for various kinds of economic output. Some are direct measures of demand, such as the number of new building permits issued for residential houses or the dollar value of new orders for industrial plants and equipment (in constant dollars, of course). Others are indirect measures of demand, such as the average work week of manufacturing workers. This is a leading indicator because manufacturers increase or reduce overtime quickly when demand for their products changes. They hire or lay off workers more slowly; measures of employment and unemployment are coincident indicators.

The Department of Commerce has spared us from trying to make sense of 11 different variables at once by publishing the *Composite Index of Leading Indicators* (CLI) each month. This is a fixed-weight index number of the kind we have met already. It is a weighted average of the 11 leading indicators. Since combining several leading indicators usu-

"Have you ever thought of adding an indicator of how people feel about having their opinions asked every other day?"

ally results in better forecasts of the future, the CLI is not far behind the CPI in the attention it gets from the news media.

How well does the CLI predict future economic activity? In the past, not too badly. Take the change in the CLI over a quarter (three-month period) as the explanatory variable. Let the percent change in the Gross National Product for the following quarter be the response variable. (The Gross National Product is the total value of all goods and services produced in the economy. It is the usual measure of overall economic activity.) For the period 1953 to 1970 these variables had $r^2 = 0.37$. That's pretty good by social science standards and suggests that the CLI has some ability to predict the future course of the GNP.[3] Unfortunately, the recent record is less impressive. In both 1984 and 1987 the CLI decreased for three or more months in a row, the traditional sign that a recession lies ahead. Yet the real GNP continued to rise. We can expect some changes in the CLI as economists try to restore its ability to predict. Such attempts may be futile; a more complex economy in which much information moves instantly may move too quickly for the indicators to catch.

Despite the recent lag in the performance of the CLI, a question should occur to any properly greedy reader: Can I use the CLI to forecast what will happen to the stock market and make a killing? Sorry. Stock prices are one of the 11 leading indicators in the CLI; they tend to move before the economy as a whole.

Social indicators

The national economic indicators are well-established with the government, the media, and the public. Social scientists have urged the adoption of a similarly authoritative series of *social indicators*. Social indicators are series of data collected regularly over time and intended to provide statistical measures of social values and well-being. Such indicators might provide information for policy makers and for the public similar to the official economic information already available. The government showed some interest in this idea prior to the 1980s. The Census Bureau published volumes entitled *Social Indicators* in 1973, 1976, and 1980.

Many statistical measures of health, housing, crime, and other social concerns already are published by the government. But these are much less complete than economic data. We have good data about how much money is spent on food, but little information about how many people are poorly nourished. Social data are also far less carefully collected than are our economic data. Economic statistics are more precise, are compiled more regularly, and are published with a shorter time lag. It is noteworthy that economic indicators such as the unemployment rate are based on regular large sample surveys of citizens. The government does not expend a comparable effort on social statistics. Official social indicators would treat social variables with the level of statistical care now reserved for economic variables.

The suggestion that social indicators be added to economic indicators brings up one new idea and several difficulties. The new idea is to add "subjective" or opinion information to improved versions of the "objective" social statistics now available. This is an attractive suggestion. It would be interesting to follow changes in the values Americans hold, their degree of satisfaction with their jobs, how much they are afraid of crime, and so on. Politicians and others now operate on anecdotal evidence in these areas. A well-designed sample survey, regularly repeated, would provide fascinating data.

Some of the difficulties in the proposal for official social indicators are directly related to the inclusion of subjective responses. Measuring even so objective a variable as unemployment is not entirely easy. How then shall we measure satisfaction with the quality of life? Any measure placed on an official list of social indicators will receive much attention and so should be well-chosen. To measure a property, you recall, we must have a clear concept of what the property is. For example, economists have a clear concept and definition of money supply. They agree that money supply has an important influence on the economy. Several versions of the money supply are regularly measured, and one of them is among the leading economic indicators. Social scientists have no such agreement over the concept of quality of life. Many different measures of the quality of life, both objective and subjective, have been proposed, none of which commands the kind of respect that money supply has among economists. The measurement and conceptualization problems so common in the social sciences appear once again. It is probably necessary to begin with several measures of the quality of life, with the hope that regular data collection will clarify which are most useful. Certainly there is interesting information to be gained. One-time surveys suggest, for example, that satisfaction with the quality of life goes *down* as level of education increases. Has this dissatisfaction grown in recent years? Is there really a turning away from material things to things of the spirit as sources of satisfaction? I'd like to know.

Social indicators have problems other than conceptualization and measurement. The economic indicators serve as a basis for making short-term economic policy and for transferring large sums in government payments. That justifies the expense needed to compile the indicators at frequent intervals. Social indicators do not appear to have the same direct relation to policy making by either public or private bodies. I'd like to know if more Americans are looking for satisfaction in things of the spirit this year than last, but I'm not sure my curiosity justifies the kind of massive effort described for the CPI in Section 2. A considerable effort would be required to collect accurate social indicators. At least this would employ more statisticians. But perhaps by now you feel that there are statisticians enough already.

Section 3 exercises

6.21. Choose one of the following economic indicators. Write a short essay describing what the indicator measures, why it is economically important, and what statistical procedures are used to compile the indicator. To help you locate material in the library, the government agency responsible for each indicator is listed. Each agency publishes material describing the indicators for which it is responsible.

    (a) Producer Price Indexes (Bureau of Labor Statistics, U.S. Department of Labor).

    (b) Gross National Product (Office of Business Economics, U.S. Department of Commerce).

    (c) Index of Industrial Production (Federal Reserve Board).

6.22. Changing economic conditions afflict economic indicators with some of the same conceptualization and measurement problems that trouble social indicators. For example, the quantity of a manufacturing firm's output may now be less important than the quality of that output. Quantity is easy to measure; quality is much tougher. Consider automobiles: Suggest how you might measure the quality of cars in order to compare different makers and to track changes in quality over time.

6.23. We wish to include as part of a set of social indicators measures of the amount of crime and of the impact of crime on people's attitudes and activities. Suggest some possible indicators in each of the following categories:

    (a) Statistics to be compiled from official sources.

    (b) "Objective" information to be collected by a sample survey of citizens.

    (c) "Subjective" information on opinions and attitudes to be collected by a sample survey.

6.24. The two primary collections of official statistics on crime in America are the FBI's annual *Crime in the United States* and the Bureau of Justice Statistics' annual *Criminal Victimization in the United States*. You can find selected data from both sources in the *Statistical Abstract*. Describe these sources in a short essay. What kinds of data do they contain? How are the data collected? How frequently are they collected? (Basic information is given in the appendix of the *Statistical Abstract*. If possible, look at the two publications themselves.)

6.25. The closest approach to social indicators in the United States is provided by the General Social Survey conducted by the National

Opinion Research Center at the University of Chicago. Write a brief essay describing the conduct of this survey and the types of information that it collects. [A detailed description of the findings of this survey appears in Charles H. Russell and Inger Megaard, *The General Social Survey, 1972–1986: The State of the American People*, (New York: Springer-Verlag, 1988).]

4.   Interpreting time series

The monthly Consumer Price Index, the monthly unemployment rate, and most other economic indicators are produced at regular time intervals. Indeed frequently the *change* in prices or unemployment from month to month is the center of attention rather than the actual value of the CPI or the unemployment rate.

> **A sequence of measurements of the same variable made at different times is called a *time series*. Usually (but not always) the variable is measured at regular intervals of time, such as monthly.**

The variable measured in a time series may be any of the many kinds we have studied—counts (such as the number of persons employed), rates (such as the unemployment rate), index numbers (such as the CPI and CLI), and other kinds of variables as well. The essential idea of a time series is that each observation records both the value of the variable and the time when the observation was made. Time series data are often used to predict future values of the variable in question. The goal of the machinery behind the time series used as economic indicators is in part to tell us where we are but also to suggest where are going. Predicting the future is always risky business, and statistical time series do not remove the risk. The interpretation of time series is complicated by the fact that several types of movement are going on together. Let us look at each type separately.

Seasonal variations are regular changes that recur in periods of less than a year. Consider Figure 6-1, which displays the quarterly count of civil disturbances in the troubled period of the late 1960s and early 1970s. The line graph shows a large seasonal variation, with disturbances regularly increasing in the summer and decreasing in winter. Riots are less frequent in the cold and snow. We must remove the effect of seasonal variation to see the underlying trend in the time series. In Figure 6-1 we can do this visually, and we find that underneath the seasonal variation the frequency of civil disturbances was decreasing during these years.

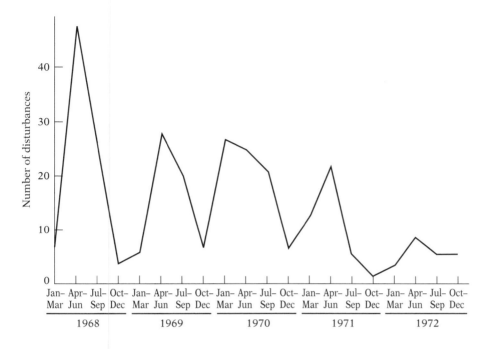

Figure 6-1. Count of civil disturbances by quarter, 1968–1972.

In the case of many official statistics, a calculation is made to adjust for the expected seasonal variation and so allow month-to-month comparisons. A rise in unemployment in January, for example, need not mean that the economy is in trouble. Rising unemployment is expected in January as Christmas employees are laid off and construction employment drops owing to bad weather. A *seasonally adjusted* unemployment rate for January takes this expected increase into account and can be compared directly with December's rate. If no seasonal adjustment is made, we must be aware of the seasonal variation and take it into account in interpreting the published data.

Example 6. President Reagan, eager to publicize an upturn in the economy, claimed in an April 1982 speech that unemployment had fallen by 88,000 in March. But the Bureau of Labor Statistics had just announced an *increase* of 98,000 in the number of persons out of work. The President used unadjusted figures, while the Bureau's unemployment data were seasonally adjusted to take account of the fact that employment usually increases from February to March. "The statisticians in Washington have funny ways of counting," said Mr. Reagan.

Once seasonal variation is removed, we can follow our usual strategy for examining data by looking for the overall pattern in a time series. *Trends and cycles* are long-term features of a time series. A trend is a persistent long-term rise or fall, while cycles are up-and-down movements of irregular strength and duration. The terms are not precise, since a trend may be part of a cycle when the time series is followed for a longer period. For example, the level of prices has been generally increasing since about 1940, a trend that has affected most people. But in the longer view, this inflationary period may be followed by a deflationary period and so take its place as one cycle (though the strongest to date) in the rise and fall of prices over several centuries past.

The goal of the analysis of a time series is usually to study trends or cycles or both. The Composite Index of Leading Indicators has shown a long-term increasing trend with strong up-and-down cycles. The trend is probably a result of inflation and is of no particular interest. The cycles, however, reflect the up-and-down business cycle of the economy as a whole (in advance, since the CLI is composed of leading indicators).

"Isn't it fascinating that you were ruined by a business cycle, while I was ruined by an erratic fluctuation?"

Analysis of the CLI therefore seeks to decide where we are in the current business cycle so that proper economic policies can be adopted. Even with seasonal variation removed, this is not easy to do. First, business cycles are irregular in length and strength, so they cannot be systematically predicted as seasonal variation can be. Even isolating past business cycles is quite subjective. Second, superimposed on trend, cycles, and seasonal variations in a time series are movements of a final type: *erratic fluctuations*. Storms, strikes, oil embargoes, and other chance occurrences large and small affect the economy and the CLI. It is extremely difficult to tell if a three-month flat spot in the CLI means that the business cycle has peaked and is about to turn down. Perhaps it is only a "pause" (to quote the president's chief economic adviser on one such occasion) caused by lack of rain in Nebraska, a strike at Ford Motor Company, and an upcoming presidential election. We cannot see the future clearly, even with the aid of a statistical record of the past.

A poor record of success does not prevent the foolhardy from continued attempts to predict the future. Economists often fail despite sophisticated statistics, foiled by the complexity of the economy and by erratic outside shocks. More simple-minded predictions overlook the prevalence of coincidence, which we noted in our discussion of causation in Chapter 5. Here is an example.

> **Example 7.** A writer in the early 1960s remarked on the great success of a simple method for predicting the outcome of presidential elections: just choose the candidate with the longer name.[4] In the 22 elections from 1876 to 1960, this method failed only once. (In two cases the longer name correctly predicted the winner in the popular vote, but the winner then lost in the Electoral College. In 1916, both names were the same length so the method did not apply. Nonetheless, the record is impressive.) Let us hope that the writer didn't bet the family silver in later elections after coming up with this clever idea. The seven elections from 1964 to 1988 presented six tests of the long-name-wins method (the 1980 candidates, Reagan and Carter, had names of the same length). The longer name lost five of the six.

Section 4 exercises

6.26. The price of fresh oranges is collected monthly as part of the data for the CPI. Figure 4-3 on page 182 displays this time series over a period of six years. What causes can you think of that might account for the erratic fluctuations, seasonal variation, and trend that the data display?

6.27. The sales at your new gift shop in December are double the November value. Should you conclude that your shop is growing more popular and will soon make you rich? Explain your answer.

6.28. The BLS publishes the CPI both unadjusted and seasonally adjusted. There is some seasonal variation owing to weather, holidays, and so forth, so these two versions of the CPI often differ slightly.

    (a) If you want to follow general price trends in the economy, would you use the seasonally adjusted or unadjusted version of the CPI? Why?

    (b) If you have a labor contract with an escalation clause tied to the CPI, you want your wages to keep pace with the actual prices you must pay. Which version of the CPI would you use for this purpose, and why?

6.29. Figure 6-2 is a plot of the annual average number of sunspots on the sun's visible face from 1610 to 1989. Since there are those who think that sunspots influence all manner of earthly happenings, study of this time series should be interesting.

    (a) The regular sunspot cycle is the most obvious feature of this time series. About how long is the sunspot cycle from maximum to maximum? Does the length of the cycle remain constant over time, or are there significant variations in the length?

    (b) By tracing the curve of the sunspot maxima over many cycles, I think I can see a longer cycle superimposed on the sunspot cycle. Comment on this suggestion.

    (c) Does this time series show any striking noncyclical phenomena? Describe any you notice.

6.30. A rich eccentric once founded an institute to seek out cycles in natural and human affairs and use the regularity of cycles for prediction. The institute noted after studying thousands of time series that there had been, for example, a quite regular 5.9-year cycle in both cotton prices and the abundance of grouse. Would you be willing to risk your money in speculation on cotton prices on the basis of this cycle? What do you think explains the similar behavior of cotton prices and abundance of grouse?

6.31. One of the best-known though least-serious methods of predicting the stock market is the Super Bowl indicator. If the winner of the Super Bowl, played in January each year, is a team from the old National Football League, stock prices (measured by the Dow Jones Industrial Average) will rise that year. If a team that came out of the old American Football League wins, stocks will drop. In the 23 Super Bowls played through 1989, the indicator was wrong

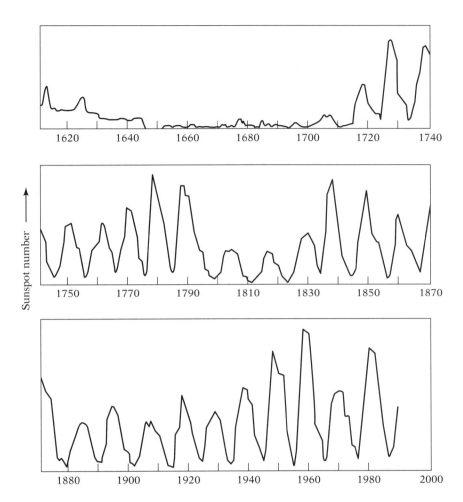

Figure 6-2. Average annual sunspot number. [Adapted from Peter V. Foukal, "The Variable Sun," *Scientific American*, February 1990, page 36.]

only twice. That's a better record than most. How has this indicator performed in 1990 and later years?

NOTES

1. Detailed information about the CPI and the surveys that sustain it, as well as about the Current Population Survey and other important data sources, can be found in the Bureau of Labor Statistics' *Handbook of Methods*, 1988.

2. More information on these topics can be found in P. Cagan and G. H. Moore, "Some Proposals to Improve the Consumer Price Index," *Monthly Labor Review*, September 1981, pp. 20–25. The *Monthly Labor Review*, published by the BLS, is a source of current information on many of the topics in this chapter.

3. The r^2 result is from Maury N. Harris and Deborah Jamroz, "Evaluating the Leading Indicators," *Federal Reserve Band of New York Monthly Review*, June 1976, p. 170.

4. The writer is Dmitri A. Borgman, *Language on Vacation* (New York: Scribner's, 1965).

Review exercises

6.32. As part of a report on trends in recreation, you decide to present index numbers to compare changes in different recreational activities. Here are data on participation in golf and bowling, from the *Statistical Abstract*. Give the index numbers (1975 = 100) for both sports for 1980, 1985, and 1988. Which sport has grown faster? Do the index numbers tell you which sport has more participants?

Sport	1975	1980	1985	1988
Bowlers (millions)	62.5	72.0	67.0	68.0
Golfers (thousands)	13,036	15,112	17,520	23,400

6.33. A food faddist eats only steak, pepper to flavor his steak, and swiss chard. In 1985, his purchases are as follows:

Food	Amount	Price
Steak	200 pounds	$4.49 per pound
Pepper	3 pounds	$1.50 per pound
Swiss chard	100 pounds	$0.40 per pound

By 1990, the food faddist likes a bit less pepper on his steak and has begun to take vitamin pills after a bout of malnutrition. Here are his 1990 purchases.

Food	Amount	Price
Steak	220 pounds	$4.99 per pound
Pepper	1 pound	$1.75 per pound
Swiss chard	120 pounds	$0.50 per pound
Vitamins	2 bottles	$8.50 per bottle

Find the Food Faddist Price Index (a fixed market basket price index) for 1990, with 1985 as the base period.

6.34. A three-minute telephone call to Japan cost $30 in 1934, when AT&T opened the first radio channel across the Pacific. The same call cost $6.34 in 1964 on the first sea-bottom copper cable. By 1989, you could talk to Japan for three minutes for $3.78 via communications satellite. Restate the 1934 and 1964 prices in 1989 dollars to see the price decrease in real terms. (Use the 1935 and 1965 entries in Table 6.1.)

6.35. Here are the mean and minimum salaries for major league baseball players over a number of past years. (The Major League Baseball Players Association compiles, and the news media publish, mean salaries. Because a few players are paid enormous amounts, the mean salary is about double the median salary. You can find the full distribution of 1989 salaries for one team in Exercise 4.80.)

Year	1970	1975	1980	1985	1989
Mean	$29,303	$44,676	$143,756	$371,571	$497,254
Minimum	$12,000	$16,000	$ 30,000	$ 60,000	$ 68,000

Write a brief discussion of the increase in major league salaries, taking into account the changing buying power of the dollar. In 1930, Babe Ruth earned $80,000, an astounding amount in those days. How does the Babe's salary compare in earning power with the 1989 average for major leaguers?

6.36. The federal government sets the minimum hourly wage that employers can pay a worker. Labor wants a high minimum wage, but some economists argue that too high a minimum makes employers reluctant to hire workers with low skills and so increases unemployment. The minimum wage was $1.60 an hour in 1970, $2.10 an hour in 1975, $3.10 an hour in 1980, and remained fixed at $3.35 an hour from 1981 to 1989. In 1989, Congress agreed to raise the minimum wage to $3.80 in 1990 and $4.25 in 1991.
  (a) Restate the 1970, 1975, 1980, 1985, and 1989 amounts in 1970 dollars to show the real change in the minimum wage. If you have the 1990 and 1991 CPI, convert the minimum wages for those years into 1970 dollars as well.
  (b) Draw line graphs of the actual minimum wage and the minimum wage in constant 1970 dollars for these years on the same graph. Briefly describe the real change in the minimum wage over these years.

Part III

Drawing conclusions from data

The experiment was designed, the protocol tested, the subjects assembled, the myriad details dealt with, double-blind enforced, the diagnoses completed. The raw data have been graphed and tabled, averaged and correlated. They seem to point to a clear conclusion. But how strong is the evidence that the data offer for that conclusion? Will your evidence stand the scrutiny of nit-picking journal editors or unhappy special interests?

> *It is a truism to say that a "good" experiment is precisely that which spares us the exertion of thinking: the better it is, the less we have to worry about its interpretation, about what it "really" means.**

In recognition of this wisdom, your experiment was carefully controlled and randomized, so that any effect that appears must be due to your treatment. But wait. You assigned subjects to treatments at random, and although this eliminates systematic bias, there will still be differences between the groups due to the luck of the draw. Will someone charge that the good performance of your favored treatment is just an accidental result of the randomization? Alas, it isn't true that any effect that appears must be due to the treatment. Unless the effect you observed is

---

*Sir Peter Medawar, *Induction and Intuition in Scientific Thought* (Philadelphia: American Philosophical Society, 1969), p. 15.

very large, it might be due to an unlucky randomization that happened to produce very unlike groups of subjects.

The best experiments, as Medawar observes, hardly need interpretation; their results are immediately convincing. The first studies of penicillin needed no elaborate statistics to see that lives were being saved. Even good experiments, however, are rarely so obviously convincing. Patients receiving the new treatment live a little longer; students taught by the new method learn a little more; the new variety of corn gives a little higher yield. Could that "little" just be chance variation? You need more statistics to argue convincingly that your effects really are due to the treatment. What you need is *statistical inference:* formal methods for drawing conclusions from data taking into account the effects of randomization and other chance variation.

Formal statistical inference is based on *probability theory,* the mathematics of randomness. Probability is a subject of interest to those who wish to understand such worldly pursuits as roulette and state lotteries as well as to students of the lofty subject of statistics. It is the subject of Chapter 7. Chapter 8 presents some of the concepts behind statistical inference. Our emphasis is on the reasoning of inference. We will leave details of the methods of inference to more traditional introductions to statistics.

Probability: The study of randomness

Even the rules of football agree that tossing a coin avoids favoritism. Favoritism in choosing subjects for a sample survey or allotting patients to treatment and placebo groups is as undesirable as it is in awarding first possession of the ball in football. Statisticians therefore recommend probability samples and randomized experiments, which are fancy-dress versions of tossing a coin. The central idea of statistical data collection is the deliberate introduction of randomness into the choice or assignment of units. Both tossing a coin and choosing an SRS are *random* in the sense that:

- The exact outcome is not predictable in advance.
- Nonetheless, a predictable long-term pattern exists and can be expressed by a relative frequency distribution of the outcomes after many trials.

The inventors of probability samples and randomized experiments in this century were not the first to notice that some phenomena are random in this sense. They were drawing upon a long history of the study of randomness and applying the results of that study to statistics.

Randomness is most easily noticed in many repetitions of games of chance — rolling dice, dealing shuffled cards, spinning a roulette wheel. Chance devices similar to these have been used from remote antiquity to discover the will of the gods. The most common method of randomization in ancient times was "rolling the bones," tossing several *astragali*.

The astragalus is a solid, quite regular bone from the heel of animals that, when thrown, will come to rest on any of four sides. (The other two sides are rounded.) Cubical dice, made of pottery or bone, came later, but even dice existed before 2000 B.C. Gambling on the throw of astragali or dice is, in contrast to divination, almost a modern development; there is no clear record of this vice before about 300 B.C. Gambling reached flood tide in Roman times, then temporarily receded (along with divination) in the face of Christian displeasure.[1]

Unpredictable outcomes have been noticed and used from the beginning of recorded history. But none of the great mathematicians of antiquity considered that the outcomes of throwing bones or dice have a clear pattern in many trials. Perhaps this is because astragali and most ancient dice were so irregular that each had a different pattern of outcomes. Or perhaps the reasons lie deeper, in the classical reluctance to engage in systematic experimentation. Professional gamblers, not so inhibited as philosophers and mathematicians, must have long known something of the regular pattern of outcomes of dice or cards and adjusted their bets to the "odds" of success. These odds the gamblers often could not guess correctly from experience alone. The systematic study of randomness began (I oversimplify, but not too much) when seventeenth-century French gamblers asked French mathematicians for help in figuring out the "fair value" of bets on games of chance. The mathematical study of randomness, *probability theory*, originated with Pierre de Fermat and Blaise Pascal in the seventeenth century and was well-developed by the

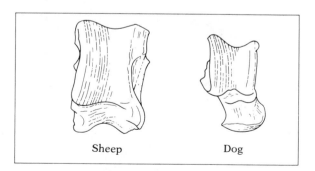

Sheep   Dog

Animal astragali, actual size. [Reproduced by permission of the publishers, Charles Griffin & Company, Ltd., of London and High Wycombe. From F. N. David, *Games, Gods, and Gambling*, 1962.]

time statisticians took it over in the twentieth century. In this chapter we examine probability, but without mention of the actual mathematics that has grown from the first attempt of the French mathematicians to aid their gambler friends.

Probability is now important for many reasons having little to do with either gambling or collecting data. Many natural and artificial phenomena are random in the sense that they are not predictable in advance but have long-term patterns. For example, the science of genetics is based on Gregor Mendel's observation that for given parents, the characteristics of offspring are random with long-run patterns that he began to uncover. And the emission of particles from a radioactive source occurs randomly over time, with a pattern that helped to suggest the cause of radioactivity. Probability theory is used to describe phenomena in genetics, physics, and many other fields of study. Although we will not meet such applications here, the ideas of this chapter shed light on these fields as well.

1.   What is probability?

Probability begins with the observed fact that some phenomena are random. "Random" is not a synonym for "haphazard," but a description of a kind of order that emerges only in the long run. We often encounter the unpredictable side of randomness in our everyday experience, but we rarely see enough repetitions of the same random phenomenon to

"What kind of childish nonsense are you working on now?"

observe the long-term regularity that probability describes. The first five exercises for this section are intended to give you some experience with the somewhat mysterious phenomenon of randomness. Careful study of games of chance will also do the job, but at some risk to your financial health. It's safer to study random sampling, as we did in Chapter 1. Let's look first at a very simple random phenomenon.

> **Example 1.** Toss a coin in the air. Will it land heads or tails? Sometimes it lands heads and sometimes tails. We cannot say what the next outcome will be. But toss a coin many times and a pattern emerges.
>
> - The eighteenth-century French naturalist Buffon tossed a coin 4040 times. Result: 2048 heads, or relative frequency for heads of 2048/4040 = 0.5069 for heads.
> - Around 1900, the English statistician Karl Pearson heroically tossed a coin 24,000 times. Result: 12,012 heads, a relative frequency of 0.5005.
> - The English mathematician John Kerrich, while imprisoned by the Germans during World War II, tossed a coin 10,000 times. Result: 5067 heads, a relative frequency of 0.5067.

The example of coin tossing suggests how to use numbers to describe the long-run regularity of random phenomena. We describe regularity by the proportion of many trials on which an outcome occurs, that is, by the relative frequency of the outcome. *Probability* is succinctly defined as long-term relative frequency.

> **If in a long sequence of repetitions of a random phenomenon the relative frequency of an outcome approaches a fixed number, that number is the probability of the outcome. A probability is always a number between 0 (the outcome never occurs) and 1 (the outcome always occurs).**

You may feel that it is obvious from the balance of a coin that the probability of a head is about 1/2. Such opinions are not always correct. Take a penny and, instead of tossing it, hold it on end on a hard surface with the index finger of one hand and snap it with the other index finger. The coin will spin for several seconds and then fall with either heads or tails showing. A long series of trials reveals that the probability of a head in this random experiment is clearly different from 1/2. Moral: We defined probability *empirically*, that is, in terms of observations. Only by observation can we be reasonably sure of the approximate value of the probability of an outcome.

When we toss a coin there are only two possible outcomes, heads and tails. Most random phenomena have more than two possible outcomes. We are then interested in assigning probabilities not just to individual outcomes but to collections of outcomes. Any collection of the individual outcomes is called an *event*. Probability as long-term relative frequency applies to events as well as to outcomes.

> **Example 2.** Return once more to the bead-sampling experiment summarized in Figure 1-2 on page 23. This experiment consisted of 200 trials of the random phenomenon of drawing an SRS of size 25 from a large box of beads, 20% of which were dark. The outcome "5 dark beads in the sample" had frequency 47 and relative frequency $47/200 = 0.24$. The *event* "4, 5, or 6 dark beads in the sample" had frequency 111 and relative frequency $111/200 = 0.56$ because one of the outcomes 4, 5, or 6 occurred on 111 of the 200 trials. The number 0.56 is not exactly the probability of the event "4, 5, or 6 dark beads," but it is reasonably close because it is the relative frequency in 200 trials. A much larger number of trials would give a relative frequency very close indeed to the probability.

The probability of an event is a measure of how likely the event is to occur, but does not make sense unless we can at least imagine many repetitions of the random phenomenon. When we say that a coin has probability 1/2 of coming up heads *on this toss*, we are applying to a single toss a measure of the chance of a head based on what would happen in a long series of tosses. This "definition" of probability as long-term relative frequency is not intended to satisfy either the mathematician or the philosopher. It merely builds on the observation that long-term relative frequencies often do settle down to fixed numbers, and it provides the terminology to describe this situation. I do not know why the world is made so that randomness exists, but it is an observed fact that it is so made.

Of course, not all phenomena are random. Some, such as dropping a coin from a fixed height and measuring the time it takes to fall, are predictable; that is, repetitions give the same result time after time. Such phenomena we call *deterministic*. They are the subject of much of the older physical sciences.

Other phenomena are unpredictable but display no long-term pattern. This is often because someone can intervene in the sequence of outcomes to disrupt the regularity on which randomness depends. If, for example, the operator of a roulette wheel has a brake that she can apply at will, she can prevent the relative frequency of "red" from settling down to a fixed probability in the long run. Thinking of the roulette operator watching the play and hitting the brake from time to time reminds us that probability as long-run relative frequency requires more than imagining a long series of repetitions. The repetitions must be *independent*. Independence means that the outcome of one trial gives no information about the outcome of any other trial. Spins of an honest roulette wheel are independent because the wheel doesn't remember where it stopped on previous spins. The operator of a dishonest wheel can remember previous outcomes and intervene to destroy independence.

Although many phenomena are not random, randomness is nonetheless widespread. The behavior of many individuals is often as random — and so as well described by the laws of probability — as the behavior of many coin tosses or many random samples. Life insurance, for example, is based on the fact that deaths occur at random among many individuals.

> **Example 3.** If all males in the United States are observed during their twenty-second year, some will die during that year. Whether a specific 21-year-old male will die is not predictable, but observation of several million such men shows that about 0.18% of them (that's 0.0018 in decimal form) die each year. So if an insurance company sells many policies to 21-year-old men, it knows that it will have to pay off on about 0.18% of them and sets the premium high enough to cover this cost. The number 0.0018 is the *probability* that an American male will die in his twenty-second year.

Some fine points

The idea of probability is a bit subtle. Here are some fine points to set you thinking.

The "Law of Averages." If in tossing a fair coin we get 10 straight heads, is tails more likely on the next toss? Because the probability of tails is 1/2, many people think that the "Law of Averages" demands that some tails now appear to balance the 10 heads. Are they right?

No. The coin has no memory, so it does not know that it has come up heads 10 straight times. Put more formally, because tosses are independent, the probability of a tail on the next toss is still 1/2.

"So the law of averages doesn't guarantee me a girl after seven straight boys, but can't I at least get a group discount on the delivery fee?"

That answer is correct but perhaps not convincing. It can be checked empirically if you have several years to waste: Toss a penny many times. Every time you get 10 straight heads, record whether the next toss gives heads or tails. In the long run you will find tails occurring about half the time after 10 heads. A better tactic is to seek to understand how the long-run relative frequency of tails can be 1/2 *without* making up for the 10 straight heads with some extra tails. To understand this, let us suppose that the next 10,000 tosses are evenly divided, giving 5000 heads and 5000 tails. Then the relative frequency of tails after 10,010 tosses is

$$\frac{5000}{10,010} = 0.4995$$

This is very close to 1/2. The 10 straight heads are swamped by later tosses and need not be made up for by extra tails. (Of course, the next 10,000 tosses will not yield exactly 5000 tails, but the point is that if the fraction of tails in the long run is about 1/2, 10 straight heads will not affect this at all.)

Belief in this phony "Law of Averages" can lead to consequences close to disastrous. A few years ago, Dear Abby published in her advice column a letter from a distraught mother of eight girls. It seems that she and her husband had planned to limit their family to four children. But when all four were girls, they tried again. And again, and again. After seven straight girls, even her doctor had assured her that "the law of averages was in our favor 100 to 1." Unfortunately for this couple, having children is like tossing coins: Eight girls in a row is highly unlikely, but once seven girls have been born it is not at all unlikely that the next child will be a girl. And it was.

What Probability Doesn't Say. There is, you may think, little difference between the statements:

(a) "In many tosses of a fair coin, the fraction of heads will be close to one-half" and

(b) "In many tosses of a fair coin, the number of heads will be close to one-half the number of tosses."

Alas, Statement (a) is true and is what we mean by saying that the probability of a head is 1/2. Statement (b) is false; in many tosses of a fair coin, the number of heads is certain to deviate more and more from one-half the number of tosses. This is, as Pooh would say, mystigious. To see why it is true, consider the following example.

Number of tosses	Number of heads	Fraction of heads	Difference between number of heads and 1/2 number of tosses
100	51	0.51	1
1000	505	0.505	5
10,000	5025	0.5025	25
100,000	50,125	0.50125	125

There it is: The *fraction* of heads gets closer to 1/2 while the *number* of heads departs more and more from one-half the number of tosses. (Again, this exact outcome is unlikely, but it is typical of what happens in many repetitions.)

The myth of short-run regularity. The idea of probability is that randomness is regular in the long run. Unfortunately, our intuition about randomness tries to tell us that random phenomena should also be regular in the short run. When they aren't, we look for some explanation other than chance variation. One kind of departure from regularity is a *run* of identical outcomes, such as several consecutive heads in tossing a coin. Runs of moderate length are much more likely to occur by chance than our intuition suggests. Try this experiment: ask several people to write down a sequence of heads and tails that they think imitates 10 tosses of a balanced coin. How long was the longest run of consecutive heads or consecutive tails in the tosses? Most people will write a sequence with no runs of more than two consecutive heads or tails. But in fact the probability of a run of three or more consecutive heads or tails in 10 tosses is greater than 0.8, and the probability of *both* a run of three or more heads and a run of three or more tails is almost 0.2.

The belief that runs must be due to some special cause influences behavior. If a basketball player makes several consecutive shots, both the fans and his teammates believe that he has a "hot hand" and is more likely to make the next shot. This is wrong. Careful study has shown that runs of baskets made or missed are no more frequent in basketball than would be expected if each shot is independent of the player's previous shots. Players perform consistently, not in streaks. (Of course, some players make a higher percent of their shots in the long run than do others.) Our perception of hot or cold streaks simply shows that we don't perceive random behavior very well.

Probability models

Enough of conceptualizing about probability. In practice, we often wish to describe a random phenomenon by listing the possible outcomes and assigning a probability to each outcome. This is called giving a *probability model* for the phenomenon. A probability model is often based partly on observation and partly on our feeling about what the probability of the outcomes should be. These models are useful for thinking about random phenomena and for computing the probabilities of complicated events from probabilities of simple events. We shall soon meet some examples. But the correctness of the model must always be judged by comparing it with observations of the random phenomenon it is supposed to describe, because probabilities are defined empirically.

The mathematics of probability begins by describing properties of all legitimate probability models. Because we intend to avoid most of the mathematics, we need only two such properties.

> **A. The probability of any event must be a number between 0 and 1.**
>
> **B. If we assign a probability to every possible outcome of a random phenomenon, the sum of these probabilities must be 1.**

Properties A and B of any probability model follow from our understanding of probability as long-run relative frequency. Any relative frequency is a number between 0 and 1, and this is property A. Because some outcome must occur on each repetition, the sum of the relative frequencies of all possible outcomes must be 1, and this is property B. Once we have assigned probabilities to each individual outcome, we find the probability of any event by simply adding together the probabilities of the outcomes that make up the event. This rule also follows from thinking about how relative frequencies behave. Here are some simple examples of probability models.

> **Example 4.** The *New York Times* of August 21, 1989 reported a poll that interviewed a probability sample of 1025 women. The married women in the sample were asked whether their husband did his fair share of household chores. Here are the results.
>
> | Outcome | Probability |
> |---|---|
> | Does more than his fair share | 0.12 |
> | Does his fair share | 0.61 |
> | Does less than his fair share | 0.27 |
>
> This example would be described more naturally as relative frequencies in a sample. Instead we have thought of it as giving (from sample observations) a probability model for the random phenomenon of choosing a married woman at random and asking her

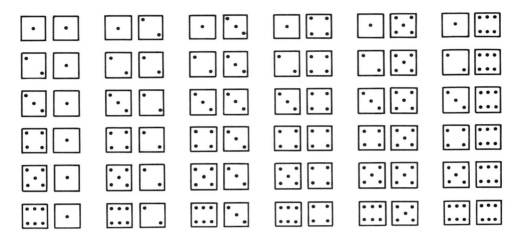

Figure 7-1. The 36 possible outcomes when two dice are rolled.

opinion. Notice that all three outcomes have probabilities between 0 and 1 (property A), and that these probabilities add to 1 (property B). The *event* "I think my husband does at least his fair share" contains the first two outcomes. Its probability is therefore $0.12 + 0.61 = 0.73$.

Example 5. Rolling two dice is a common way to lose money in casinos. There are 36 possible outcomes when we roll two dice and record the up faces in order (first die, second die). Figure 7-1 displays these outcomes. What probabilities should we assign to these outcomes? Casino dice are carefully made. Their spots are not hollowed out, which would give the faces different weights, but are filled with white plastic of the same density as the red plastic of the body. For casino dice it is reasonable to assign the same probability to each of the 36 outcomes in Figure 7-1. Because these 36 probabilities must have sum 1 (property B), each outcome must have probability 1/36.

We are interested in the sum of the spots on the up faces of the dice. What is the probability that this sum is 5? The event "roll a 5" contains these four outcomes:

Because each outcome has probability 1/36, the probability of rolling a 5 is

$$\frac{1}{36} + \frac{1}{36} + \frac{1}{36} + \frac{1}{36} = 0.111$$

This probability model was based on assuming that the two dice are perfectly balanced and are rolled independently. Long experience shows that the model is accurate for casino dice. Inexpensive dice with hollowed-out spots aren't balanced and this assignment of probabilities doesn't describe their behavior.

Probability versus odds

Probability was born in a gambling hall, from whence it climbed to more respectable status. Back in the gambling hall, the chance of an event often is stated in terms of odds rather than probability. At the risk of encouraging you to misapply your knowledge of probability, let us learn to translate odds into probabilities.

Example 6. You are rolling two dice, as in Example 5. You would very much like to roll a 7, and you have heard that the odds against this are 5 to 1. What's the probability of a 7? Odds of 5 to 1 means that failure to roll a 7 happens five times as often as success. In the long run, then, five of every six tries will fail and one will succeed in rolling a 7. The probability of a 7 is now clear. It is one out of six, or

$$\frac{1}{6} = 0.167.$$

You can follow the method of Example 5 to check from Figure 7-1 that this is the correct probability. Notice that odds of 5 to 1 are not the same as a 1-in-5 chance. The latter means probability 1/5, while the former gives the probability as 1/6.

The recipe illustrated in Example 6 is

Odds of *A* to *B* against an outcome means that the probability of that outcome is *B/(A + B).*

Thus if the odds against the favorite in a horse race are 3 to 2, this is equivalent to that horse having probability 2/5 of winning.

Summing up: What is probability?

I have stressed that long-term stability of relative frequencies is an observed fact in some circumstances. This observation provides a way to think about probability numerically and leads to properties A and B, which characterize all legitimate probability models. But long-term relative frequency is not the only important interpretation of probability. We could instead define the probability of an event as a number between 0 and 1 that represents my personal assessment of the chance that the event will occur. ("I think the Rams have a 30% chance of going to the Super Bowl this year.") Probabilities interpreted this way are called *personal probabilities* or *subjective probabilities*, because they express personal opinion. Your personal probability for the event that the Rams will play in this year's Super Bowl is no doubt not the same as mine.

Personal probability has several advantages over the relative-frequency interpretation. The probability that the Rams will go to the Super Bowl this year, or the probability that your firm's bid will be high enough to win a certain contract, cannot be thought of as long-term relative frequencies because these probabilities refer to a single, unrepeatable chance phenomenon. So personal probability has wider scope than long-term relative frequency. The other advantage of a subjective interpretation of probability is that it provides a meaningful definition of probability, not just a way of thinking about randomness. Long-term relative frequency is not truly a definition of probability because we can never watch an endless series of trials to be certain that the relative frequency settles down to a probability. So "long-term relative frequency" refers to an ideal never quite attained. Personal probability, on the other hand, is exactly what it pretends to be: a subjective assessment of chance.

The disadvantage of personal probability is that it is personal. When long-term relative frequencies do exist, my personal assignment of probabilities may bear no relation to the pattern of outcomes actually observed. Which interpretation of probability is favored in applications varies with the weight accorded to these advantages and disadvantages. Personal probability predominates when the insight and partial information of a decision maker are important, as in business decisions and gambling on horses. Frequency ideas rule when repeatable events are in question, as in quality control of mass-produced items and gambling on roulette wheels.

I prefer to keep the frequency interpretation of probability foremost in your mind, because I want to emphasize that randomness is something we can observe and that probability gives a mathematical description of randomness. But we will meet both interpretations, not always sharply distinguished, through the rest of this book. The clash of interpretations can be played down because any assignment of probabilities must have properties A and B, whether interpreted as personal assessment of chance or as long-term relative frequency. Users of probability in science, business, gambling, and statistics begin by assigning probabilities to events. However you interpret the notion of probability, the rules describing legitimate assignments of probabilities stand forever firm.

Section 1 exercises

7.1. Toss a thumbtack on a hard surface 100 times. How many times did it land with the point up? What is the approximate probability of landing point up?

7.2. The table of random digits (Table A) was produced by a random mechanism that gives each digit probability 0.1 of being a 0. What proportion of the first 200 digits in the table are 0s? This proportion is an estimate of the true probability, which in this case is known to be 0.1.

7.3. Hold a penny on edge on a flat surface with the index finger of one hand and snap it with your other index finger so that it spins rapidly until finally falling with either heads or tails upward. Repeat this 50 times and record the number of heads. What is your estimate of the probability of a head in this random experiment? (In doing this experiment, disregard any trial in which the penny does not spin for several seconds or in which it hits an obstacle.)

7.4. When we toss a penny, experience shows that the probability (long-term relative frequency) of a head is close to 1/2. Suppose now that we toss the penny repeatedly until we get a head. What is the probability that the first head comes up in an *odd* number of tosses (1, 3, 5, and so on)? To find out, repeat this experiment 50 times, and keep a record of the number of tosses needed to get a head on each of your 50 trials.

    (a) From your experiment, estimate the probability of a head on the first toss. What value should we expect this probability to have?

    (b) Use your empirical results to estimate the probability that the first head appears on an odd-numbered toss.

7.5. Roll a pair of dice 100 times and record the sum of the spots on the upward faces in each trial. What is the relative frequency of 8 among these 100 trials? Compare your result with the probability calculated from the probability model in Example 5.

7.6. Probability is a measure of how likely an event is to occur. Match one of the probabilities that follow with each statement of likelihood given. (The probability is usually a much more exact measure of likelihood than is the verbal statement.)

$$0 \quad 0.01 \quad 0.3 \quad 0.6 \quad 0.99 \quad 1$$

(a) This event is impossible. It can never occur.
(b) This event is certain; it will occur on every trial of the random phenomenon.
(c) This event is very unlikely, but it will occur once in a while in a long sequence of trials.
(d) This event will occur more often than not.

7.7. (a) A gambler knows that red and black are equally likely to occur on each spin of a roulette wheel. He observes five consecutive reds occur and bets heavily on black at the next spin. Asked why, he explains that black is "due by the law of averages." Explain to the gambler what is wrong with this reasoning.

(b) After hearing you explain why red and black are still equally likely after five reds on the roulette wheel, the gambler moves to a poker game. He is dealt five straight red cards. He remembers what you said, and assumes that the next card dealt in the same hand is equally likely to be red or black. Is the gambler right or wrong, and why?

7.8. Buy a bag of M&M candies and draw a candy at random from the bag. The candy can have any one of six colors. The probability of drawing each color depends on the proportion of each color among all candies made. The table below gives the probability that a randomly chosen M&M has each color. What must the probability for tan candies be to make this a legitimate assignment of probabilities? What is the probability that the candy drawn is red, yellow, or orange?

Color	Brown	Red	Yellow	Green	Orange	Tan
Probability	0.3	0.2	0.2	0.1	0.1	?

7.9. Choose at random an American woman aged 25 to 29 years. The probability that you choose a single woman is just the proportion of single women in this age group. With help from the Census Bureau, we can make the assignment of probabilities given below. If this is to be a legitimate assignment of probabilities, what must be the probability that a woman in this age group is married? What is the probability that a woman in this age group is not married?

Outcome	Single	Married	Widowed	Divorced
Probability	0.288	?	0.003	0.076

7.10. Exactly one of Brown, Chavez, and Williams will be promoted to partner in the law firm that employs them all. Brown thinks that she has probability 0.25 of winning the promotion and that Williams has probability 0.2. What probability does Brown assign to the outcome that Chavez is the one promoted?

7.11. Example 5 gives a probability model for rolling two dice. According to this model, what is the probability that the sum of the spots on the up faces is either 7 or 11?

7.12. An American roulette wheel contains compartments numbered 1 through 36 plus 0 and 00. Of the 38 compartments, 0 and 00 are colored green, 18 of the others are red, and 18 are black. A ball is spun in the direction opposite to the wheel's motion, and bets are made on the number where the ball comes to rest. A simple wager is *red or black*, in which you bet that the ball will stop in, say, a red compartment. If the wheel is fair, all 38 compartments are equally likely.
   (a) What is the probability of a red?
   (b) What are the odds against a red?

7.13. Figure 7-2 displays four assignments of probability to the outcomes of rolling a single die. Which, if any, is *correct* for this die can be discovered only by rolling the die many times. But some of the models are not *legitimate* assignments of probability. Which are legitimate and which are not, and why?

7.14. In each of the following cases, state whether or not the given assignment of probabilities to individual outcomes is legitimate in the sense of satisfying properties A and B. If not, give specific reasons why not.
   (a) When a coin is spun, $P(H) = 0.55$ and $P(T) = 0.45$.
   (b) When two coins are tossed, $P(HH) = 0.4$, $P(HT) = 0.4$, $P(TH) = 0.4$, and $P(TT) = 0.4$.
   (c) The mixture of colors for M&M's given in Exercise 7.8 is about to be replaced by a new mixture. Now there will be no tan candies, while the other five colors will have the same probabilities that are given in Exercise 7.8.

7.15. A six-sided die is rolled and comes to rest with 1, 2, 3, 4, 5, or 6 spots on the up face. Make an assignment of probabilities to the six outcomes that satisfies properties A and B and that you think is close to correct for casino dice. (Assume that the die is balanced so that all faces are equally likely to come up. Probability models are

Probability

Outcome	Model 1	Model 2	Model 3	Model 4
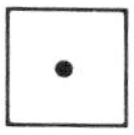 ⚀	1/7	1/3	1/3	1
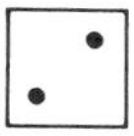 ⚁	1/7	1/6	1/6	1
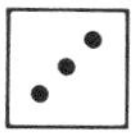 ⚂	1/7	1/6	1/6	2
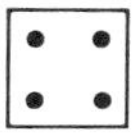 ⚃	1/7	0	1/6	1
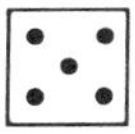 ⚄	1/7	1/6	1/6	1
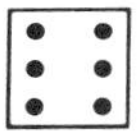 ⚅	1/7	1/6	1/6	2

Figure 7-2.

often based on assuming balance or symmetry. In this case, observation supports the assumption.)

7.16. When you toss two coins, the four possible outcomes are (head, head), (head, tail), (tail, head), and (tail, tail). Make an assignment of probabilities to these outcomes that you think is correct for balanced coins. What probability does your model give for the event that exactly one of the coins shows a head?

7.17. Using the interpretation of probability as long-run relative frequency, explain carefully why the following rule is true in any legitimate probability model:

The probability that an event does not occur is one minus the probability that the event does occur.

7.18. You are gambling with a fair coin, which has probability 1/2 of coming up heads on each toss. You are allowed to choose either 10 or 100 tosses.

    (a) On the first bet, you win if the relative frequency of head is between 0.4 and 0.6. Should you choose 10 tosses or 100 tosses?

(b) On the second bet, you win if exactly half the tosses are heads. Should you choose 10 tosses or 100 tosses?

7.19. The odds against being dealt three of a kind in a five-card poker hand are about 49 to 1. What is the probability of being dealt three of a kind?

2.   Finding probabilities by simulation

Suppose that a couple plans to have children until they have a girl or until they have four children, whichever comes first. What is the probability that they will have a girl among their children? To answer probability questions such as this, we first construct a probability model. In this case, it seems reasonable to assume that:

(a) Each child has probability 1/2 of being a girl and 1/2 of being a boy; and

(b) The sexes of successive children are independent.

"I've had it! Simulated wood, simulated leather, simulated coffee, and now simulated probabilities!"

How can we compute the probability of a somewhat complex event (having a girl in four tries) from this simple model?

There are two ways of finding the probability of complex events from known probabilities of simple events. One is to master the mathematics of probability theory, a worthwhile endeavor that we wish to avoid. The other is to *simulate* (imitate or run a small scale model of) the random phenomenon by using our trusty companion, the table of random digits. We can use the table to simulate many repetitions of the phenomenon. The relative frequency of any event eventually will be close to its probability, so many repetitions give a good estimate of probability. We will simulate the childbearing strategy of the couple discussed above.

Example 7. Step 1. A single random digit simulates the sex of a single child as follows:

$$0, 1, 2, 3, 4 \quad \text{the child is a girl}$$
$$5, 6, 7, 8, 9 \quad \text{the child is a boy}$$

Step 2. To simulate one repetition of the childbearing experiment, use successive random digits until either a girl or four children are obtained. Using line 130 of Table A, we find

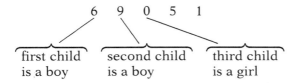

and the couple stops at three children, having obtained a girl.

Step 3. Simulate many repetitions and use the relative frequency of the event "the couple has a girl" to estimate its probability. Here is the result of using line 130. To interpret the digits, we have written G for girl and B for boy under them, have added space to separate repetitions, and under each repetition have written + if a girl was born and − if not.

690	51	64	81	7871	74	0	951	784	53	4	0	64	8987
BBG	BG	BG	BG	BBBG	BG	G	BBG	BBG	BG	G	G	BG	BBBB
+	+	+	+	+	+	+	+	+	+	+	+	+	−

In 14 repetitions, a girl was born 13 times. So our estimate of the probability that this strategy will produce a girl is

$$\frac{13}{14} = 0.93$$

This example of simulation deserves careful discussion. Step 1 simulates part (a) of the probability model because we assigned 5 of the 10 possible digits to the outcome "girl." Because all 10 digits are equally likely to occur at any point in the table, any one table entry has probability 5/10 or 1/2 of indicating "girl." Here are further examples of this idea.

Example 8. Simulate an event that has probability 3/10 as follows:

0, 1, 2   the event occurs

3, 4, 5, 6, 7, 8, 9   the event does not occur

Example 9. To simulate an event that has probability 0.33, use *two* digits to simulate one repetition, with

00, 01, 02, . . ., 31, 32   the event occurs

33, 34, 35, . . ., 98, 99   the event does not occur

Example 10. Simulate a random trial having *three* possible outcomes with probabilities 2/10, 3/10, and 5/10 as follows:

0, 1   outcome A occurs

2, 3, 4   outcome B occurs

5, 6, 7, 8, 9   outcome C occurs

Step 2 in Example 7 simulates one repetition of the childbearing process, because successive random digits are independent and therefore simulate successive independent births. In Step 3, 14 repetitions were made, and the relative frequency of bearing a girl was found. The relative frequency 0.93 is not a very precise estimate of the probability, because only 14 repetitions were made. With mathematics we could show that if the probability model given is correct then the true probability of having a girl is 0.938. Our simulated answer came quite close.

Now that we have seen a first illustration of the mechanics of simulation, let's pause for a longer look at some of the ideas.

The probability model is the foundation of any computation of probability, whether by simulation or by mathematics. It may appear a bit shady to begin the process of finding a probability by assuming that we already know some other probabilities, but not even mathematics can give us something for nothing. The idea is to state the basic structure of the random phenomenon and then use simulation to move from the basics to the probabilities of more complicated events. The model is based on opinion and past experience; if it does not correctly describe the random phenomenon, the probabilities derived from it by simulation also will be incorrect. So will the probabilities a mathematician might obtain by his black arts, like that 0.938 in the childbearing example. That is the "true" probability of a girl only if the sexes of successive children are independent and each child is equally likely to be male or female.

The probability models we will meet have two parts.

(a) A simple random phenomenon with a small number of possible outcomes. These are assigned probabilities with properties A and B as discussed in Section 1.

(b) Independent trials of phenomena such as part (a) describes, sometimes repetitions of the same random phenomenon (like the successive children in Example 7), and sometimes independent trials of different random phenomena.

The random digits play their usual role as substitutes for physical randomization. We could physically simulate the behavior of our probability model. The childbearing strategy might be simulated by an urn* with a number of balls, half red and half white. Red balls are female children and white balls are males. Draw a ball to represent the sex of the first child; you are equally likely to get a female or a male. Replace the ball (to maintain the half-red and half-white mixture), stir the urn well, and draw again; that's the second child. And so on. Urns and balls could always be used to do our simulations. But random digits have the same advantages in simulation as in random sampling: They are faster and more accurate than physical randomization. Thinking of physical models like urns full of balls does help to take the mystery out of random

---

*I would like to call this container a pot, but protocol forbids. Pots full of balls of varying colors are officially called urn models in probability theory.

To find the probability of an event by simulation

(a) Specify a probability model for the random phenomenon by assigning probabilities to individual outcomes and assuming independence where appropriate.

(b) Decide how to simulate the basic outcomes of the phenomenon, using assignments of digits to match the assignment of probabilities (step 1).

(c) Decide how to simulate a single repetition of the random phenomenon by combining simulations of the basic outcomes (step 2).

(d) Estimate the probability of an event by the relative frequency of the event in many repetitions of the simulated phenomenon (step 3).

---

digits in simulation. Both the urn and the random digits make a manageable copy of the random phenomenon under study.

The precision of a simulation for estimating probabilities increases as more repetitions are used. This is simply a restatement of the "definition" of probability as long-term relative frequency. But there is a close connection with the precision of sample statistics from probability samples. Both statistics from large samples and probabilities simulated from many trials are highly precise in the sense that repeating the whole process would give nearly the same answer. Indeed, estimating an unknown probability p by a relative frequency $\hat{p}$ is substantially the same as estimating a population proportion p by a relative frequency $\hat{p}$.

More extensive simulations

Simulation is a common procedure for finding probabilities too difficult to compute, even for those who know probability theory. Such simulation is always done by a computer, which can be programmed to do thousands of repetitions in a short time. Large numbers of repetitions give very precise results, but the ideas remain those we have met. Simulations in science and engineering are usually accompanied by a statement of precision in the form of the standard deviation of the relative frequency used to estimate the unknown probability. Recall from Section 5 of Chapter 4 that the standard deviation of the sample statistic is a common way of stating precision in sampling as well. The next example illustrates the close connection between sampling and simulation by asking a probability question about a sampling procedure.

> **Example 11.** Suppose that 80% of all consumers prefer Brand A instant coffee to Brand B. If an SRS of 10 consumers is chosen, what is the probability that seven or more of them prefer Brand A? (This is a question about the long-term pattern of results of an SRS. That long-term pattern, which has been with us since Chapter 1, can be described now in terms of the probability of various outcomes.)
>
> We first need a probability model. Here is the model we will simulate:
>
> (a) Each consumer has probability 8/10 of preferring Brand A.
>
> (b) The preferences of successive consumers are independent.
>
> This is a good model for an SRS of size 10 from a population of which 80% (8/10) favor Brand A *if* the population is large.* Now for the simulation.
>
> Step 1. One digit simulates one consumer's preference.
>
> | | |
> |---|---|
> | 0, 1, 2, 3, 4, 5, 6, 7 | Brand A |
> | 8, 9 | Brand B |
>
> Step 2. To simulate one repetition of the experiment, use 10 random digits. Count how many of these digits are 0, 1, . . ., 7. This is the number of consumers in a sample of size 10 who prefer Brand A.
>
> Step 3. We do this 10 times, starting at line 110 of Table A. Here are the 10 repetitions.

---

*That's a fine point. Think of an urn containing 80% red balls (Brand A) and 20% white balls (Brand B). We begin to draw our SRS, and the first ball is red. Now the urn contains *less* than 80% red balls. So the preferences of the 10 consumers in our SRS are *not* independent, because each consumer drawn changes the makeup of the remaining population. When the population is large, the dependence is so small we can neglect it. Drawing a red ball from an urn containing 80,000 reds and 20,000 whites does not noticeably change the chance of a red on the second draw.

38448	48789	5 prefer Brand A
ABAAB	ABABB	
18338	24697	7 prefer Brand A
ABAAB	AAABA	
39364	42006	9 prefer Brand A
ABAAA	AAAAA	
76688	08708	6 prefer Brand A
AAABB	ABAAB	
81486	69487	6 prefer Brand A
BAABA	ABABA	
60513	09297	8 prefer Brand A
AAAAA	ABABA	
00412	71238	9 prefer Brand A
AAAAA	AAAAB	
27649	39950	7 prefer Brand A
AAAAB	ABBAA	
59636	88804	6 prefer Brand A
ABAAA	BBBAA	
04634	71197	9 prefer Brand A
AAAAA	AAABA	

The event "7 or more prefer Brand A" occurs in 6 of the 10 repetitions, so we estimate its probability to be

$$\frac{6}{10} = 0.6.$$

Ten repetitions gives quite poor precision. By mathematics or more extensive simulation we can find that the true probability that 7 or more of an SRS of 10 consumers prefer Brand A is 0.88. I did a set of 50 trials and got a relative frequency of $41/50 = 0.82$, which is closer to home.

The probability we estimated by simulation in Example 11 depends on the assumption that 80% of the population prefer Brand A. If that population proportion is different, the probabilities of various outcomes in the SRS are different. A complete study of probabilities for an SRS would include a description of how these probabilities change when the population proportion changes.

The building and simulation of random models is a powerful tool of contemporary science, yet a tool that can be understood in substance without advanced mathematics. What is more, several attempts to simulate simple random phenomena will increase your understanding of probability more than many pages of my prose. Having in mind these two goals of understanding simulation for itself and understanding simulation to understand probability, let us study a more extensive example.

Example 12. We are studying the Asian stochastic beetle,[2] and we observe that females of this insect have the following pattern of reproduction:

> 20% of females die without female offspring
>
> 30% have one female offspring
>
> 50% have two female offspring

What will happen to the population of Asian stochastic beetles: Will they increase rapidly, barely hold their own, or die out? (Notice that we can ignore the male beetles in studying reproduction, as long as there are some around for certain essential purposes. Notice also that we are studying only a single population. It is common for ecologists to use probability models and simulation in their study of the interaction of several populations, such as predators and prey.)

The reproduction of a single female is simulated as follows:

> 0, 1   dies without female offspring
>
> 2, 3, 4   has one female offspring
>
> 5, 6, 7, 8, 9 has two female offspring

Moreover, we will assume that female beetles reproduce independently of each other.

To answer the question, "What is the future of the Asian stochastic beetle?", we will simulate the female descendants of several female beetles until they either die out or reach the fifth generation. Beginning at line 122 of the table of random digits,

13873   81598   95052   90908   73592

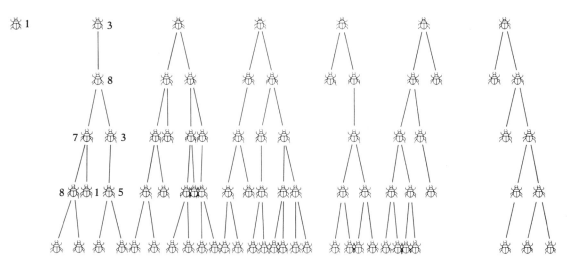

Figure 7-3. Simulation of the descendants of seven female Asian stochastic beetles. 20% die without female offspring; 30% have one female offspring; 50% have two female offspring.

the first beetle dies without offspring (1). The second has one female offspring (3); she in turn has two female offspring (8); the first of these has two (7) and the second has one (3) female offspring. So the fourth generation of this family contains three female beetles.

We need a better way to record this simulation. Figure 7-3 records the female descendants of seven female Asian stochastic beetles. Each family is followed to the fifth generation. The two families on the left are those we just met, and the random digits beside each beetle in these families remind you how line 122 was used from left to right in each generation of offspring. The fifth generation has 29 female beetles from the original 7. It is clear that the population of Asian stochastic beetles will increase rapidly until crowding, shortage of food, or increased predator populations change the reproductive pattern we have simulated.

Section 2 exercises

7.20. An opinion poll selects Americans over 18 at random and asks them, "Which political party, Democratic or Republican, do you think is better able to manage the economy?" Explain carefully how you would use Table A to simulate the response of one person in each of the following situations.

  (a) Of all Americans over 18, 50% would choose the Democrats and 50% the Republicans.

  (b) Of all Americans over 18, 60% would choose the Democrats and 40% the Republicans.

  (c) Of all Americans over 18, 40% would choose the Democrats, 40% would choose the Republicans, and 20% are undecided.

  (d) Of all Americans over 18, 53% would choose the Democrats and 47% the Republicans.

7.21. Use Table A to simulate the responses of 10 independently chosen Americans over 18 in each of the four situations of Exercise 7.20.

  For situation (a), use line 110.
  For situation (b), use line 111.
  For situation (c), use line 112.
  For situation (d), use line 113.

7.22. Probabilities involving runs of outcomes are quite hard to calculate by mathematics, but are easy to simulate. Let's look at runs of heads or tails in 10 tosses of a balanced coin.

  (a) Describe how to simulate a single toss of a balanced coin. Then describe how to simulate 10 independent tosses of a balanced coin.

  (b) Use Table A beginning at line 115 to simulate 50 repetitions of 10 coin tosses. Watch for runs of 3 or more consecutive heads or consecutive tails in the 10 tosses and keep a record of "run of heads," "run of tails," or "no run of 3" for each repetition. What is the approximate probability of a run of at least 3 heads in 10 tosses of a balanced coin? What is the approximate probability of both a run of at least 3 heads and a run of at least 3 tails in 10 tosses?

7.23. A basketball player makes 70% of her free throws in a long season. In a tournament game she shoots five free throws late in the game and misses three of them. The fans think she was nervous, but the misses may simply be chance. You will shed some light by finding a probability.

  (a) Describe how to simulate a single shot if the probability of making each shot is 0.7. The describe how to simulate five independent shots.

  (b) Simulate 50 repetitions of the five shots and record the number missed on each repetition. Use Table A starting at line 125. What is the approximate probability that the player will miss at least three of the five shots?

7.24. Elaine is enrolled in a self-paced course that allows three attempts to take an examination on the material. To pass the course, she must pass the examination. She does not study and has probability 2/10 of passing on any one attempt by luck. What is Elaine's probability of passing on at least one of the three attempts? (Assume the attempts are independent because a different examination is given on each attempt.)

   (a) Explain how you would simulate Elaine's three tries by using random digits.

   (b) Simulate this experiment 50 times by using Table A beginning at line 120. What is your estimate of the Elaine's probability of passing the course?

   (c) Do you think the assumption that the probability of passing the exam is the same on each trial is realistic? Why or why not?

7.25. A more realistic probability model for Elaine's three attempts to pass an exam in the previous exercise is as follows: On the first try she has probability 0.2 of passing. If she fails on the first try, her probability on the second try increases to 0.3 because she learned something from her first attempt. If she fails on two attempts, the probability of passing on a third attempt is 0.4. She will stop as soon as she passes and the course rules force her to stop after three attempts in any case.

   (a) Explain how to simulate one repetition of Elaine's tries at the exam.

   (b) Simulate 50 repetitions and estimate the probability that Elaine eventually passes the exam and so passes the course. Use Table A starting at line 130.

7.26. A wildcat oil driller estimates that the probability of finding a producing well when he drills a hole is 0.1. If he drills 10 holes without finding oil, he will be broke. What is the probability that he will go broke? Answer this question as follows:

   (a) State a simple probability model for drilling 10 holes.

   (b) Explain how you will use random digits to simulate drilling 1 hole and then explain how to simulate drilling 10 holes.

   (c) Use Table A beginning at line 140 to simulate 20 repetitions of drilling 10 holes. Estimate from this simulation the probability that the wildcatter will go broke.

7.27. Tossing four astragali was the most popular game of chance in Roman times. The scoring for the four possible outcomes of a single bone was

broad convex side of bone	4
broad concave side of bone	3

narrow flat side of bone	1
narrow hollow side of bone	6

Many throws of a present-day sheep's astragalus show that the approximate probability distribution for these scores is

Score	Probability
1	1/10
3	4/10
4	4/10
6	1/10

The best throw of four astragali was the Venus, when all the four uppermost sides were different. [From Florence N. David, *Games, Gods and Gambling* (London: Griffin, 1962), p. 7.]

  (a) Explain how to simulate the throw of a single astragalus and of four independent astragali.
  (b) Simulate 25 throws of four astragali. Estimate the probability of throwing a Venus.

7.28. The original Pennsylvania State Lottery (state lotteries were much simpler when first introduced) worked as follows: Each lottery ticket bore a six-digit number. Suppose that you had ticket 123456. A weekly winning number was drawn at random, so every six-digit number had the same chance to be drawn. You won

$50,000	if the winning number was	123456
$2000	if the winning number was	X23456 or 12345X
$200	if the winning number was	XX3456 or 1234XX
$40	if the winning number	XXX456 or 123XXX

where X stands for any nonmatching number.

  (a) Explain how to simulate one play of this lottery by using Table A to draw the weekly winning number.
  (b) Simulate 20 plays of the lottery and record the weekly winning numbers drawn.
  (c) On which plays did your ticket 123456 win a prize, and what were the values of the prizes won?
  (d) Based on your 20 trials, what is your estimate of the probability of winning a prize if you hold one ticket in this lottery?

7.29. From your experience with random digits, you can find the exact value of the probability of winning the Pennsylvania lottery as it is given in Exercise 7.28.

(a) How many different six-digit numbers are there?

(b) How many of these will pay you each of

$$\$50,000 \quad \$2000 \quad \$200 \quad \$40$$

if drawn as the weekly winning number?

(c) Now find the total number of winning numbers that will pay you a prize. From this and part (a), find the probability of winning a prize.

7.30. Females of the benign boiler beetle have the following reproductive pattern:

40% die without female offspring

40% have one female offspring

20% have two female offspring

(a) Explain how you would use random digits to simulate the number of offspring of a single female benign boiler beetle.

(b) Simulate the family trees to the fifth generation of enough of these beetles to decide whether the population will definitely die out, will definitely grow rapidly, or appears barely to hold its own. (Simulate the offspring of at least five beetles.)

7.31. A nuclear reactor is equipped with two independent automatic systems to shut down the reactor when the core temperature reaches the danger level. Neither system is perfect: System A shuts down the reactor 90% of the time when the danger level is reached; system B does so 80% of the time. The reactor is shut down if *either* system works.

(a) Explain how to simulate the response of system A to a dangerous temperature level.

(b) Explain how to simulate the response of system B to a dangerous temperature level.

(c) Both systems are in operation simultaneously. Combine your answers to (a) and (b) to simulate the response of both systems to a dangerous temperature level. Explain why you cannot use the same random digit to simulate both responses.

(d) Now simulate 100 trials of the reactor's response to an emergency of this kind. Estimate the probability that it will shut down.

7.32. The game of craps is played with two dice. The player rolls both dice and wins immediately if the outcome (the sum of the faces) is 7 or 11. If the outcome is 2, 3, or 12, the player loses immediately.

If he rolls any other outcome, he continues to throw the dice until he either wins by repeating the first outcome or loses by rolling a 7.

  (a) Explain how to simulate the roll of a single fair die. (It is easiest to use one digit and skip those not needed to represent outcomes.) Then explain how to simulate a roll of two fair dice.

  (b) Use Table A beginning at line 114 to simulate three plays of craps. Explain at each throw of the dice what the result was.

  (c) Now that you understand craps, simulate 25 plays and estimate the probability that the player wins.

7.33. In Example 11, I commented that the responses of successive units drawn in an SRS are not truly independent but that the dependence is negligible if the population is large. Let's examine this comment.

  (a) An urn contains 10 balls, of which 8 are red and 2 are white. One ball is drawn at random. What is the probability that it is red? That it is white?

  (b) The first ball is the first unit in our SRS from this population of 10 balls. We set it aside, and draw a second ball from the urn at random. If the first ball was red, what is the probability that the second is red? (*Hint:* What are the colors of the 9 balls left in the urn?)

  (c) If the first ball was white, what is the probability that the second is red?

Because the probability that the second ball drawn is red changes when we know the color of the first ball drawn, the first and second responses in our SRS are *not* independent.

  (d) Now answer the questions of parts (a), (b), and (c) again, this time for an urn containing 100,000 balls, of which 80,000 are red and 20,000 white.

The probability of a red on the second draw still changes with the color of the first draw. But the difference is now so small that we can ignore it. That leaves us with the probability model of Example 11. No probability model exactly describes a real-world random phenomenon, so we are satisfied with this one.

3.   State lotteries and expected values

Both public and private lotteries were common in the early years of the United States. After disappearing for a century or so, government-run gambling reappeared in 1964, when New Hampshire caused a furor by

*"I think the lottery is a great idea. If they raised taxes instead, we'd
have to pay them."*

introducing a lottery to raise public revenue without raising taxes. The
furor subsided quickly as larger states adopted the idea, until almost all
states outside the south now sponsor lotteries. State lotteries sold over
$15 billion worth of tickets in 1988 and continue to expand rapidly.
Their growth is fed by constant advertising and by new games designed
to hold the interest of the public. The most popular game in most states
is Lotto, in which players choose (for example) 6 numbers out of 54 in
the hope of matching the randomly chosen winning number.

A lottery uses random selection to distribute prizes among those who
have bought tickets. How can we measure the value of a lottery ticket or
compare the return we can expect from buying lottery tickets with that
from gambling in Las Vegas or Atlantic City? It is not enough to know the
probability of winning. The amount won is also important, because one
game of chance might offer many small prizes and another might award
a few large prizes. The larger prizes can compensate for a lower probability of winning and so make the second lottery a better bet. We need a
way to take into account both the probability of winning and the amount
won to compute the *expected value* of a lottery.

State lotteries have introduced ever more gimmicks and special prizes
in an attempt to keep public interest high, and these make computation

of the expected value of a ticket difficult. So let us go back a few years and study the original, uncluttered New York State Lottery.

> **Example 13.** The New York State Lottery awarded, for each 1 million tickets sold,
>
> | | |
> |---:|:---|
> | 1 | $50,000 prize |
> | 9 | $5000 prizes |
> | 90 | $500 prizes |
> | 900 | $50 prizes |
>
> The winning tickets were drawn at random from those sold. The total value of these prizes is
>
> $$(\$50{,}000) + (9)(\$5000) + (90)(\$500) + (900)(\$50) = \$185{,}000$$
>
> Since this amount is divided among 1,000,000 tickets, the average winnings per ticket is
>
> $$\frac{\$185{,}000}{1{,}000{,}000} = \$0.185$$
>
> or 18½ cents. This is the *expected value* of a ticket—the average value of an individual ticket.

Because lottery tickets cost 50 cents, New York State paid out a bit less than 40% (18½ over 50 is 0.37, or 37%) of the amount wagered by ticket buyers and kept over 60%. Competition has since sweetened the pot a bit. State lotteries typically pay out about half the dollars bet; another 15% goes for advertising and expenses and the remaining 35% flows into the state's treasury. The states also pay grand prize winners over time, usually 20 years. Because money earns interest over time, a jackpot advertised as $10 million actually costs the state only about $4.8 million. By way of comparison, casinos in Las Vegas or Atlantic City pay out 85% to 95% of the amount wagered, depending on which game you choose.

State lotteries are a poor bet. Professional gamblers avoid them, not wanting to waste money on so bad a bargain. Politicians avoid them for reasons nicely put by Nelson Rockefeller when he was Governor of New York: "I'm afraid I might win." Almost everyone else in lottery states plays. Surveys usually show that about 80% of the adult residents have purchased at least one lottery ticket. In Massachusetts, the leader in enticing its citizens to gamble, the 1988 take was $235 for every person in the state. Why is such a poor bet so popular? Lack of knowledge of how poor the bet is plays a role. So does skillful advertising by the state. But the major attraction is probably the lure of possible wealth, no matter how unlikely the jackpot is. Many people find 50 cents a week a fair price for the entertainment value of imagining themselves rich. As one Carmen Brutto of Harrisburg, Pennsylvania, said in a newspaper interview, "My chances of winning a million are better than my chances of earning a million."[3]

There is another way of calculating the expected value of a New York State lottery ticket that can be applied more generally than the method of Example 13. (It gives the same answer; I just wish to organize the arithmetic differently.)

Example 14. Buying a ticket in the old New York State Lottery and observing how much you win is a random phenomenon. Because winning tickets are drawn at random, each ticket has 1 chance in 1,000,000 to win $50,000, 9 chances in 1,000,000 to win $5000, and so on. Here is a probability model for the amount won by one ticket.

Amount	Probability
$50,000	1/1,000,000
$5000	9/1,000,000
$500	90/1,000,000
$50	900/1,000,000
$0	999,000/1,000,000

The total probability of winning anything is 1/1000; that is, 1000 out of 1,000,000 tickets win something. The expected value can be found by multiplying each possible outcome by its probability and summing.

$$(\$50,000)\left(\frac{1}{1,000,000}\right) + (\$5000)\left(\frac{9}{1,000,000}\right)$$
$$+ (\$500)\left(\frac{90}{1,000,000}\right) + (\$50)\left(\frac{900}{1,000,000}\right) + (\$0)\left(\frac{999,000}{1,000,000}\right)$$
$$= \$0.185 \text{ (just as before)}$$

We have arrived at the following definition.

If a random phenomenon has numerical outcomes $a_1, a_2, \ldots, a_k$ that have probabilities $p_1, p_2, \ldots, p_k$, the *expected value* is found by multiplying each outcome by its probability and then summing over all possible outcomes. In symbols,

$$\text{Expected value} = a_1 p_1 + a_2 p_2 + \ldots + a_k p_k$$

The law of large numbers

The expected value is an average of the possible outcomes, but an average in which outcomes with higher probability count more. The expected value is the average outcome in another sense as well:

If a random phenomenon with numerical outcomes is repeated many times independently, the mean value of the actually observed outcomes approaches the expected value.

Statisticians call this fact the *law of large numbers*. The law of large numbers explains why gambling, which is a recreation or an addiction for individuals, can be a business for a casino. The house in a gambling operation is not gambling at all. The average winnings of a large number of customers will be quite close to the expected value. The house has calculated the expected value ahead of time and knows what its take will be in the long run. There is no need to load the dice or stack the cards to guarantee a profit. Casinos concentrate on inexpensive entertainment and cheap bus trips to keep the customers flowing in. If enough bets are placed, the law of large numbers guarantees the house a profit.

The law of large numbers is closely related to our definition of probability: In many independent repetitions, the relative frequencies of the possible outcomes will be close to their probabilities, and the average outcome obtained will be close to the expected value. Expected values, because they give the average outcome in two senses, are widely used. Some nongambling examples are given in the exercises. Here is another gambling example.

> **Example 15.** The numbers racket is a well-entrenched illegal gambling operation in the poorer areas of most large cities. The New York City version works as follows: You choose any one of the 1000

three-digit numbers 000 to 999 and pay your friendly local numbers runner 50 cents to enter your bet. Each day, one three-digit number is chosen at random and pays off $300.

Amount	Probability
$300	1/1000
$0	999/1000

$$\text{Expected value} = (\$300) \left(\frac{1}{1000}\right) + (\$0) \left(\frac{999}{1000}\right) = \$0.30$$

This 30-cent expected value is considerably higher than the 18½-cent expected value of a 50-cent old New York State Lottery ticket and somewhat higher than the roughly 25-cent expected value of current state lottery tickets. Crime pays.

Because the criminal organization receives many thousands of bets through its runners each day, by the law of large numbers it will have to pay out very close to 30 cents per ticket. The rest is profit. This guaranteed profit might be endangered if bets were dependent, that is, if many customers chose to bet on the same number. Some numbers (such as 333) are so popular that they are "cut numbers"; a bet on such a number pays off less than $300 if it is the winning number. For a time in the good old days, Willie Mays' batting average was a cut number. The cut numbers protect the racket against very large payouts when a popular number wins.

You might trust the state lottery to conduct an honest random drawing, but perhaps you would not trust the local branch of organized crime.* How, then, does the numbers racket choose a three-digit number each day in a way its customers can trust? The winning number is the last three digits of a large number published every day in the newspaper. In New York, this number is the total amount bet at a local race track. If $2,454,123 was bet at the track, 123 is the winning number. The race track handle is printed in newspapers, so the winning number is publicly available. And the last three digits of a large number such as this are close to being random digits. (Can you see why the first three digits are

---

*Maybe you shouldn't trust anyone. The New York State Lottery was closed down for almost a year in 1975 and 1976 after it was discovered that ticket holders were being cheated by the inclusion of unsold tickets in the prize drawings.

not close to random?) There was a time when the numbers racket in Jersey City, New Jersey, paid off the winning New York State Lottery number. That's adding insult to the injury of a higher payout. In Cleveland, Pittsburgh, and Indianapolis the numbers-game payoff is based on certain stock market tables printed in the regional edition of the *Wall Street Journal*. This edition is printed in Cleveland, and gamblers appear at the printing plant at midnight to learn the winning number. A *Journal* official told a reporter, "The cars line up bumper to bumper. You should see some of them—they're solid chrome."[4] Those, I suppose, are the racketeers, not the players. The law of large numbers guarantees the house a steady income and regular players a steady deficit.

As with probability, it is worth exploring a few fine points about expected values and the law of large numbers.

How large is a large number? The law of large numbers says that the actual average outcome of many trials gets closer to the expected value as more trials are made. It doesn't say how many trials are needed to guarantee an average outcome close to the expected value. That depends on the *variability* of the random phenomenon. The more variable the outcomes, the more trials are needed to ensure that the mean outcome is close to the expected value. Games of chance must be quite variable if they are to hold the interest of gamblers. Even a long evening in the casino has an unpredictable outcome. Gambles with extremely variable outcomes, like state lottos with their large but very improbable jackpots, require impossibly large numbers of trials to ensure that the average outcome is close to the expected value. Though most forms of gambling are less variable than lotto, the layman's answer to the applicability of the law of large numbers is usually that the house plays often enough to rely on it, but you don't. Much of the psychological allure of gambling is its unpredictability for the player; the business of gambling rests on the fact that the result is not unpredictable for the house.

Is there a winning system? Serious gamblers often follow a system of betting in which the amount bet on each play depends on the outcome of previous plays. You might, for example, double your bet on each spin of the roulette wheel until you win—or, of course, until your fortune is exhausted. Such a system tries to take advantage of the fact that you have a memory even though the roulette wheel doesn't. Can you beat the odds with a system? No. Mathematicians have established a stronger version of the law of large numbers that says that your average winnings (the expected value) remains the same so long as successive trials of the game (such as spins of the roulette wheel) are independent and you do not have an infinite fortune to gamble with. Sorry.

Finding expected values by simulation

How can we find expected values in practice? You know the mathematical recipe, but that requires that you start with the probability of each outcome. Expected values too difficult to compute in this way can be found by simulation. The procedure is as before: Give a probability model and simulate many repetitions. By the law of large numbers, the average outcome of these repetitions will be close to the expected value.

> **Example 16.** If a fair coin is tossed repeatedly, what is the expected number of trials required to obtain the first head? (Here is the same question in a different guise: What is the expected number of children a couple must have in order to have a girl?) Each random digit simulates one toss, with odd meaning head and even meaning tail. One repetition is simulated by as many digits as are required to obtain the first odd digit. Here are simulated trials using line 131 of the table of random digits, with vertical bars dividing the repetitions:
>
> 05|007| 1|663|2   81|1|9|4   1|487|3| 041|9|7|
>
> There are 13 trials, which required the following numbers of tosses to the first head:
>
> 2, 3, 1, 3, 3, 1, 1, 2, 3, 1, 3, 1, 1
>
> The average number of tosses required was
>
> $$(2 + 3 + 1 + . . . + 1)/13 = 25/13 = 1.9$$
>
> and this is our estimate of the expected value. (It is a good estimate; the true expected value is 2.)

A summing up

Probability and expected values give us a language to describe randomness. Random phenomena are not haphazard or chaotic any more than random sampling is haphazard. Randomness is instead a kind of order, a long-run regularity as opposed to either chaos or a determinism that fixes events in advance. When randomness is present, probability answers the question "How often in the long run?" and expected value answers the question "How much in the long run?" The two answers are tied together by the definition of expected value in terms of probabilities.

It appears more and more that randomness is embedded in the way the world is made. Albert Einstein reacted to the growing emphasis on randomness in physics by saying, "I cannot believe that God plays dice with the universe." Lest you have similar qualms, I remind you again that randomness is not chaos but a kind of order. Our immediate concern, however, is man-made randomness—not God's dice, but Reno's. In particular, statistical designs for data collection are founded on deliberate randomizing. The order thus introduced into the data is the basis for statistical inference, as we have noticed repeatedly and will study more thoroughly in the next chapter. If you understand probability, statistical inference is stripped of mystery. That may console you as you contemplate the remark of the great economist John Maynard Keynes on long-term orderliness: "In the long run, we are all dead."

Section 3 exercises

7.34. The Connecticut State Lottery (in its original simple form and ignoring some gimmicks that raise the payout slightly) awarded at random, for each 100,000 50-cent tickets sold,

1	$5000 prize
18	$200 prizes
120	$25 prizes
270	$20 prizes

What is the expected value of the winnings of one ticket in this lottery?

7.35. A household is a group of people living together, regardless of their relationship to each other. Sample surveys such as the Current Population Survey select a random sample of households. Here is the probability model for the number of people living in a randomly chosen American household.

Household size	1	2	3	4	5	6	7
Probability	0.236	0.320	0.181	0.156	0.069	0.024	0.014

These probabilities are the proportions for all households in the country and so give the probabilities that a single household chosen at random will have each size. (The very few households with more than 7 members are placed in the 7 group.) Check that this is a legitimate assignment of probabilities by verifying that properties

A and B of Section 1 are satisfied. Then find the expected size of a randomly chosen household.

7.36.

(a) What is the expected number of female offspring produced by a female Asian stochastic beetle? (See Example 12 in Section 2 for this insect's reproductive pattern.)

(b) What is the expected number of female offspring produced by a female benign boiler beetle? (See Exercise 7.30 in Section 2 for the reproductive pattern.)

(c) Use the law of large numbers to explain why the population should grow if the expected number of female offspring is greater than 1 and die out if this expected value is less than 1. Do your expected values in parts (a) and (b) confirm the results of the simulations of these populations done in Section 2?

7.37. A life insurance company sells a term insurance policy to a 21-year-old male that pays $20,000 if the insured dies within the next five years. The probability that a randomly chosen male will die each year can be found in mortality tables. The company collects a premium of $100 each year as payment for the insurance. The amount that the company earns on this policy is $100 per year, less the $20,000 that it must pay if the insured dies. Here are the probabilities for the possible amounts the company can earn. Calculate the expected value of the earnings.

Age at death	21	22	23	24	25	≥26
Payout	−$19,900	−$19,800	−$19,700	−$19,600	−$19,500	$500
Probability	0.0018	0.0019	0.0019	0.0019	0.0019	0.9906

7.38. It would be quite risky for you to insure the life of a 21-year-old friend under the terms of the previous exercise. With high probability your friend would live and you would gain $500 in premiums. But if he were to die, you would lose almost $20,000. Explain carefully why selling insurance is not risky for an insurance company that insures many thousands of 21-year-old men.

7.39. We play a game by reading a pair of random digits from Table A. If the two digits are the same (for example, 2 and 2), you win $10. If they differ by 1 (for example, 1 and 0, 7 and 8, or 9 and 0 — note that we say 0 and 9 differ by one), you win $5. Otherwise, you lose $3. What is your expected outcome in this game?

7.40. Green Mountain Numbers, the Vermont state lottery, offers a choice of several bets. In each case, the player chooses a three-digit number. The lottery commission announces the winning number, chosen at random, at the end of the day. Find the expected winnings for a $1 bet for each of the following options in this lottery.

  (a) The "triple" pays $500 if your chosen number exactly matches the winning number.

  (b) The "box" pays $83.33 if your chosen number has the same digits as the winning number, in any order. (Assume that you chose a number having three distinct digits.)

7.41. A psychic runs the following ad in a magazine:

Expecting a baby? Renowned psychic will tell you the sex of the unborn child from any photograph of the mother. Cost $10. Moneyback guarantee.

This may be a profitable con game. Suppose that the psychic simply replies "girl" to each inquiry. In the worst case, everyone who has a boy will ask for her money back. Find the expected value of the psychic's profit by filling in the table below.

Sex of child	Probability	Profit in this case
Male		
Female		

7.42. Use the probability distribution of prizes for the Pennsylvania State Lottery (Exercise 7.28 in Section 2) to compute the expected value of the winnings from one ticket in this lottery.

7.43. Simulate the offspring (one generation only) of 100 female Asian stochastic beetles. What is your estimate of the expected number of offspring of one such beetle, based on this simulation? Compare the simulated value with the exact expected value you found in Exercise 7.36(a). Explain how your results illustrate the law of large numbers.

7.44. Section 2 opened with a discussion of a couple who plan to have children until they have either a girl or four children. What is the expected number of children that such a couple will have? (We don't know enough mathematics to find this expected value from the definition, so we must use simulation. Do the simulation as outlined in Section 2, and make 30 repetitions.)

7.45. OK, friends, I've got a little deal for you. We have a fair coin (heads and tails each have probability 1/2). Toss it twice. If two heads come up, you win right there. If you get any result other than two heads, I'll give you another chance: Toss the coin twice more, and if you get two heads you win. (Of course, if you fail to get two heads on the second try, I win.) Pay me a dollar to play. If you win I'll give you your dollar back plus another dollar.

   (a) Explain how to simulate one play of this game. Start by simulating two tosses of a fair coin.

   (b) You have only two possible monetary outcomes, −$1 if you lose and $1 if you win. Simulate 50 plays, using Table A starting at line 125. Use your simulation to estimate the expected value of the game.

7.46. If your state has a lottery, find out what percent of the money bet is returned to the bettors in the form of prizes. What percent of the money bet is used by the state to pay lottery expenses, and what percent is net revenue to the state that can be used for other purposes?

7.47. Write a brief essay giving arguments for and against state-run lotteries as a means of financing state government. Conclude the essay by explaining why you support or oppose such lotteries.

NOTES

1. More detail can be found in the opening chapters of Florence N. David, *Games, Gods and Gambling* (London: Griffin, 1962). The historical information given here comes from this excellent and entertaining book.

2. Stochastic beetles are well known in the folklore of simulation, if not in entomology. They are said to be the invention of Arthur Engle of the School Mathematics Study Group.

3. Quoted in an article by Frank J. Prial in the *New York Times* of February 17, 1976.

4. Quoted in an Associated Press dispatch appearing in the *Lafayette Journal and Courier* of March 7, 1976.

Review exercises

7.48. In the game of *heads or tails*, Betty and Bob toss a coin four times. Betty wins a dollar from Bob for each head, and pays Bob a dollar for each tail. That is, Betty wins or loses the difference between the number of heads and the number of tails. For example, if there are 1 head and 3 tails, Betty loses $2. You can check that Betty's possible outcomes are

$$-4, -2, 0, 2, 4$$

Assign probabilities to these outcomes by playing the game 20 times and using the relative frequencies of the outcomes as estimates of the probabilities. If possible, combine your trials with those of other students to obtain long-run relative frequencies that are closer to the probabilities.

7.49. All human blood can be typed as one of O, A, B, or AB, but the distribution of the types varies a bit with race. Here is the distribution of the blood type of a randomly chosen black American. If this is to be a legitimate assignment of probabilities, what must be the probability of type AB blood?

Blood type	O	A	B	AB
Probability	0.49	0.27	0.20	?

7.50. A bridge deck contains 52 cards, 4 of each of the 13 face values — ace, king, queen, jack, ten, nine, . . ., two. You deal a single card from such a deck and record the face value of the card dealt. Give an assignment of probabilities to these outcomes that should be correct if the deck is thoroughly shuffled. Give a second assignment of probabilities that is legitimate (that is, has properties A and B) but differs from your first choice. Then give a third assignment of probabilities that is *not* legitimate, and explain what is wrong with this choice.

7.51. Las Vegas Zeke, when asked to predict the Atlantic Coast Conference basketball champion, follows the modern practice of giving probabilistic predictions. He says, "North Carolina's probability of winning is twice Duke's. North Carolina State and Virginia each have probability 0.1 of winning, but Duke's probability is three times that. Nobody else has a chance." Has Zeke given a legitimate assignment of probabilities to the eight teams in the conference?

7.52. A study selected a sample of fifth-grade pupils and recorded how many years of school they eventually completed. Based on this study we can give the following assignment of probabilities for the years of school that will be completed by a randomly chosen fifth grader.

Years	4	5	6	7	8	9	10	11	12
Probability	0.010	0.007	0.007	0.013	0.032	0.068	0.070	0.041	0.752

(a) Verify that this is a legitimate assignment of probabilities.
(b) What outcomes make up the event "the student completed at least one year of high school"? (High school begins with the ninth grade.) What is the probability of this event?
(c) What is the expected number of years of school completed?

7.53. Sociologists have studied how children do or do not move out of their parents' occupational class. The overall result can be expressed in terms of probabilities. If a father has a white-collar occupation, here are the probabilities for the occupational class of his adult son.

professional	0.2
white collar	0.5
blue collar	0.2
no steady occupation	0.1

(a) Explain how to simulate the occupational class of a randomly selected son of a white-collar father.
(b) If we follow five sons of white-collar fathers, and these sons progress independently of each other, what is the probability that at least two of the five will have professional occupations? Use Table A, beginning at line 101, to simulate 50 repetitions of the five sons' occupations and estimate this probability.

7.54. A grocery chain runs a prize game based on giving each customer a ticket that may award a prize when a box on the ticket is scratched. The chain is required to reveal the probabilities of winning. For customers who visit the store once each week during the game, these probabilities are:

Amount won	Probability
$250	0.01
$100	0.05
$10	0.25

(a) What is the probability of winning nothing?
(b) What is the expected value of the amount won?
(c) Describe how to simulate the winnings of a single customer

who visits the store once each week and so has the stated probabilities.

(d) Three friends agree to pool their winnings. Explain how you would simulate their three outcomes.

(e) Simulate 50 repetitions of the three friends' winnings, using Table A starting at line 140. From your results, estimate their expected total winnings and also the probability that they will win at least $100.

7.55. A famous example in probability theory shows that the probability that at least two people in a room have the same birthday is already greater than 1/2 when 23 people are in the room. The probability model for this situation is:

(a) The birth date of a randomly chosen person is equally likely to be any of the 365 dates of the year.

(b) The birth dates of different people in the room are independent.

Explain carefully how you would simulate the birth dates of 23 people to see if any two have the same birthday. Do the simulation *once*, using line 139 of Table A. (*Comment*: This simulation is most easily done by letting three-digit groups stand for the birth dates of successive people, so that 001 is January 1 and 365 is December 31. Ignore leap years. Some groups must be skipped in doing this. The simulation is too lengthy to ask you to repeat it many times, but in principle you can find the probability of matching birthdays by routine repetition. The birthday problem is too hard for most of your math-major friends to solve, so it shows the power of simulation.)

7.56. Example 5 in Section 1 gives a probability model for rolling two casino dice and recording the number of spots on each of the two up faces. Suppose that instead we just count the spots on the two up faces and record the sum. The possible outcomes are now 2, 3, 4, . . ., 12. Follow the method of Example 5 to find the probability model for this random phenomenon. Then use the probabilities to find the expected value.

7.57. Keno is a common casino game. In Keno, 20 numbers between 1 and 80 are chosen at random and gamblers attempt to guess some of the numbers in advance. A bewildering variety of Keno bets are available. Here are two of the simpler bets. Give the expected winnings for each.

(a) A $1 bet on "Mark 1 number" pays $3 if the single number you mark is one of the 20 chosen; otherwise you lose your dollar.

(b) A $1 bet on "Mark 2 numbers" pays $12 if both your numbers are among the 20 chosen. The probability of this is about 0.06. Is Mark 2 a more or a less favorable bet than Mark 1?

7.58. The behavior of common stock prices is often described by probability models. The stock of Random Enterprises behaves as follows: Each day it either gains $1 or loses $1, independent of earlier price movements. Mr. Smart buys a share for $10; he plans to sell as soon as it goes up to $11. After five days, he must sell anyway because he needs to use the money. Mr. Smart can gain $1 or lose as much as $5, depending on the stock's price changes over the five-day period. Describe how to simulate his gain or loss, then simulate 50 repetitions. Based on your simulation, estimate the probability that Mr. Smart will finish with a gain. Also estimate the expected value of his gain. (Take losses to be negative gains; for example, a loss of $2 is a gain of −$2.)

Chapter 8

Formal statistical reasoning

How shall we come to sound conclusions from empirical evidence? How shall we decide, for example, whether large doses of vitamin C reduce the incidence of colds and flu? Or how shall we decide whether the 1970 draft lottery was biased in favor of men born early in the year? It may happen that the evidence is so clear that no reasonable person would argue with our conclusion. If in a randomized and controlled experiment 80% of the placebo group caught colds during the winter while only 30% of the vitamin C group did so, who would hesitate to recommend vitamin C? If in 1970 the 31 birth dates in December had all got draft numbers less than 50, the unfairness of the lottery would have been obvious. So it may happen that simply describing the data by graphs or descriptive statistics points to the conclusion. But it happens at least as often that the data only hint at the proper conclusion. In a large experiment on the effect of vitamin C (Example 7 on page 88), 18% of the placebo group and 26% of the vitamin C group were free of illness. Is this good evidence that vitamin C prevents colds and flu? Or is the difference (26% versus 18%) so small that we should instead conclude that vitamin C is not noticeably more effective than a placebo? Common sense is not enough to answer such questions. We need systematic methods for drawing conclusions from data. Such methods make up the subject of *statistical inference*.

Drawing conclusions in mathematics is a matter of starting from a hypothesis and using accepted methods of logical argument to prove without doubt that the conclusion follows. Empirical science argues in almost the reverse order. If vitamin C prevents colds, we would expect the vitamin C group to have fewer colds than the placebo group; the vitamin C group did have fewer colds, so this is evidence in favor of vitamin C's effect. This is an *inductive argument* (from consequences back to a hypothesis) as opposed to the *deductive arguments* of mathematics (from hypothesis to consequences). Inductive arguments do not produce proof. The good health of the vitamin C group *might* be due to something other than the vitamin C they took. You have no doubt heard that "statistics cannot prove anything." True enough. Neither can any other kind of inductive argument. Outside mathematics there is no proof. But inductive arguments can be quite convincing, and statistical arguments are sometimes among the most convincing.

In Chapter 7 I sidestepped the heated philosophical argument over "What is probability?" by explaining probability theory simply as the vocabulary used to describe the observed phenomenon of randomness. Now we again face an area of controversy—controversy that this time we cannot quite ignore. Since inductive arguments seem in general harder to grasp than deductive arguments, it is not surprising that statisticians disagree over the proper kind of reasoning for drawing conclusions from data. I have chosen to give most of this chapter over to the kinds of statistical reasoning most favored by users of statistics. Even theoreticians who believe that the users have poor taste cannot fault me for a choice solidly based on such empirical grounds. The final section of the chapter will bring to light some of the controversies surrounding statistical inference by introducing a way of thinking about the subject quite different from those appearing in earlier sections.

One principle is agreed upon by all sides in the discussion about statistical inference: *Formal statistical reasoning is based on the laws of probability.* The views of statistics presented in this chapter are based on the approach to probability emphasized in Chapter 7. Probability there was long-term relative frequency.* So formal statistical reasoning is based on considering what would happen in many repetitions of the experiment or survey. This entire chapter is a working out of that idea.

Here is a distinction that you first met in Chapter 1 and that you absolutely must keep in mind when thinking about statistical inference.

A *parameter* is a number describing the population. For example, the proportion of the population with some special property is a

---

*Another school of thought holds that statistical reasoning should begin with the personal or subjective idea of probability as a personal assessment of chance. Of this I will say nothing.

parameter that we call p. In a statistical inference problem, the population parameters are fixed numbers, but we do not know their values.

A *statistic* is a number describing the sample data. For example, the proportion of the sample with some special property is a statistic that we call $\hat{p}$. Statistics change from sample to sample. We use the observed statistics to get information about the unknown parameters.

1.   Estimating with confidence

Senator Bean wants very much to know what fraction of the voters in his state plan to vote for him in the election, now only a month away. (This unknown proportion of the population of voters is a *parameter p.*) He therefore commissions a poll. Being rich, he can afford a genuine SRS of 1000 registered voters. Of these voters, 570 say that they plan to vote for Bean. That's a reassuring 57% of those polled. This observed proportion of the sample is a *statistic*, $\hat{p} = 0.57$. But wait. Bean and his pollster know very well that a different sample of 1000 voters would no doubt have produced a different response—perhaps 59% or 55%. Or perhaps even 51% or (horrors) 49%. How should the senator interpret the 57% sample result?

   If you advised the senator to demand a confidence statement, you may join the chorus of the wise. A confidence statement will attach a margin of error to that 57% estimate and also state how confident we can be that the true fraction of voters who favor Bean falls within the margin of error. Confidence statements are based on the distribution of values of the sample proportion $\hat{p}$ that would occur if many independent SRSs were taken from the same population. This is the *sampling distribution* of the statistic $\hat{p}$. Armed with the language of probability, we now see that the sampling distribution is the *probability distribution* of $\hat{p}$. It gives the probability of each possible outcome for $\hat{p}$.

   A probability distribution is interpreted exactly like a relative frequency distribution. Instead of describing a particular set of data, it describes the distribution that would arise after many, many repetitions. For example (to return to the senator), suppose that in fact 55% of the several million voters in Bean's state plan to vote for him. What would be the pattern (probability distribution) of the sample proportion $\hat{p}$ favoring Bean in many, many independent SRSs of size 1000?

   We could discover this pattern by a long simulation. Fortunately, it is also possible to discover the pattern by mathematics, and I will tell you

"It was a numbers explosion."

the answer. The probability distribution of the sample proportion $\hat{p}$ favoring Bean in an SRS of size 1000 when 55% of the population favor Bean is very close to the normal distribution with mean 0.55 and standard deviation 0.015. If you don't believe me, go and simulate several thousand SRSs of size 1000 from such a population and make a histogram of the several thousand values of $\hat{p}$ you get. That histogram will look very much like the normal curve with mean 0.55 and standard deviation 0.015. (Because $\hat{p}$ is unbiased, you could guess that the mean of the distribution of $\hat{p}$ is 0.55 when the true p is 0.55.) We did a similar "simulation" in the bead-sampling experiment of Chapter 1, with results that appear in Figures 4-14, 4-15, and 4-16. That the probability distribution of a sample proportion $\hat{p}$ is close to normal is not news to you.

Figure 8-1 shows the probability distribution of the sample proportion $\hat{p}$ in Senator Bean's case. This distribution shares the properties of all normal curves, which were described in Section 5 of Chapter 4. Now we can use the language of probability instead of relative frequency. The 68-95-99.7 rule can be stated for Figure 8-1 as follows:

- The probability is 0.68 that an SRS of size 1000 from this population will have a $\hat{p}$ between 0.535 and 0.565. (This is true because 0.535 is one standard deviation below the mean and 0.565 is one standard deviation above).

- The probability is 0.95 that an SRS of size 1000 from this population will have a $\hat{p}$ between 0.52 and 0.58, that is, within two standard deviations of the mean.

- The probability is 0.997 (almost certainty) that an SRS of size 1000 from this population will have a $\hat{p}$ between 0.505 and 0.595.

The reasoning of confidence intervals

Senator Bean is becoming impatient. We are telling him how $\hat{p}$ behaves for a known population proportion p. He wants to know what he can say about p, the *unknown* fraction of voters who are for him, once he has taken a sample and observed $\hat{p} = 0.57$. Patience, Senator. We're coming to that. Chew on this line of argument.

A. When an SRS of size 1000 is chosen from a large population of which a proportion p favor Bean, the proportion $\hat{p}$ of the sample

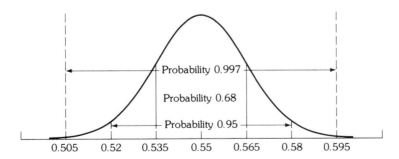

Figure 8-1. The probability distribution of a statistic. The sample proportion $\hat{p}$ of an SRS of size 1000 drawn from a population in which the population proportion is $p = 0.55$ has the normal probability distribution shown. The probability is 0.95 that $\hat{p}$ will fall within two standard deviations of p.

who favor Bean has approximately the normal distribution with mean equal to p and standard deviation 0.015*

B. By the 68-95-99.7 rule, the probability is 0.95 that $\hat{p}$ will be within 0.03 (two standard deviations) of its mean p.

C. That's exactly the same as saying that the probability is 0.95 that the unknown p is within 0.03 of the observed $\hat{p}$.

Aha! We see that 95% of all such SRSs produce a $\hat{p}$ within 0.03 of the true p. Bean's SRS had $\hat{p} = 0.57$. So he can be quite confident that the true proportion of voters who favor him lies in the interval between

$$\hat{p} - 0.03 = 0.57 - 0.03 = 0.54$$

and

$$\hat{p} + 0.03 = 0.57 + 0.03 = 0.60$$

Bean can be confident that he's favored by between 54% and 60% of the population.

Be sure you understand the ground of his confidence. There are only two possibilities. *Either* the true p lies between 0.54 and 0.60, *or* Bean's SRS was one of the few samples for which $\hat{p}$ is not within 0.03 of the true p. Only 5% of all samples give such inaccurate results, because of the 95% rule. It is possible that Bean had the bad luck to draw a sample for which $\hat{p}$ misses p by more than 0.03, but over many drawings this will happen only 5% of the time (probability 0.05). We say that Bean is "95% confident" that he is favored by between 54% and 60% of the voters.

Bean is a cautious man, and a method that will be right 95% of the time and wrong 5% of the time is not good enough for him. Very well, let's use the 99.7% rule. The probability is 0.997 that $\hat{p}$ falls within 0.045 (three standard deviations) of its mean p. So Bean can be 99.7% confident that the true p falls between

$$\hat{p} - 0.045 = 0.57 - 0.045 = 0.525$$

and
$$\hat{p} + 0.045 = 0.57 + 0.045 = 0.615$$

---

*Actually, the standard deviation changes when p changes. But for p anywhere between $p = 0.3$ and $p = 0.7$, the standard deviation of $\hat{p}$ is within 0.001 of 0.015. You can find more detailed information in Section 2.

Now he's smiling. A method that is correct 997 times in 1000 in the long run (probability 0.997) estimates that he is safely ahead.

You have just followed one of the most common lines of reasoning in formal statistical inference. Here it is in general terms.

> **A.** **A statistic computed from a sample survey or a randomized experiment has a probability distribution (a regular long-term pattern of outcomes) because of the randomization used to collect the data.**
>
> **B.** **This probability distribution changes when the population parameter changes. That is, the behavior of the sample statistic reflects the truth about the population.**
>
> **C.** **Knowing this probability distribution, we can give a recipe for finding from the sample statistic an interval that has a specified probability of covering the unknown true parameter value. The interval is called a *confidence interval* and the coverage probability is called the *confidence level*.**

Step C is the definition of a confidence interval.* A confidence interval has two parts: A recipe for finding the interval from sample data and the confidence level, which is the probability that the recipe will produce an interval that really does contain the true parameter value. Steps A and B remind us that confidence intervals are based on a knowledge of the probability distributions of sample statistics. This knowledge is obtained by mathematics and provides recipes for confidence intervals in many different settings. We saw two such recipes, and some of their background, in the case of Senator Bean. In our new vocabulary, if an SRS of size 1000 gives a sample proportion $\hat{p}$, we can make the following statements:

- A 95% confidence interval for the population proportion p is the interval from $\hat{p} - 0.03$ to $\hat{p} + 0.03$.

- A 99.7% confidence interval for the population proportion p is the interval from $\hat{p} - 0.045$ to $\hat{p} + 0.045$.

The key idea of a 95% confidence interval is that the recipe gives a correct answer (an interval that covers the true parameter value) 95% of

---

*Confidence intervals as a systematic method were invented in 1937 by Jerzy Neyman (1894–1981). Neyman was a Pole who moved to England in 1934 and spent the last half of his long life at the University of California at Berkeley. He remained active until his death, almost doubling the list of his publications after his official retirement in 1961.

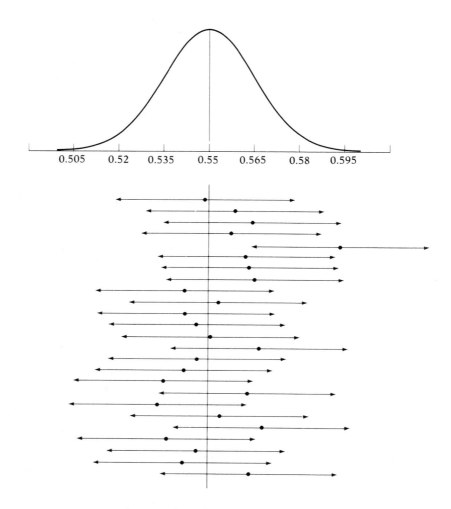

Figure 8-2. Behavior of confidence intervals in repeated sampling. The intervals above are 95% confidence intervals for p computed from 25 independent SRSs of size 1000 drawn from a population in which $p = 0.55$. The intervals vary from sample to sample, but all except one contain the true value of p.

the time in the long run. A 90% confidence interval is right 90% of the time, and a 99.7% confidence interval is right 99.7% of the time. Figure 8-2 shows the results of simulating 25 trials of the 95% confidence interval $\hat{p} \pm 0.03$ based on an SRS of size 1000 from a population with $p = 0.55$. The normal curve at the top of the figure is the probability distribution of the sample proportion $\hat{p}$, in this case having a mean equal to 0.55 (the true value of p) and standard deviation 0.015. Below are the confidence intervals resulting from 25 SRSs. The dot in the center of each interval is the observed value of $\hat{p}$ for that sample. The intervals shift from sample to sample, but only one of the 25 fails to cover the true p, represented by the vertical line. In the long run, the interval from $\hat{p} - 0.03$ to $\hat{p} + 0.03$ would cover the true p in 95% of all the samples drawn.

Some fine points

The statement that we are 95% confident that between 54% and 60% of the voters favor Senator Bean is shorthand for "We got these numbers by a method that gives correct results 95% of the time." That's the essential idea. As always, some fine points should be pondered. To these we now turn.

What confidence does not mean. For Senator Bean's poll, a method with probability 95% of being right estimated that he was favored by between 0.54 and 0.60 of the voters in this state. Be careful: We *cannot* say that the probability is 95% that the true p falls between 0.54 and 0.60. It either does or does not; we don't know which. No randomness is left after we draw a particular sample and get from it a particular interval, 0.54 to 0.60. So it makes no sense to give a probability. All we can say is that the interval 0.54 to 0.60 was obtained by a method that covers the true p in 95% of all possible samples. The language of statistical inference uses this fact about what would happen in the long run to express our confidence in the results of a single sample.

High confidence is not free. Why would anyone use a 95% confidence interval when 99.7% confidence is available? Look again at Senator Bean. His 95% confidence interval was $\hat{p} \pm 0.03$, while 99.7% confidence required a wider interval, $\hat{p} \pm 0.045$. It is generally true that there is a trade-off between the confidence level and the width of the interval. To obtain higher confidence from the same sample, you must be willing to accept a larger margin of error (a wider interval). The way to get higher confidence and still have a short interval is to take a larger sample. The precision of a sample statistic increases as the sample size increases. That means that for a fixed level of confidence, the confidence interval grows ever shorter as the sample size increases. Or, if you prefer, the confidence level for an interval of the same length grows ever higher as the sample size increases.

It's a poor cook who uses the same recipe in every meal. The Gallup Poll's probability sampling method is such that when a sample of size

1000 is taken, we can be 95% confident that the announced sample proportion is within 4 points of the true population proportion. (Table 1-1 on page 34 gives this information in more detail. That table shows that the confidence statement I just made holds whenever the true population p falls between 0.3 and 0.7. Otherwise the margin of error is even smaller.) In more formal language, a 95% confidence interval for the population proportion p based on a sample size 1000 drawn by Gallup's probability sampling method is the interval from $\hat{p} - 0.04$ to $\hat{p} + 0.04$.

Now the 95% confidence interval for an SRS of size 1000 was $\hat{p} - 0.03$ to $\hat{p} + 0.03$. That recipe is wrong for the Gallup Poll because Gallup does not use an SRS. The recipe for a confidence interval depends on how the data were collected. Section 2 gives some more-detailed recipes for use when you have an SRS. To use them when the data are not an SRS is tantamount to pouring catsup into your egg-drop soup.

You might notice that Gallup's sampling method is less precise (has a wider 95% confidence interval) than an SRS of the same size. That is the price paid for the convenience of Gallup's multistage sampling design.

Confidence intervals are used whenever statistical methods are applied, and some of the recipes are complicated indeed. But the idea of 95% confidence is always the same: The recipe employed catches the true parameter value 95% of the time when used repeatedly.

Section 1 exercises

8.1. The report of a sample survey of 1500 adults says, "With 95% confidence, between 27% and 33% of all American adults believe that drugs are the most serious problem facing our nation's public schools." Explain to someone who knows no statistics what the phrase "95% confidence" means in this report.

8.2. The last closely contested presidential election pitted Jimmy Carter against Gerald Ford in 1976. A poll taken immediately before the 1976 election showed that 51% of the sample intended to vote for Carter. The polling organization announced that they were 95% confident that the sample result was within ±2 points of the true percent of all voters who favored Carter.

    (a) Explain in plain language to someone who knows no statistics what "95% confident" means in this announcement.

    (b) The poll showed Carter leading. Yet the polling organization said the election was too close to call. Explain why.

8.3. Suppose that Senator Bean's SRS of size 1000 produced 520 voters who plan to vote for Bean.

(a) What is the 95% confidence interval for the population proportion p who plan to vote for Bean?

(b) What is the 99.7% confidence interval for p?

(c) Based on the sample results, can Bean be confident that he is leading the race?

8.4. An SRS of 1000 graduates of a university showed that 54% earned at least $40,000 a year. Give a 95% confidence interval for the proportion of all graduates of the university who earn at least $40,000 a year.

8.5. Table 1-1 on page 34 is a table of margins of error for samples of several sizes drawn by the Gallup Poll's probability sampling procedure. We now can use that table to give 95% confidence intervals.

A Gallup Poll survey of 1028 adults finds that 678 favor a constitutional amendment that would permit organized prayer in public schools. Give a 95% confidence interval for the proportion of all American adults who support such an amendment.

8.6. A Gallup Poll survey of 1540 adults finds that 15% jog regularly. Use Table 1-1 on page 34 to give a 95% confidence interval for the proportion p of all adults who jog regularly.

8.7. If Senator Bean took an SRS of only 25 voters, the probability distribution of the sample proportion $\hat{p}$ who favor him would be (very roughly) normal with mean equal to the population proportion p and standard deviation 0.1.

(a) What property of this distribution shows that $\hat{p}$ is an unbiased estimate of p?

(b) A sample of size 25 gives less precision than a sample of size 1000. How is this reflected in the distributions of $\hat{p}$ for the two sample sizes?

(c) For the SRS of size 25, what is the number m such that the interval from $\hat{p} - m$ to $\hat{p} + m$ is a 95% confidence interval for p? Explain your answer.

(d) If Senator Bean found 17 of 25 voters in an SRS favoring him, give a 95% confidence interval for the proportion of all voters who favor Bean.

8.8. In Senator Bean's SRS of 1000 voters, what is the number m such that the interval from $\hat{p} - m$ to $\hat{p} + m$ is a 68% confidence interval for p? Explain your answer.

8.9. We are going to simulate the performance of a 68% confidence interval for p based on an SRS of size 25.

(a) Using the information given in Exercise 8.7, explain why the interval from $\hat{p} - 0.1$ to $\hat{p} + 0.1$ is a 68% confidence interval for p.

(b) Suppose that in fact $p = 0.6$ (that is, 60% of all the voters favor Bean). Explain how to use Table A to simulate drawing an SRS of size 25.

(c) Starting at line 101 of Table A, simulate drawing 10 SRSs of size 25 from this population. For each sample, compute the sample proportion $\hat{p}$ who favor Bean. Then compute the 68% confidence interval $\hat{p} - 0.1$ to $\hat{p} + 0.1$ for each sample. How many of the intervals covered the true $p = 0.6$? How many failed to cover p?

8.10. The recipes for confidence intervals depend on the probability distribution of the sample statistic used. In the text we considered only $\hat{p}$, the sample proportion. Here is a fact about the probability distribution of the sample mean $\bar{x}$.

If an SRS of size 100 is chosen from a population that has standard deviation 1 but unknown mean μ, the sample mean $\bar{x}$ has approximately the normal distribution with mean μ and standard deviation 0.1.

From this fact, follow the argument given in the text for $\hat{p}$ step by step to derive the recipe for a 95% confidence interval for the unknown population mean μ.

8.11. A scientist measures the length in microns of a small object. (A micron is one-millionth of a meter.) Her measuring procedure is not perfectly accurate. In fact, repeated measurements follow the normal distribution with mean equal to the unknown true length μ (the measurement process is unbiased). The standard deviation of the distribution of measurements is 1 micron. (This is a statement of the reliability of the measuring process.)

(a) The results of 100 independent measurements can be thought of as an SRS of size 100 from the population of all possible measurements. If the mean of 100 measurements of the object is $\bar{x} = 21.3$ microns, use the result of Exercise 8.10 to give a 95% confidence interval for the true length μ.

(b) Suppose that only one measurement were taken and gave the result $x = 21.3$ microns. Give a 95% confidence interval for μ based on this single observation.

8.12. On hearing of the poll mentioned in Exercise 8.2, a nervous politician asked, "What is the probability that over half the voters prefer Carter?" A statistician said in reply that this question not only can't be answered from the poll results, it doesn't even make sense to talk about such a probability. Can you explain why?

2.   Confidence intervals for proportions and means*

Although the idea of a confidence interval remains ever the same, specific recipes vary greatly. The form of a confidence interval depends first on the parameter you wish to estimate — a population proportion or mean or median or whatever. The second influence is the design of the sample or experiment; estimating a population proportion from a stratified sample requires a different recipe than if the data come from an SRS. The sampling design and the parameter to be estimated usually determine the form of the confidence interval. The final details depend on the sample size and the confidence level you choose. The two recipes in this section are quite useful, but in comparison with the statistician's full array of confidence intervals for all occasions, these two resemble a tool kit containing only a chisel and a roofing square. These are useful tools, but only sometimes.

Confidence intervals for a population proportion

When a large SRS is selected from a population, the sampling distribution of the sample proportion $\hat{p}$ is close to a normal distribution. This normal sampling distribution has mean equal to the population proportion p because $\hat{p}$ is unbiased as an estimator of p. When the sample size n is 1000 and p is between 0.3 and 0.7, the standard deviation of $\hat{p}$ is close to 0.015. So by the 68-95-99.7 rule, $\hat{p} \pm (2)(0.015)$ is a 95% confidence interval for p. The same reasoning leads to the conclusion that whenever we know the standard deviation of the sampling distribution of $\hat{p}$, a 95% confidence interval for p is

$$\hat{p} \pm (2)(\text{standard deviation of } \hat{p})$$

because 95% of the probability in the normal distribution of $\hat{p}$ falls within two standard deviations of the mean p.

   By mathematics we can discover the standard deviation of the normal sampling distribution of $\hat{p}$. Here is the full story.

Sampling distribution of a sample proportion

Suppose that an SRS of size n is drawn from a population in which the proportion p of the units have some special property. The

*This section is optional. It contains material more technical than the rest of the book.

proportion of units in the sample having this property is the statistic $\hat{p}$. When n is large, the sampling distribution of $\hat{p}$ is approximately normal with mean p and standard deviation

$$\sqrt{\frac{p(1-p)}{n}}$$

Figure 8-3 illustrates this sampling distribution. The standard deviation of $\hat{p}$ depends on the true p and on the sample size n. For example, when $n = 1000$ and $p = 0.5$, the standard deviation of $\hat{p}$ is

$$\sqrt{\frac{(0.5)\,(0.5)}{1000}} = (0.00025)^{1/2} = 0.0158$$

And when $p = 0.3$ (or 0.7), the standard deviation is

$$\sqrt{\frac{(0.3)\,(0.7)}{1000}} = (0.00021)^{1/2} = 0.0145$$

These more exact results lie behind the statement in Section 1 that for p between 0.3 and 0.7, the standard error of $\hat{p}$ in an SRS of size 1000 is close to 0.015.

We now know that a more accurate recipe for a 95% confidence interval for p, taking p and n into account, is

$$\hat{p} \pm 2\sqrt{\frac{p(1-p)}{n}}$$

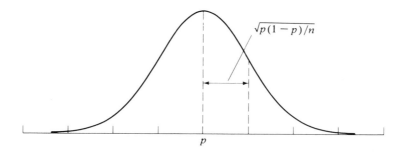

Figure 8-3. The distribution of $\hat{p}$ in a large SRS. The probability distribution of $\hat{p}$ is approximately normal with mean equal to the population proportion p. The standard deviation is $\sqrt{p(1-p)/n}$, which decreases as larger sample sizes n are chosen.

But this formula is unusable because we don't know p. (If we did know p, we would not need to settle for 95% confidence!) When n is large, $\hat{p}$ is quite close to p. So at the cost of further approximation, the estimated standard deviation formed by replacing p by $\hat{p}$ can be employed in the recipe. So the final version of a 95% confidence interval for p is

$$\hat{p} \pm 2\sqrt{\frac{\hat{p}(1 - \hat{p})}{n}}$$

All this started with the fact that in any normal distribution there is probability 0.95 within two standard deviations of the mean. What if we want a 90% confidence interval or a 75% confidence interval? For any probability C between 0 and 1 ($0 < C < 1$), there is a number z^* such that any normal distribution has probability C within z^* standard deviations of the mean. Figure 8-4 illustrates the situation, and Table C-1 in the back of the book lists the numbers z^* for various choices of C. These numbers are often called *critical values* for the normal distributions. For example, any normal distribution has probability 0.90 within ± 1.64 standard deviations of its mean. And any normal distribution has probability 0.95 within ± 1.96 standard deviations of its mean. (The number $z^* = 1.96$ is rounded off to 2 in the 68-95-99.7 rule.)

Now for the final assault on p. With probability C, the sample proportion $\hat{p}$ is within z^* standard deviations of p. Otherwise said, the unknown p is within z^* standard deviations of the observed $\hat{p}$ with probability C.

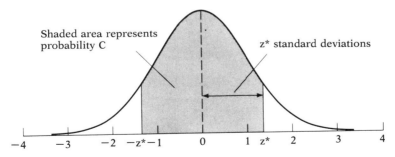

Figure 8-4. Critical points of the normal distributions. The critical point z^* is the number such that any normal distribution assigns probability C to the interval from z^* standard deviations below the mean to z^* standard deviations above it.

Using the estimated standard deviation of $\hat{p}$ produces at last the following recipe.

Confidence interval for a population proportion

Suppose that an SRS of size n is drawn from a population of units of which the proportion p have some special characteristic. When n is large, an approximate level C confidence interval for p is

$$\hat{p} \pm z^* \sqrt{\frac{\hat{p}(1-\hat{p})}{n}}$$

where z^* is the two-sided critical value for C.

Please note that this recipe is valid only when an SRS is chosen. Even then it is only approximately correct, for two reasons. First, the sampling distribution of $\hat{p}$ is only approximately normal. Second, the estimated standard deviation of $\hat{p}$ is only approximately equal to the exact standard deviation $\sqrt{p(1-p)/n}$. Both of these approximations improve as the sample size n increases. For samples of size 100 and larger, the recipe given is quite accurate. It is often used for sample sizes as small as 25 or 30. If you have a small sample, or if your sample is not an SRS, please visit your friendly local statistician for advice.

Example 1. Senator Bean, our acquaintance from Section 1, took an SRS of 1000 registered voters. Of these, 570 supported the Senator in his bid for reelection. So

$$\hat{p} = \frac{570}{1000} = 0.57$$

and the estimated standard deviation of $\hat{p}$ is

$$\sqrt{\frac{\hat{p}(1-\hat{p})}{n}} = \sqrt{\frac{(0.57)(0.43)}{1000}} = 0.0157$$

A 95% confidence interval for the proportion p of all registered voters who support Bean is therefore

$$\hat{p} \pm z^* \sqrt{\frac{\hat{p}(1-\hat{p})}{n}} = 0.57 \pm (1.96)(0.0157)$$

$$= 0.57 \pm 0.03$$

or 0.54 to 0.60. This is the same result we found in Section 1. If Bean insists on 99% confidence, the interval is

$$\hat{p} \pm z^* \sqrt{\frac{\hat{p}(1-\hat{p})}{n}} = 0.57 \pm (2.58)\,(0.0157)$$

$$= 0.57 \pm 0.04$$

or 0.53 to 0.61. As usual, higher confidence exacts its price in the form of a larger margin of error.

That $\sqrt{n}$ in the recipe for our confidence interval tells us exactly how the sample size n affects the margin of error. Because n appears in the denominator, the margin of error gets smaller as n gets larger. Because it is the square root of n that appears, a sample four times larger is needed to give a margin of error half as large. Senator Bean's SRS of 1000 voters gave a margin of error of about ± 0.04 with 99% confidence; to reduce the margin of error to ± 0.02, the senator would need an SRS of 4000 voters. The square root of n increases much more slowly than n itself. This mathematical fact means that small margins of error require large samples and therefore lots of money. Sorry about that.

Confidence intervals for a population mean

Suppose now that we wish to estimate the mean of a population. To distinguish the population mean (a parameter) from the sample mean $\bar{x}$, we denote it by μ, the Greek letter mu. The unknown μ is estimated by the mean $\bar{x}$ of an SRS. Like the sample proportion $\hat{p}$, the sample mean $\bar{x}$ from a large SRS has a sampling distribution that is close to normal. Since the sample mean of an SRS is an unbiased estimator of μ, the sampling distribution of $\bar{x}$ has μ as its mean. The standard deviation of $\bar{x}$ depends on the standard deviation of the population, which is usually denoted by σ (the Greek letter sigma). By mathematics we can discover the following fact.

Sampling distribution of the sample mean

Suppose that an SRS of size n is drawn from a population having mean μ and standard deviation σ. The mean of the sample is the statistic $\bar{x}$. When n is large, the sampling distribution of $\bar{x}$ is approximately normal with mean μ and standard deviation $\sigma/\sqrt{n}$.

The standard deviation of $\bar{x}$ depends on both σ and the sample size n. We know n, but not σ. But when n is large, the sample standard deviation s is close to σ and can be used to estimate it. So the estimated standard deviation of $\bar{x}$ is $s/\sqrt{n}$. Now confidence intervals for μ can be found just as for p.

Confidence interval for a population mean

Suppose that an SRS of size n is drawn from a population of units having mean μ. When n is large, an approximate level C confidence interval for μ is

$$\bar{x} \pm z^* \frac{s}{\sqrt{n}}$$

where z^* is the two-sided critical value for probability C.

The cautions cited in estimating p apply here as well. The recipe is valid only when an SRS is drawn and the sample size n is reasonably large. And the margin of error again decreases only at a rate proportional to $\sqrt{n}$ as the sample size n increases.

Example 2. Exercise 4.17 on page 199 reports the count of coliform bacteria per milliliter in each of 100 specimens of milk. Suppose that these specimens can be considered to be an SRS of the milk sold in a certain region. Give a 90% confidence interval for the mean coliform count of all milk sold in that region.

We first compute the sample mean and standard deviation for this set of data. The results are

$$\bar{x} = 5.88 \quad \text{and} \quad s = 2.02$$

The 90% confidence interval is therefore

$$\bar{x} \pm z^* \frac{s}{\sqrt{n}} = 5.88 \pm (1.64)\frac{2.02}{\sqrt{100}}$$
$$= 5.88 \pm 0.33$$

We are 90% confident that the mean coliform count in the population falls between 5.55 and 6.21 per milliliter.

Section 2 exercises

8.13. You are the polling consultant to a member of Congress. An SRS of 500 registered voters finds that 28% name environmental problems as the most important issue facing the nation. Give a 90% confidence interval for the proportion of all voters who hold this opinion.

8.14. In the setting of the previous exercise:
- **(a)** Give a 75% confidence interval and a 99% confidence interval. Note how the confidence level affects the width of the interval.
- **(b)** Suppose that the sample result $\hat{p} = 0.28$ had come from an SRS of 100 persons or an SRS of 4000 persons. Give a 90% confidence interval in both cases. Note how the sample size affects the width of the interval.

8.15. The member of Congress you advise receives 1310 pieces of mail on pending gun control legislation. Of these, 86% oppose the legislation. Having learned from you about estimating with confidence, he asks you for an analysis of these opinions. What will you tell him?

8.16. The Forest Service is considering additional restrictions on the number of vehicles allowed to enter Yellowstone National Park. To assess public reaction, the Service asks a simple random sample of 150 visitors if they favor the proposal. Of these, 89 say "Yes." Give a 95% confidence interval for the proportion of all visitors to Yellowstone who favor the restrictions. Are you 95% confident that more than half are in favor? Explain your answer.

8.17. An agricultural extension agent is concerned about hornworms infesting the tomatoes in the home gardens of her district. She checks a sample of 128 hornworms and is pleased to find 67 of them being parasitized by wasp larvae. Assuming these 128 can be regarded as an SRS, give an 80% confidence interval for the proportion of hornworms in her district that are parasitized.

8.18. Some people use mayonnaise jars, rather than specially made canning jars, to can garden vegetables. *Organic Gardening* magazine (August 1983, p. 86) tested mayonnaise jars for canning to estimate how likely they are to break. Said the magazine, "The mayonnaise jars didn't do badly—only three out of 100 broke. Statistically this means you'd expect between 0 and 6.4% to break." Verify this statistical statement by giving a 95% confidence interval.

8.19. The member of Congress you advise knows from preliminary polls that about half the registered voters in his district favor his reelection. He wants to commission a poll that will estimate this proportion accurately. You decide to take an SRS large enough to get a 95% confidence interval with margin of error ± 0.02. How large a sample must you take? (*Hint*: The margin of error here is $\pm z^* \sqrt{p(1-p)/n}$, and p is close to 0.5. You must solve for n.)

8.20. Scores on the American College Testing (ACT) college admissions examination for the reference population used to develop the test vary normally with mean $\mu = 18$ and standard deviation $\sigma = 6$. The range of reported scores is 1 to 36.

(a) What range of scores contains the middle 95% of all students in the reference population?

(b) If the ACT scores of 25 randomly selected students are averaged, what range contains the middle 95% of the averages $\bar{x}$?

8.21. An SRS of 120 farmers in north central Indiana is selected and asked their corn yields last year. The sample mean of the replies is $\bar{x} = 125$ bushels per acre, and the sample standard deviation is $s = 11$ bushels per acre. Give a 90% confidence interval for the mean corn yield of all north central Indiana farmers.

8.22. In a randomized comparative experiment on the effect of diet on blood pressure, 54 healthy white males were divided at random into two groups. One group received a calcium supplement, the other a placebo. At the beginning of the study, the researchers measured many variables on the subjects. The paper reporting the study gives $\bar{x} = 114.9$ and $s = 9.3$ for the seated systolic blood pressure of the 27 members of the placebo group.

(a) Give a 95% confidence interval for the mean blood pressure of the population from which the subjects were recruited.

(b) The recipe you used in part (a) requires an important assumption about the 27 men who provided the data; what is this assumption?

8.23. The income of the householder was one of the items included on only 17% of the forms in the 1990 census. Suppose (alas, it is too simple to be true) that the households who answer this question are an SRS of the households in each district. In Middletown, a city of 40,000 persons, the 1990 census asked 2621 householders their 1989 income. The mean of the responses was $\bar{x} = \$23,453$ and the standard deviation was $\$8721$. Give a 99% confidence interval for the 1989 mean income of Middletown householders.

8.24. A laboratory scale is known to have errors of measurement normally distributed with mean 0 and standard deviation $\sigma = 0.0001$ gram. So repeated measurements of the same quantity are normally distributed with mean equal to the true weight and standard deviation $\sigma = 0.0001$ gram. A series of 25 weighings gives $\bar{x} = 2.32143$ grams and sample standard deviation $s = 0.00013$ gram.

    (a) Give a recipe for a level C confidence interval for the mean μ of a population having *known* standard deviation σ. Explain why you should use σ and ignore the sample standard deviation s.

    (b) Use your recipe to give a 75% confidence interval for the true weight of the quantity weighed 25 times above.

8.25. The estimated standard deviation of a statistic is often called a *standard error.* For example, the standard error of $\bar{x}$ from an SRS of size n is $s/\sqrt{n}$. When the standard error is given, you can compute confidence intervals for means and proportions from complex sample designs without knowing the formula that led to the standard error.

    A report based on the Current Population Survey gives the 1987 mean household income as \$32,144 with a standard error of \$161. Give a 99% confidence interval for the mean 1987 income of all U.S. households.

3.   Statistical significance

Confidence intervals are one of the two most common types of statistical inference. They are appropriate when our goal is to estimate a population parameter. The second common type of inference is directed at a quite different goal: To assess the evidence provided by the data in favor of a statement. An example will illustrate the reasoning used.

> **Example 3.** When the correlation coefficient between birth date (1 to 366) and draft number (1 to 366) for the 1970 draft lottery is calculated, we get $r = -0.226$. Is this correlation good evidence that the lottery was not random?
>
> *Probability Question:* Suppose for the sake of argument that the lottery were truly random. What is the probability that a random lottery would produce an r at least as far from 0 as the observed $r = -0.226$?
>
> *Answer* (from mathematics or simulation): The probability that a random draft lottery will have an r this far from 0 is less than 0.001 (one in a thousand).

Conclusion: Since an *r* as far from 0 as that observed in 1970 would almost never occur in a random lottery, we have strong evidence that the 1970 draft lottery was not random.

In a random assignment of draft numbers to birth dates, we would expect the correlation to be close to 0. The correlation for the 1970 lottery was −0.226, showing that men born later in the year tended to get lower draft numbers. This is not a large correlation. The scatterplot (Figure 8-5) shows little association. Common sense is not enough to decide if *r* = −0.226 means the lottery was not random. After all, the correlation will almost never be exactly 0, and perhaps *r* = −0.226 is within the range that a random lottery would be expected to produce. So as an aid to answering the informal question ''Is this good evidence of a

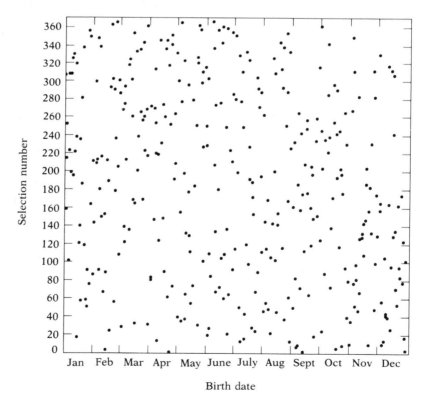

Figure 8-5. The 1970 draft lottery. The scatterplot shows the association between birth dates and draft numbers in the 1970 draft lottery. When birth dates are labeled from 1 (January 1) to 366 (December 31), their correlation with draft numbers 1 to 366 is *r* = −0.226.

nonrandom lottery?" we stated a formal question about probabilities. We asked just how often a random lottery would produce an r as far from 0 as the r observed in 1970. The answer could be obtained by simulating many repetitions of a random lottery, but I obtained it by mathematics. If a random draft lottery were run each year, a correlation as strong as that observed in 1970 would occur less than once in a thousand years! This convinces us that the 1970 lottery was biased.

Be sure you understand why this evidence is convincing. There are two possible explanations of that notorious $r = -0.226$.

(a) The lottery was random, and by bad luck a very unlikely outcome occurred.

(b) The lottery was biased, so the outcome is about what would be expected from such a lottery.

We cannot be certain that explanation (a) is untrue. The 1970 results *could* be due to chance alone. But the probability that such results would occur by chance in a random lottery is so small (0.001) that we are quite confident that explanation (b) is right. Here is a second example of this reasoning.

> **Example 4.** The Toronto vitamin C experiment (Example 7 on page 88) was a randomized double-blind experiment with about 400 subjects in each of two groups. In this study, 26% of the vitamin C group and 18% of the placebo group were free of illness through the winter. Is this good evidence that vitamin C prevents illness better than a placebo?
>
> *Probability Question:* Suppose for the sake of argument that there is no difference between the effects of vitamin C and the placebo. That is, suppose that the only difference between the two groups of volunteers is due to the random allocation in the experimental design. What is the probability of observing an outcome favoring vitamin C by 26% versus 18% or more?
>
> *Answer:* The probability that a difference this large would occur because of the random allocation alone is less than 0.01 (one in a hundred).
>
> *Conclusion:* Since a difference as large as the one actually observed would occur only one time in a hundred if vitamin C has no more effect than a placebo, there is good evidence that vitamin C is more effective than a placebo.

Tests of significance

The reasoning used in these examples is codified in *tests of significance*. In both examples, we hoped to show that an effect was present — that the draft lottery was biased in Example 3 and that vitamin C outperformed the placebo in Example 4. To do this, we began by supposing for the sake of argument that the effect we sought was *not* present. In Example 3, we supposed that the lottery was random, not biased. In Example 4, we supposed that a volunteer given vitamin C was no more likely to be free of illness than one given a placebo. We then looked for evidence against the supposition we made. Such evidence is also evidence in favor of the effect we hoped to find. The first step in tests of significance is to state a claim ("Suppose for the sake of argument that . . .") that we will try to find evidence against.

> **The statement being tested in a test of significance is called the**
> ***null hypothesis*. The test of significance is designed to assess the**
> **strength of the evidence against the null hypothesis. Usually the**
> **null hypothesis is a statement of "no difference" or "no effect."**

The term "null hypothesis" is abbreviated as H_0, which is read "H nought." It is a statement about the population and so is stated in terms of a population parameter or parameters. For example, suppose that p_1 is the proportion of the whole population of North American males who would be free of illness if they were to take a gram of vitamin C each day, and let p_2 stand for the illness-free proportion if a placebo were given instead. Then the null hypothesis of Example 4 is

$$H_0: \ p_1 = p_2$$

This hypothesis says that vitamin C has the same effectiveness as a placebo. It is useful to give a name to the statement we hope or suspect is true instead of H_0. This is called the *alternative hypothesis* and is abbreviated by H_a. In Example 4, the alternative hypothesis is that vitamin C is more effective than a placebo. In terms of population parameters this is

$$H_a: \ p_1 > p_2$$

A test of significance assesses the strength of the evidence against the null hypothesis in terms of probability. If the observed outcome is unlikely under the supposition that the null hypothesis is true, but is more probable if the alternative hypothesis is true, that outcome is evidence against H_0 in favor of H_a. The less probable the outcome is, the stronger is the evidence that H_0 is false. Now, usually any individual outcome has low probability. It is unlikely that a random draft lottery would give exactly $r = 0$, but if we observed $r = 0$, we would certainly not have evidence against the null hypothesis that the lottery is random. Our procedure is therefore to say what kinds of outcomes would count as evidence against H_0 and in favor of H_a. In the draft lottery case, correlations r away from 0 (either positive or negative) count against the hypothesis of a random lottery. The farther from 0 the observed r is, the stronger the evidence. The probability that measures the strength of the evidence that the 1970 lottery is nonrandom is therefore the probability that a random lottery would produce an r *at least as far from* 0 as the 1970 lottery did.

In general, we find the probability of getting an outcome at least as far as the actually observed outcome from what we would expect when H_0 is true. What counts as "far from what we would expect" depends on H_a as well as H_0. In Example 3, an observed r away from 0 *in either direction* is evidence of a nonrandom lottery. In Example 4, we wanted to know if vitamin C was more effective than a placebo (that's H_a). So evidence against H_0 is measured by the probability that the percent free of illness in the vitamin C group *exceeds* that percent in the placebo group by as much as 26% versus 18%.

> **The probability of getting an outcome at least as far from what we would expect if H_0 were true as was the actually observed outcome is called the *P-value*. The smaller the *P*-value is, the stronger is the evidence against H_0 provided by the data.**

Significance level

One final step is sometimes taken in assessing the evidence against H_0. We can compare the evidence we obtained with a fixed level of evidence that we regard as decisive. Because the strength of the evidence provided by the data is measured by the *P*-value, we need only say how small a *P*-value we insist on. This decisive value is called the *significance level*. It is always denoted by α, the Greek letter alpha. If we take $\alpha = 0.05$, we are requiring that the data give evidence against H_0 so strong that it would happen no more than 5% of the time (one time in twenty) when H_0 is really true. If we take $\alpha = 0.01$, we are insisting on stronger evidence against H_0, evidence so strong that it would appear only 1% of the time (one time in a hundred) if H_0 is really true. If the *P*-value is as small or smaller than α, we say that the data are *statistically significant at level α*. This is just a way of saying that the evidence against the null hypothesis reached the standard set by the level α. A common abbreviation for significance at, let us say, level 0.01 is "The results were significant ($P < 0.01$)." Here P stands for the P-value.

A recipe for testing the significance of the evidence against a null hypothesis H_0 is called a *test of significance* (or a *test of hypotheses*, but I will save that language for a slightly different kind of reasoning, which we will meet in Section 6). Courses on statistical methods teach many such recipes for different hypotheses H_0 and H_a and for different data-collection designs. An outline of what such a recipe must include appears in the box below. We are concerned not with any specific recipe but rather with the reasoning that lies behind all such recipes. You are now well-prepared to understand the meaning of conclusions stated in terms of statistical significance. "The results of the Toronto vitamin C experiment were significant at level $\alpha = 0.01$" summarizes in one sentence the long chain of reasoning we followed in discussing Example 4. It means that the experimental evidence favoring vitamin C was so strong that it would appear in less than 1% of a long series of experiments if the only difference between the experimental and control groups were due to random allocation of subjects.

Steps in a test of significance

(a) Choose the *null hypothesis H_0* and the *alternative hypothesis H_a*. The test is designed to assess the strength of the evidence against H_0. H_a is a statement of the alternative we will accept if the evidence against H_0 is sufficiently strong.

(b) (Optional) Choose the *significance level α*. This states how much evidence against H_0 we will regard as decisive.

(c) Choose the *test statistic* on which the test will be based. This is a statistic that measures how well the data conform to H_0. In Example 3, the correlation coefficient was used because in a random lottery there should be little or no correlation between birth date and draft number.

(d) Find the *P-value* for the observed data. This is the probability that the test statistic would weigh against H_0 at least as strongly as it does for these data if H_0 were in fact true. If the P-value is less than or equal to α, the test is *statistically significant at level α*.

To review and solidify this introduction to tests of significance, here is a concluding example.

> **Example 5.** Is a new method of teaching reading to first graders (method B) more effective than the method now in use (method A)? An experiment is called for. We will use a *matched-pairs experimental design:* 20 pairs of first graders are available; the two children in each pair are carefully matched on IQ, socioeconomic status, reading-readiness score, and other variables that may influence their reading performance. One student from each pair is randomly assigned to method A, while the other student in the pair is taught by method B. At the end of first grade, all the children take a test of reading skill.
>
> (a) We want to know if the new method B is superior to the old method A. Our null hypothesis is that method B and method A are equally effective. We will examine the experimental evidence against this H_0 in favor of the alternative that method B is more effective. To state H_0 and H_a in terms of parameters, and later to do the necessary simulation, we must give a probability model for the experiment. The children in each pair are (as closely as possible) exactly identical as reading students. So if the two teaching methods are equally effective, each child has the same chance to score higher on the test. Let p stand for the proportion of all possible matched pairs of children for which the child taught by method B will have the higher score. The null hypothesis is
>
> $$H_0:\ p = 1/2$$
>
> and the alternative is
>
> $$H_a:\ p > 1/2$$
>
> We can think of p as the probability that the method-B child in any one pair will do better.
>
> (b) We choose $\alpha = 0.10$. That is, we are willing to say that method B is better if it shows a superiority in the experiment that would occur no more than 10% of the time if the two methods were equally effective.
>
> (c) The test statistic is the number of pairs out of 20 in which the child taught by method B has a higher score. The larger this number is, the stronger is the evidence that method B is better.

(d) Now we take a year off to do the experiment. The result is that method B gave the higher score in 12 of the 20 pairs of students. The P-value for these data is the probability that 12 *or more* pairs out of 20 favor method B when H_0 is true. This is exactly the same as the probability of 12 or more heads in 20 tosses of a coin with heads and tails equally likely. We can find this probability by simulation as follows:

Step 1. Each random digit simulates one pair of students.

odd   method A scores higher
even   method B scores higher

Step 2. One repetition requires 20 digits representing 20 pairs of students.

Step 3. Begin simulating at line 110 of Table A, recording for each repetition how many times B scored higher.

38448	48789	18338	24697	
ABBBB	BBABA	ABAAB	BBBAA	12 Bs
39364	42006	76688	08708	
AAABB	BBBBB	ABBBB	BBABB	15 Bs
81486	69487	60513	09297	
BABBB	BABBA	BBAAA	BABAA	11 Bs
00412	71238	27649	39950	
BBBAB	AABAB	BABBA	AAAAB	10 Bs

and so on. In these four trials the relative frequency of 12 or more B's was 1/2. Many more trials are needed to estimate the probability precisely. When they are made, the probability of 12 or more B's in 20 pairs of students is 0.25. This is the P-value. The P-value did not reach the required level of significance. The experimental results were not statistically significant at the $\alpha = 0.10$ level. We do not have adequate evidence that method B is superior to method A.

Section 3 exercises

8.26. A social psychologist reports that, "In our sample, ethnocentrism was significantly higher ($P < 0.05$) among church attenders than among nonattenders." Explain to someone who knows no statistics what this means.

8.27. The financial aid office of a university asks a sample of students about their employment and earnings. The report says that, "For academic year earnings, a significant difference ($P = 0.038$) was found between the sexes, with men earning more on the average. No difference ($P = 0.476$) was found between the earnings of black and white students." Explain both of these conclusions, for the effects of sex and of race on mean earnings, in language understandable to someone who knows no statistics. (From a study by M. R. Schlatter et al., Division of Financial Aid, Purdue University.)

8.28. A sociologist asks a large sample of high school students which academic subject is their favorite. She suspects that a lower percentage of females than of males will say that mathematics is their favorite subject. State in words the sociologist's null hypothesis H_0 and alternative hypothesis H_a.

8.29. An educational researcher randomly divides sixth-grade students into two groups for gym class. He teaches both groups basketball skills with the same methods of instruction. He encourages Group A with compliments and other positive behavior but acts cool and neutral toward Group B. He hopes to show that Group A does better (on the average) than Group B on a test of basketball skills at the end of the instructional unit. State in words the researcher's null hypothesis H_0 and alternative hypothesis H_a.

8.30. A political scientist hypothesizes that among registered voters there is a negative correlation between age and the percent who actually vote. To test this, she draws a random sample from public records on registration and voting. State in words the null hypothesis H_0 and alternative hypothesis H_a.

8.31. Exercise 2.25 on page 95 describes an experiment that used a matched-pairs design like that of Example 5. Out of six pairs of plots, the experimental plot had a higher yield in five cases. We will assess the significance of this evidence against the null hypothesis that each plot in a pair is equally likely to have the higher yield. (This null hypothesis says that the experimental treatment — praying for the soybeans — has no effect on the yield.)

    (a) Supposing that the null hypothesis is true, explain how to simulate for one pair whether the experimental or control plot has higher yield. Then explain how to simulate one repetition of the experiment with six independent pairs of plots.

    (b) Simulate 20 repetitions of the experiment; begin at line 113 of Table A.

(c) The *P*-value is the probability that five or more (that is, either five or six) out of six pairs favor the experimental plot. Estimate the *P*-value from your simulation. (Of course, 20 trials will not give a precise estimate.)

(d) About how low would the *P*-value have to be for you to conclude that prayer does increase yields? That is, what level of significance would you insist on to believe the result suggested by the experiment? Were the experimental results significant at that level?

8.32. A classic experiment to detect ESP uses a shuffled deck of cards containing five suits (waves, stars, circles, squares, and crosses). As the experimenter turns over each card and concentrates on it, the subject guesses the suit of the card. A subject who lacks ESP has probability 1/5 of being right by luck on each guess. A subject who has ESP will be right more often. Julie is right in 5 of 10 tries. (Actual experiments naturally use much longer series of guesses so that weak ESP could be spotted. No one has ever been right half the time in a long experiment!)

(a) Give H_0 and H_a for a test to see if this result is significant evidence that Julie has ESP.

(b) Explain how to simulate the experiment; assume for the sake of argument that H_0 is true.

(c) Simulate 20 repetitions of the experiment; begin at line 121 of Table A.

(d) The actual experimental result was 5 right in 10 tries. What is the event whose probability is the *P*-value for this experimental result? Give a (not very precise) estimate of the *P*-value based on your simulation. How convincing was Julie's performance?

8.33. An old farmer claims to be able to detect the presence of water with a bent stick. To test this claim, he is presented with five identical barrels, some containing water and some not. He is correct in four out of five cases. Assess the strength of this result as evidence of the farmer's special ability. (You must formulate the hypotheses and do a simulation to estimate the *P*-value.)

8.34. Read the article by Sandy L. Zabell, "Statistical proof of employment discrimination" in *Statistics: A Guide to the Unknown*. State in words the null and alternative hypotheses in the Hazelwood discrimination case. What was the *P*-value for the significance test used in this court case?

8.35. A study on predicting job performance reports that

An important predictor variable for later job performance was the score X on a screening test given to potential employees.

The variable being predicted was the employee's score Y on an evaluation made after a year on the job. In a sample of 70 employees the correlation between X and Y was $r = 0.4$, which is statistically significant at the 1% level.

(a) Explain to someone who knows no statistics what information "$r = 0.4$" carries about the connection between screening test and later evaluation score.

(b) The null hypothesis in the test reported above is that there is *no association* between X and Y when the population of all employees is considered. What value of the correlation for the entire population does this null hypothesis correspond to?

(c) In order to use the screening test, the employer must show that screening scores are positively associated with on-the-job performance. What is the alternative hypothesis?

(d) Explain to someone who knows no statistics why "statistically significant at the 1% level" means there is good reason to think that there is a positive association between the two scores.

4.   Significance tests for proportions and means*

As is the case for confidence intervals, statisticians have developed tests of significance for use in many different circumstances. The recipes for carrying out these tests generally have two parts: A statistic that measures the effect being sought and a probability distribution for the statistic that gives the corresponding *P*-value. Tests are now often implemented by computer routines that compute the test statistic and its *P*-value starting from the raw data. The computer will not explain to you what statistical significance means, however. Your understanding of the reasoning common to all tests is therefore more valuable than knowledge of the specific procedures discussed in this section.

Tests for a population proportion

Inspired by Senator Bean's use of polls and statistics, his colleague Senator Caucus commissions an SRS of 1200 registered voters in her state. The poll finds that 53% of the sample plan to vote for Caucus. We

---

*This section is optional. It contains material more technical than the rest of the book.

know how to use this sample result $\hat{p} = 0.53$, to give a confidence interval for the unknown proportion p of all voters who favor Caucus. But the senator just wants to know if she's ahead. Does the sample provide strong evidence that a majority of the population plan to vote for Caucus? We wish to test the null hypothesis

$$H_0: p = 0.5$$

against the alternative hypothesis

$$H_a: p > 0.5$$

As usual, H_a states that the effect we seek (a majority for Caucus) is present, and H_0 that the effect is absent.

The statistic is $\hat{p}$. We recorded the sampling distribution of $\hat{p}$ on page 389. One point to emphasize is that for computing the P-value, we must use the sampling distribution *assuming that the null hypothesis is true*. Since H_0 states that $p = 0.5$, the distribution of $\hat{p}$ is then approximately normal with mean 0.5 and standard deviation

$$\sqrt{\frac{p(1-p)}{n}} = \sqrt{\frac{(0.5)(0.5)}{1200}} = 0.0144$$

The P-value is the probability that a statistic having this distribution will take a value at least as large as 0.53, the actual observed value of $\hat{p}$. Figure 8-6 illustrates this probability as an area under a normal curve.

To find this probability, recall again that all normal distributions are the same when measured in standard deviation units from the mean. So convert $\hat{p} = 0.53$ to a standard score:

$$z = \frac{0.53 - 0.5}{0.0144} = 2.1$$

Table B shows that a standard score of 2.1 corresponds to the 98.2 percentile. The P-value is therefore $P = 0.018$, since if 98.2% of the distribution lies below 2.1, the remaining 1.8% lies above. Senator Caucus is pleased. If the race were even, there is only probability 0.018 that the poll would show her as far ahead as it did. This is good evidence that she actually is ahead.

If a particular degree of evidence, stated as a significance level α, is required, the test procedure is simpler. Compute the standard score as before, and compare it with the number z^* such that any normal distri-

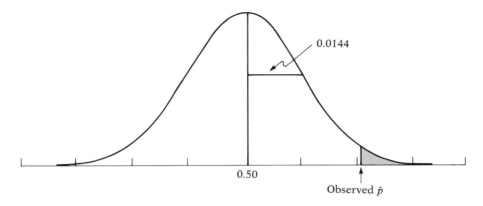

Figure 8-6. The *P*-value for Senator Caucus's poll. The normal curve is the sampling distribution of $\hat{p}$ assuming that the null hypothesis is true. The shaded area represents the *P*-value.

bution has probability α falling more than z^* standard deviations above the mean. If the standard score is greater than z^*, the observed $\hat{p}$ is statistically significant at level α. The numbers z^* are *one-sided critical values* for the normal distribution. They are not the same as the two-sided critical values used in Section 2. Table C-2 at the back of the book lists the numbers z^* for various choices of α. We can now give the following recipe.

Suppose that an SRS of size n is drawn from a population of units of which proportion p have some special characteristic. When n is large, a test of

$$H_0\!: p = p_0 \qquad \text{vs.} \qquad H_a\!: p > p_0$$

for a specified value p_0 is obtained by computing the standard score

$$z = \frac{\hat{p} - p_o}{\sqrt{\dfrac{p_o(1 - p_o)}{n}}}$$

The test result is significant at level α if $z \geq z^*$, where z^* is the one-sided critical value for α.

I recommend that you give the P-value rather than simply stating "significant at $\alpha = 0.05$." To do this, just use Table B to find the normal probability lying above z.

The same reasoning that led to this recipe shows how to test $H_0: p = p_0$ against other alternatives.

To test

$$H_0: p = p_0 \qquad \text{vs.} \qquad H_a: p < p_0$$

compute z and declare the result significant at level α if $z \le -z^*$, where z^* is the one-sided critical value for α.

To test

$$H_0: p = p_0 \qquad \text{vs.} \qquad H_a: p \ne p_0$$

compute z and declare the result significant at level α if $z \ge z^*$ or $z \le -z^*$, where z^* is the two-sided critical value for $C = 1 - \alpha$.

In all three cases, the direction of the alternative hypothesis H_a determines which deviations of z from zero count as evidence against H_0. If H_a is two-sided, so that deviations in either direction count against H_0, we use two-sided critical values. That table is arranged for confidence intervals, but entering it with $C = 1 - \alpha$ finds the z^* with probability C within z^* of the mean and therefore probability α more than z^* from the mean.

Example 6. We suspect that spinning a penny, unlike tossing it, does not give heads and tails equal probabilities. I spun a penny 200 times and got 83 heads. How significant is this evidence against equal probabilities?

We can think of this as an SRS of size $n = 200$ from the population of all spins of the penny. If p is the probability of a head, we must test

$$H_0: p = 0.5 \qquad \text{vs.} \qquad H_a: p \ne 0.5$$

since we are seeking evidence of imbalance in either direction. The sample gave $\hat{p} = 83/200 = 0.415$, so

$$z = \frac{0.415 - 0.5}{\sqrt{\dfrac{(0.5)(0.5)}{200}}} = -2.40$$

Now compare z with two-sided critical values in Table C-1. Our result is significant at $\alpha = 0.05$, because $2.40 > 1.96$, the table entry for $C = 0.95$. It is not significant at $\alpha = 0.01$, because $2.40 < 2.58$.

We now know that $0.01 < P < 0.05$. If we need the exact P-value, we must use Table B and think a bit. Figure 8-7 illustrates the calculation. The standard score -2.40 is the 0.82 percentile. So probability 0.0082 lies below -2.40 and another 0.0082 above 2.40. Adding these, $P = 0.0164$ is the probability that z takes a value farther from zero in either direction than that observed.

Tests for a population mean

The reasoning that led us to the test statistic z for a population proportion is remarkably general. It applies without change whenever a parameter is estimated by a statistic having a normal distribution with known standard deviation when the null hypothesis is true. When we wish to test a hypothesis about the mean μ of a population, the natural statistic is the sample mean $\bar{x}$. When the sample size n is large and the population has standard deviation σ, we know (see page 393) that $\bar{x}$ is approximately normal with mean μ and standard deviation $\sigma/\sqrt{n}$. This leads to tests for hypotheses about μ.

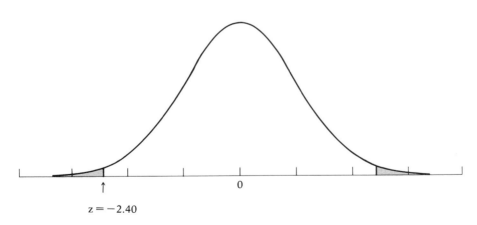

$z = -2.40$

Figure 8-7. The P-value for Example 6. The standard normal curve that is the sampling distribution of the standard score z is pictured. The shaded area represents the P-value. This area is two-tailed because H_a is two-sided.

Suppose that an **SRS** of size n is drawn from a population with unknown mean μ and known standard deviation σ. When n is large, a test of

$$H_0: \mu = \mu_0 \qquad \text{vs.} \qquad H_a: \mu > \mu_0$$

is obtained by computing the standard score

$$z = \frac{\bar{x} - \mu_0}{\sigma / \sqrt{n}}$$

The test result is significant at level α if $z \geq z^*$, where z^* is the one-sided critical value for α.

The variations for the other one-sided alternative $H_a: \mu < \mu_0$ and the two-sided alternative $H_a: \mu \neq \mu_0$ are the same as in the case of proportions. So too is the process of finding the P-value. The common notation z for a standard score emphasizes the similarity of these tests.

Example 7. A camera company buys a machined metal part from a supplier. The length of a slot in the part is supposed to be 0.550 inch. No manufacturing process is perfectly precise, so the actual lengths vary according to a distribution with $\sigma = 0.001$ inch. This standard deviation measures the precision of the supplier's work and is known from long experience. But the mean μ of the current shipment of 20,000 parts may have moved away from the desired value $\mu_0 = 0.550$ inch because of improper adjustment, tool wear, or other reasons. The camera company measures a sample of $n = 80$ parts and finds that the mean slot length is $\bar{x} = 0.5501$ inch. Is the mean of the shipment correct or not?

We must test against a two-sided alternative, since slot lengths either too short or too long are undesirable. The hypotheses are

$$H_0: \mu = 0.550 \qquad \text{vs.} \qquad H_a: \mu \neq 0.550$$

The test statistic is

$$z = \frac{0.5501 - 0.550}{0.001/\sqrt{80}} = 0.89$$

Comparing z with the two-sided critical values in Table C-1, we see that this result is significant at $\alpha = 0.40$ but not at $\alpha = 0.35$. (In fact, we can compute from Table B that $P = 0.37$.) So over 35% of all samples would have an $\bar{x}$ at least this far from 0.550 even if μ were exactly equal to 0.550. There is no suspicion that μ is incorrect, and the lot should be accepted.

If the population standard deviation σ is unknown, we can estimate it by the sample standard deviation s and replace σ by s in the formula for z. When n is large, the test remains approximately correct with this substitution. Small samples (say n less than 30) are a more complex matter. Now the distribution of z depends on the distribution of the population, and replacing σ by s changes the critical values of z. Tables B and C may not be accurate in this case.

Section 4 exercises

8.36. A preelection poll of 1500 registered voters finds that 781 would vote for Senator Caucus if the election were held today. Is this convincing evidence that a majority of all voters would vote for Caucus? Answer this question by stating H_0 and H_a and using Table B to give the P-value.

8.37. A wholesale supplier of tree and shrub seedlings advertises that 90% of its seedlings will survive a year when given proper care. A retail nursery buys 250 dogwood seedlings to grow for later sale. After a year, 207 survive. Is this significant evidence that less than 90% of the wholesaler's seedlings will survive? State H_0 and H_a and use Table C to assess significance.

8.38. The probability of rolling a 7 or 11 as the sum of the faces of two fair dice is 8/36, or 0.222. You decide to test this by watching a casino craps table. Of the first 300 rolls, 61 give 7 or 11. Is this good evidence against the hypothesis that 0.222 is the correct probability? State H_0 and H_a and use Table C to assess the significance of the result.

8.39. An environmentalist group collects a liter of water from each of 45 locations along a stream and measures the amount of dissolved oxygen in each specimen. The mean is 4.62 milligrams and the standard deviation 0.92 milligram. Is this strong evidence that the stream has a mean oxygen content of less than 5 milligrams per liter? State the hypotheses and use Table B to give the P-value.

8.40. Scores on the SAT are approximately normally distributed with standard deviation $\sigma = 100$. When the population itself has a normal distribution, our recipe for tests about the mean μ is correct even for small samples (as long as σ is known). A random sample of 20 students from a large school system is given a training course before the SAT. Their SAT math scores are as follows:

452	438	577	421
498	450	396	743
514	520	483	328
508	398	429	450
449	547	788	593

Is there evidence that the training has raised the mean above the system mean of $\mu_0 = 483$?

8.41. Packages of broccoli are supposed to weigh 12 ounces. Packages are automatically weighed at the end of the production line. For the last 50, $\bar{x} = 11.92$ ounces and $s = 0.11$ ounce. Is this convincing evidence that the mean weight being produced is not equal to the desired 12 ounces? (Use Table C.)

8.42. A maker of light bulbs knows that the lifetime in service of the bulbs is normally distributed with mean 1500 hours and standard deviation 50 hours. Use Table C to fill in the blank in this advertising claim: "Ninety-five percent of our bulbs last _____ hours or more."

8.43. The weight of tomato juice in mechanically filled cans varies from can to can according to a normal distribution with standard deviation 8 grams. This describes the precision of the filling machinery. The mean of the distribution can be set by adjusting the machine. Use Table C to answer the following questions:
  (a) A penalty is charged if more than 10% of the cans contain less than 454 grams. What mean should be set in order to have exactly 10% of the cans weigh less than 454 grams?
  (b) Suppose instead that the penalty applies to the sample average weight $\bar{x}$ of a case of 12 cans. That is, no more than 10% of all cases may have $\bar{x}$ below 454 grams. To what mean weight should the filling machine be set to produce just 10% of cases with average can weight below 454 grams? (*Hint:* What is the standard deviation of the average weight $\bar{x}$ of the 12 cans in a case?)

5. Use and abuse of tests of significance

Tests of statistical significance are routinely used to assess the results of research in agriculture, education, medicine, psychology, and sociology and increasingly in other fields as well. Any tool used routinely is often used unthinkingly. This section therefore offers some comments for the thinking researcher on the use and abuse of this tool. Thinking consumers of research findings (such as students) might also ponder these comments.

Choosing a level of significance

The spirit of a test of significance is to give a clear statement of the degree of evidence against the null hypothesis obtained from the sample. This is best done by the P-value. But sometimes some action will be taken if the evidence reaches a certain standard. Such a standard is set by giving a level of significance α. Perhaps you will announce a new scientific finding if your data are significant at the $\alpha = 0.05$ level. Or perhaps you will recommend using a new method of teaching reading if the evidence of its superiority is significant at the $\alpha = 0.01$ level. When α serves as a standard of evidence, we speak of *rejecting H_0 at level α* if the data are significant at that level.

The idea of using α as a criterion for taking some decision or action goes beyond the idea of a test of significance. But when you want to give such a rule, α is chosen by deciding how much evidence is required to reject H_0. This depends first on how plausible H_0 is. If H_0 represents an assumption that everyone in your field has believed for years, strong evidence (small α) will be needed to reject it. Second, the level of evidence required to reject H_0 depends on the consequences of such a decision. If rejecting H_0 in favor of H_a means making an expensive changeover from one medical therapy or instructional method to another, strong evidence is needed. Both the plausibility of H_0 and H_a and the consequences of any action that rejection may lead to are somewhat subjective. Different persons may feel that different levels of significance are appropriate. When this is the case, it is better to report the P-value, which leaves each of us to decide individually if the evidence is sufficiently strong.

When you really must make a decision with well-defined consequences, you should abandon the idea of testing significance and think about rules for making decisions. This approach to statistical inference is discussed in Section 6. It is different in spirit from testing significance, though the two are usually mixed in textbooks and are often mixed in practice. Choosing a level α in advance makes sense if you must make a decision, but not if you wish only to describe the strength of your evidence. In short: When a test of significance is what you want, don't set α in advance and do always report the P-value. This advice is easy to follow, since the computer programs used for most statistical arithmetic automatically print out the *P*-value. It is acceptable to use "significant at level $\alpha = 0.01$" or the shorthand "$P < 0.01$" to describe your results, but the actual *P*-value is more informative.

Textbooks commonly stress certain standard levels of significance, such as 10%, 5%, and 1%. The 5% level ($\alpha = 0.05$) is particularly common. Significance at that level is still a widely accepted criterion for meaningful evidence in research work. Now there is no sharp border between "significant" and "insignificant," only increasingly strong evidence as the *P*-value decreases. It makes no sense to treat $\alpha = 0.05$ as a universal rule for what is significant. There is a reason for the common use of $\alpha = 0.05$ — the great influence of Sir R. A. Fisher.* Fisher's ideas on statistical inference agreed with the reasoning behind tests of significance. Here is his opinion on level of significance:

> . . . *it is convenient to draw the line at about the level at which we can say: "Either there is something in the treatment, or a coincidence has occurred such as does not occur more than once in twenty trials. . . ."*
>
> *If one in twenty does not seem high enough odds, we may, if we prefer it, draw the line at one in fifty (the 2 percent point), or one in a hundred (the 1 percent point). Personally, the writer prefers to set a low standard of significance at the 5 percent point, and ignore entirely all results which fail to reach that level. A scientific fact should be regarded as experimentally established only if a properly designed experiment rarely fails to give this level of significance.[1]*

There you have it. Fisher thought 5% was about right, and who was to disagree with the master?

What statistical significance doesn't mean

When a null hypothesis ("no effect" or "no difference") can be rejected at the usual levels, $\alpha = 0.05$ or $\alpha = 0.01$, there is good evidence that an effect is present. But that effect may be extremely small. When large samples are available, statistical tests are very sensitive and will detect

---

*We met the great British statistician R. A. Fisher in Chapter 2 as the father of randomized experimental design. He was the father of much else in modern statistics as well, including the general use of regression (Chapter 5) and the mathematical derivation of the probability distributions of common test statistics. Fisher did not originate tests of significance. But since his writings organized statistics, especially as a tool of scientific research, his views on tests were enormously influential.

even tiny deviations from the null hypothesis. For example, suppose that we are testing the hypothesis of no correlation between two variables. With 1000 observations, an observed correlation of only $r = 0.08$ is significant evidence at the $\alpha = 0.01$ level that the correlation in the population is not zero but positive. The low significance level does not mean there is a strong association, only that there is strong evidence of some association. The true population correlation is probably quite close to the observed sample value, $r = 0.08$. We might well conclude that for practical purposes there is no association between these variables, even though we are confident (at the 1% level) that this is not literally true.

Remember the wise saying: *Statistical significance is not the same thing as practical significance.* I am tempted to interpret the results of the Toronto vitamin C experiment in this light. Since the observed difference (that 26% versus 18% again) was significant at the $\alpha = 0.01$ level, it does appear that vitamin C prevented colds better than a placebo. But not much better. The difference between an 18% chance of avoiding colds and a 26% chance is nothing to fuss about.

The remedy for attaching too much importance to statistical significance is to pay attention to the actual experimental results as well as to the *P*-value. It is usually wise to give a confidence interval for the population parameter you are interested in. Confidence intervals are not used as often as they should be, while tests of significance are perhaps overused.

Don't ignore lack of significance

Researchers typically have in mind the research hypothesis that some effect exists. Following the peculiar logic of tests of significance, they set up as H_0 the null hypothesis that no such effect exists and try their best to get evidence against H_0. Now a perverse legacy of Fisher's opinion on $\alpha = 0.05$ is that research in some fields has rarely been published unless significance at that level is attained. A survey of four journals of the American Psychological Association once showed that of 294 articles using statistical tests, only 8 did not attain the 5% significance level.[2]

Such a publication policy impedes the spread of knowledge. If a researcher has good reason to suspect that an effect is present and then fails to find significant evidence of it, that may be interesting news — perhaps more interesting than if evidence in favor of the effect at the 5% level had been found. If you follow the history of science, you will recall examples such as the Michelson-Morley experiment, which changed the course of physics by *not* detecting a change in the speed of light that everyone expected to find. (Of course, an experiment that fails only causes a stir if it is clear that the experiment would have detected the effect if it were really there. Such experiments are much rarer in psychology than in physics.) Keeping silent about negative results may condemn other researchers to repeat the attempt to find an effect that isn't there. Witness this parable.

> . . .*There's this desert prison, see, with an old prisoner, resigned to his life, and a young one just arrived. The young one talks constantly of escape, and, after a few months, he makes a break. He's gone a week, and then he's brought back by the guards. He's half dead, crazy with hunger and thirst. He describes how awful it was to the old prisoner. The endless stretches of sand, no oasis, no signs of life anywhere. The old prisoner listens for a while, then says, "Yep, I know. I tried to escape myself, twenty years ago." The young prisoner says, "You did? Why didn't you tell me, all these months I was planning my escape? Why didn't you let me know it was impossible?" And the old prisoner shrugs, and says, "So who publishes negative results?"[3]*

Statistical inference is not valid for all sets of data

We learned long ago that badly designed surveys or experiments often produce invalid results. Formal statistical inference cannot correct basic flaws in the design. There is no doubt a significant difference in English vocabulary scores between high school seniors who have studied Latin and those who have not. (Recall Example 3 on page 79.) But as long as the effect of actually studying Latin is confounded with the differences between students who choose Latin and those who do not, this statistical significance has little meaning. It does indicate that the difference in English scores is greater than would often arise by chance alone. That leaves unsettled the issue of *what* other than chance caused the difference.

Both tests of significance and confidence intervals are based on the laws of probability. Randomization in sampling or experimentation assures that these laws apply. When these statistical strategies for collecting data cannot be used, statistical inference from the data obtained should be done only by caution. Many data in the social sciences by necessity are collected without randomization. It is universal practice to use tests of significance on such data. It can be argued that significance at least points to the presence of an effect greater than would be likely by chance. But that indication alone is little evidence against H_0 and in favor of the research hypothesis H_a. Do not allow the wonders of this chapter to obscure the common sense of Chapters 1 and 2.

Beware of searching for significance

Statistical significance is a commodity much sought after by researchers. It means (or ought to mean) that you have found something you were looking for. The reasoning behind statistical significance works well if you decide what effect you are seeking, design an experiment or sample to search for it, and use a test of significance to weigh the evidence you get. But because a successful search for a new scientific phenomenon often ends with statistical significance, it is all too easy to make significance itself the object of the search. There are several ways to do this, none of them acceptable in polite scientific society.

One such tactic is to make many tests on the same data. The story is told of three psychiatrists who studied a sample of schizophrenic persons in comparison with a sample of nonschizophrenic persons. They measured 77 variables for each subject — religion, family background, childhood experiences, and so on. Their goal was to discover what distinguishes persons who later become schizophrenic. Having measured 77 variables, they made 77 separate tests of the significance of the differences between the two groups of subjects. Now pause for a moment of reflection. If you made 77 tests at the 5% level, you would expect a few of them to be significant by chance alone, right? After all, results significant at the 5% level do occur five times in a hundred in the long run even when H_0 is true. Well, our psychiatrists found 2 of their 77 tests significant at the 5% level and immediately published this exciting news.[4] Running one test and reaching the $\alpha = 0.05$ level is reasonably good evidence that you have found something; running 77 tests and reaching that level only twice is not.

The case of the 77 tests happened long ago, and such crimes are rarer now — or at least better concealed, for some common practices are not very different. The computer has freed us from the labor of doing arithmetic, and this is surely a blessing in statistics, where the arithmetic can be long and complicated. Statistical software is universally available, so a few simple commands will set the machine to work performing all manner of complicated tests and operations on your data. The result can be much like the 77 tests of old. I will state it as a law that any large set of data — even several pages of a table of random digits — contains some unusual pattern. Sufficient computer time will discover that pattern, and when you test specifically for the pattern that turned up, the test will be significant. It also will mean exactly nothing.

One lesson here is not to be overawed by the computer. It is a wondrous tool that makes possible statistical analysis of large data sets and allows ever more complicated and sensitive statistical procedures to be used. The computer has greatly extended the range of statistical inference. But it has changed the logic of inference not one bit. Doing 77 tests and finding 2 significant at the $\alpha = 0.05$ level was not evidence of a real discovery. Neither is doing multiple regression analysis followed by principal components analysis followed by factor analysis and at last discovering a pattern in the data. Fancy words, and fancy computer programs, but still bad scientific logic. It is convincing to hypothesize that an effect or pattern will be present, design a study to look for it, and find it at a low significance level. It is not convincing to search for any effect or pattern whatever and find one.

I do not mean that searching data for suggestive patterns is not proper scientific work. It certainly is. Many important discoveries have been made by accident rather than by design. New computer-based methods of searching through data are important in statistics. I do mean that the usual reasoning of statistical inference does not apply when the search is successful. You cannot legitimately test a hypothesis on the same data that first suggested that hypothesis. After all, any data set has some peculiarity, and you may have found only the peculiarity of this one set of data. The remedy is clear. Now that you have a hypothesis, design a study to search specifically for the effect you now think is there. If the result of this study is statistically significant, you have real evidence at last.

Section 5 exercises

8.44. Which of the following questions does a test of significance help answer? Explain.
   (a) Is the sample or experiment properly designed?
   (b) Is the observed effect due to chance?
   (c) Is the observed effect important?

8.45. A researcher looking for ESP tests 500 subjects. (See Exercise 8.32 for the experiment employed to detect ESP.) Four of these subjects do significantly better ($P < 0.01$) than random guessing.
   (a) Is it proper for the researcher to conclude that these four people have ESP? Explain.
   (b) Is it proper to choose these four people for more tests, conducted independently of the previous tests? Explain.

8.46. Every user of statistics should understand the distinction between statistical significance and practical significance. A sufficiently large sample will declare very small effects statistically significant. Suppose you are trying to decide whether a coin is fair when tossed. Let p be the probability of a head. If $p = 0.505$ rather than 0.500, this is of no practical significance to you. Test H_0: $p = 0.5$ vs. H_a: $p \neq 0.5$ in each of the following cases. Use Table C to assess significance.
  (a) 1,000 tosses give 505 heads ($\hat{p} = 0.505$).
  (b) 10,000 tosses give 5050 heads ($\hat{p} = 0.505$).
  (c) 100,000 tosses give 50,500 heads ($\hat{p} = 0.505$).

8.47. The previous exercise shows that simply reporting a P-value can be misleading. A confidence interval is more informative. Give a 95% confidence interval for p in each of the cases in the previous exercise. You see that for large n the confidence interval says, "Yes, p is larger than $1/2$, but it is very little larger."

8.48. Table 4-4 on page 217 gives the Nielsen ratings for all national prime-time television programs in the week of July 31 to August 6, 1989. It is *not* proper to do a test of significance on these data to decide whether NBC had a significantly higher mean rating than CBS. Why not?

8.49. Few accounts of really complex statistical methods are readable without extensive training. One that is, and that is also an excellent essay on the abuse of statistical inference, is "The Real Error of Cyril Burt," a chapter in Stephen Jay Gould's *The Mismeasure of Man* (New York: Norton, 1981). We met Cyril Burt under suspicious circumstances in Exercise 3.29 on page 165. Gould's long chapter shows Burt and others engaged in discovering dubious patterns by complex statistics. Read it, and write a brief explanation of why "factor analysis" failed to give a firm picture of the structure of mental ability.

8.50. The article by Arie Y. Lewin and Linda Duchan, "Women in Academia" [*Science*, Volume 173 (1971), pp. 892–895], reports an investigation in which applications of equally qualified male and female Ph.D.s were sent to academic department chairmen. The chairmen were asked which applicant they would hire. The number who chose the male was somewhat greater than the number who would hire the female. The authors concluded that "the results, although not statistically significant, showed definite trends that confirm our hypothesis that discrimination against women does exist at the time of the hiring decision." This conclusion was strongly attacked in letters to the editor. One irate statistician wrote, "Are the standards of *Science* the standards of science?"

Discuss the validity of the conclusion that the survey results confirm the existence of discrimination. (If possible, read the entire article first.)

6.   Inference as decision

Tests of significance were presented in Section 3 as methods for assessing the strength of evidence against the null hypothesis. This assessment is made by the P-value, which is a probability computed under the assumption that the null hypothesis is true. The alternative hypothesis (the statement we seek evidence for) enters the test only to help us see what outcomes count against the null hypothesis. Such is the reasoning of tests of significance as advocated by Fisher and as practiced by many users of statistics.

But already in Section 5, signs of another way of thinking were present. A level of significance α chosen in advance points to the outcome of the test as a *decision*. If the P-value is less than α, we reject H_0 in favor of H_a, otherwise we fail to reject H_0. The transition from measuring the strength of evidence to making a decision is not a small step. It can be argued (and is argued by followers of Fisher) that making decisions is too grand a goal, especially in scientific inference. A decision is reached only after the evidence of many experiments is weighed, and indeed the goal of research is not "decision" but a gradually evolving understanding. Better that statistical inference should content itself with confidence intervals and tests of significance. Many users of statistics are content with such methods. It is rare (outside textbooks) to set up a level α in advance as a rule for making a decision in a scientific problem. More commonly, users think of significance at level 0.05 as a description of good evidence. This is made clearer by talking about P-values.

Yet there are circumstances in which a decision or action is called for as the end result of inference. *Acceptance sampling* (Example 4 on page 6) is one such circumstance. The supplier of bearings and the consumer of the bearings agree that each carload lot shall meet certain quality standards. When a carload arrives, the consumer chooses a sample of bearings to be inspected. On the basis of the sample outcome, the consumer will either accept or reject the carload. Fisher agreed that this is a genuine decision problem. But he insisted that acceptance sampling is completely different from scientific inference. Other eminent statisticians have argued that if "decision" is given a broad meaning, almost all problems of statistical inference can be posed as problems of making decisions in the presence of uncertainty. I am not going to venture further into the arguments over how we ought to think about inference. I do want to show how a different concept—inference as decision—changes the reasoning used in tests of significance.

Tests of significance concentrate attention on H_0, the null hypothesis. If a decision is called for, however, there is no reason to single out H_0. There are simply two alternatives, and we must accept one and reject the other. It is convenient to call the two alternatives H_0 and H_a, but H_0 no longer has the special status (the statement we try to find evidence against) that it had in tests of significance. In the acceptance sampling problem, we must decide between

H_0 :   the lot of bearings meets standards

H_a :   the lot does not meet standards

on the basis of a sample of bearings. There is no reason to put the burden of proof on the consumer by accepting H_0 unless we have strong evidence against it. It is equally sensible to put the burden of proof on the producer by accepting H_a unless we have strong evidence that the lot meets standards. Producer and consumer must agree on where to place the burden of proof, but neither H_0 nor H_a has any special status.

In a decision problem, we must give a *decision rule*—a recipe based on the sample that tells us what decision to make. Decision rules are expressed in terms of sample statistics, usually the same statistics we would use in a test of significance. In fact, we have seen that a test of significance becomes a decision rule if we reject H_0 (accept H_a) when the sample statistic is statistically significant at level α and otherwise accept H_0 (reject H_a).

Suppose, then, that we use statistical significance at level α as our criterion for decision. Suppose also that the null hypothesis H_0 is really true. Then sample outcomes significant at level α will occur with probability α. (That's our definition of "significant at level α"; outcomes weighing this strongly against H_0 occur with probability α when H_0 is really true.) We make a *wrong decision* in all such outcomes, by rejecting H_0 when it is really true. That is, the significance level α now can be understood as the probability of a certain type of wrong decision.

The other hypothesis H_a requires equal attention. Just as rejecting H_0 when H_0 is really true is an error, so is accepting H_0 when H_a is really true. We can make two kinds of errors.

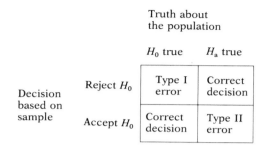

Truth about
the population

		H_0 true	H_a true
Decision based on sample	Reject H_0	Type I error	Correct decision
	Accept H_0	Correct decision	Type II error

Figure 8-8. Possible outcomes of a two-action decision problem.

If we reject H_0 (accept H_a) when in fact H_0 is true, this is a *Type I error*.

If we accept H_0 (reject H_a) when in fact H_a is true, this is a *Type II error*.

The possibilities are summed up in Figure 8-8. If H_0 is true, our decision is either correct (if we accept H_0) or is a Type I error. Only one error is possible at one time. Figure 8-9 applies these ideas to the acceptance sampling example.

So the significance level α is the probability of a Type I error. In acceptance sampling, this is the probability that a good lot will be rejected. The probability of a Type II error is the probability that a bad lot will be accepted. A Type I error hurts the producer, while a Type II error hurts the consumer. *Any decision rule is assessed in terms of the probabilities of the two types of error.* This is in keeping with the idea that

Truth about the lot

		Does meet standards	Does not meet standards
Decision based on sample	Reject the lot	Type I error	Correct decision
	Accept the lot	Correct decision	Type II error

Figure 8-9. Possible outcomes of an acceptance sampling decision.

statistical inference is based on probability. We cannot (short of inspecting the whole lot) guarantee that good lots will never be rejected and bad lots never be accepted. But by random sampling and the laws of probability, we can say what the probabilities of both kinds of error are. Because we can find out the monetary cost of accepting bad lots and of rejecting good ones, we can determine how much loss the producer and consumer each will suffer in the long run from wrong decisions.

Advocates of decision theory argue that the kind of "economic" thinking natural in acceptance sampling applies to all inference problems. Even a scientific researcher decides whether to announce results or to do another experiment or to give up the research as unproductive. Wrong decisions carry costs, though these costs are not always measured in dollars. A scientist suffers by announcing a false effect and also by failing to detect a true effect. Decision theorists maintain that the scientist should try to give numerical weights (called *utilities*) to the consequences of the two types of wrong decisions. Then the scientist can choose a decision rule with error probabilities that reflect how serious the two kinds of error are. This argument has won favor where utilities are easily expressed in money. Decision theory is widely used by business in making capital investment decisions, for example. But scientific researchers have been reluctant to take this approach to statistical inference.

To sum up, in a test of significance we focus on a single hypothesis (H_0) and a single probability (the P-value). The goal is to measure the strength of the sample evidence against H_0. If the same inference problem is thought of as a decision problem, we focus on two hypotheses and give a rule for deciding between them based on the sample evidence. We therefore must focus on two probabilities, the probabilities of the two types of error.

Such a clear distinction between the two ways of thinking is helpful for understanding. In practice, the two approaches often merge, to the dismay of partisans of one or the other. We continued to call one of the hypotheses in a decision problem H_0. In the common practice of *testing hypotheses*, we mix significance tests and decision rules as follows:

(a) Choose H_0 and H_a just as in a test of significance.

(b) Think of the problem as a decision problem, so the probabilities of Type I and Type II error are relevant.

(c) Because of step (a), Type I errors are more serious. So choose an α (significance level) and consider only tests with probability of Type I error no greater than α.

(d) Among these tests, select one that makes the probability of a Type II error as small as possible. If this probability is too large, you will have to take a larger sample to reduce the chance of an error.

Testing hypotheses may seem to be a hybrid approach, or maybe a bastard approach. It was, historically, the effective beginning of decision-oriented ideas in statistics. Hypothesis testing was developed by Jerzy Neyman and Egon S. Pearson from 1928 to 1938. The decision theory approach came later and grew out of the Neyman-Pearson ideas. Because decision theory in its pure form leaves you with two error probabilities and no simple rule on how to balance them, it has been used less often than tests of significance. Decision theory ideas have been applied in testing problems mainly by way of the Neyman-Pearson theory. That theory asks you first to choose α, and the influence of Fisher often has led users of hypothesis testing comfortably back to $\alpha = 0.05$ or $\alpha = 0.01$ (and also back to the warnings of Section 5 about this state of affairs). Fisher, who was exceedingly argumentative, violently attacked the Neyman-Pearson decision-oriented ideas, and the argument still continues.

The reasoning in statistical inference is subtle, and the principles at issue are complex. I have (believe it or not) oversimplified the ideas of all the viewpoints mentioned and omitted other important viewpoints altogether. If you are feeling that you do not fully grasp all the ideas in this chapter, you are in excellent company. Nonetheless, any user of statistics should make a serious effort to grasp the conflicting views on the nature of statistical inference. More than most other kinds of intellectual exercises, statistical inference can be done automatically, by recipe or by computer. These valuable shortcuts are of no worth without understanding. What Euclid said of his own science to King Ptolemy of Egypt long ago remains true of all knowledge: "There is no royal road to geometry."

Section 6 exercises

8.51. A criminal trial can be thought of as a decision problem, the two possible decisions being guilty and not guilty. Moreover, in a criminal trial there is a null hypothesis in the sense of an assertion that we will continue to hold until we have strong evidence against it. Criminal trials are therefore similar to hypothesis testing.

(a) What are H_0 and H_a in a criminal trial? Explain your choice of H_0.

   (b) Describe in words the meaning of Type I error and Type II error in this setting, and display the possible outcomes in a diagram like Figures 8-8 and 8-9.

   (c) Suppose that you are a jury member. Having studied statistics, you think in terms of a significance level α, the (subjective) probability of a Type I error. What considerations would affect your personal choice of α? (For example, would the difference between a charge of murder and a charge of shoplifting affect your personal α?)

8.52. You are designing a computerized medical diagnostic program that will scan the results of tests conducted by technicians (pulse rate, blood pressure, urinalysis, etc.) and either clear the patient or refer the case to the attention of a doctor. This program will be used to screen many thousands of persons who do not have specific medical complaints as part of a preventive medicine system.

   (a) What are the two hypotheses and the two types of error? Display the situation in a diagram like Figures 8-8 and 8-9.

   (b) Briefly discuss the costs of each of the two types of error. These costs are not entirely monetary.

   (c) After considering these costs, which error probability would you choose to make smaller?

8.53. You are the consumer of bearings in an acceptance sampling situation. Your acceptance sampling plan has probability 0.01 of passing a lot of bearings that does not meet quality standards. You might think the lots that pass are almost all good. Alas, it is not so.

   (a) Explain why low probabilities of error cannot ensure that lots which pass are mostly good. (*Hint:* What happens if your supplier ships all bad lots?)

   (b) The paradox that most decisions can be correct (low error probabilities) and yet most lots that pass can be bad has important analogs in areas such as medical diagnosis. Explain why most conclusions that a patient has a rare disease can be false alarms even if the diagnostic system is correct 99% of the time.

8.54. A major advantage of the decision approach to inference is that it is not restricted to the two-decision situations characteristic of hypothesis testing. For example, suppose that three decisions are possible in an acceptance sampling setting.

      Decision 1:   The lot of bearings is of high quality; accept it at full price.

      Decision 2:   The lot is of medium quality; accept it at a lower price.

Decision 3:   The lot is of low quality;
reject it.

The performance of a decision rule is still assessed in terms of the probabilities of error. By filling in the display below, count the different types of error now possible.

True quality of the lot

	High	Medium	Low
Accept at full price			
Accept at lower price			
Reject			

Decision based on sample

NOTES

1. R. A. Fisher, "The Arrangement of Field Experiments," *Journal of the Ministry of Agriculture of Great Britain*, Volume 33 (1926), p. 504. Quoted in Leonard J. Savage, "On Rereading R. A. Fisher," *The Annals of Statistics*, Volume 4 (1976), p. 471.
2. Theodore D. Sterling, "Publication Decisions and their Possible Effects on Inferences Drawn from Tests of Significance — or Vice Versa," *Journal of the American Statistical Association*, Volume 54 (1959), pp. 30–34.
3. From Jeffrey Hudson, *A Case of Need* (New York: New American Library, 1968). Quoted in G. William Walster and T. Anne Cleary, "A Proposal for a New Editorial Policy in the Social Sciences," *The American Statistician*, April 1970, p. 16.
4. This example is cited by William Feller, "Are Life Scientists Overawed by Statistics?" *Scientific Research*, February 3, 1969, p. 26.

Review exercises

8.55. A television news program conducts a call-in poll about a proposed city ban on handgun ownership. Of the 2372 calls, 1921 oppose the ban. The station, following recommended practice, makes a confidence statement: "Eighty-one percent of the Channel 13 Pulse Poll sample opposed the ban. We can be 95% confident that the true proportion of citizens opposing a handgun ban is within 1.6% of the sample result." In this conclusion justified?

8.56. A randomized comparative experiment was conducted to examine whether a calcium supplement in the diet will reduce the blood pressure of healthy men. The subjects received either a calcium supplement or a placebo for 12 weeks. The statistical analysis was quite complex, but one conclusion was that, "The calcium group had lower seated systolic blood pressure ($P = 0.008$) compared with the placebo group." Explain this conclusion, especially the P-value, as if you were speaking to a doctor who knows no statistics. [From R. M. Lyle, et al., "Blood Pressure and Metabolic Effects of Calcium Supplementation in Normotensive White and Black Men." *Journal of the American Medical Association*, Volume 257(1987), pp. 1772–1776.]

8.57. When asked to explain the meaning of "statistically significant at the $\alpha = 0.05$ level," a student says, "This means there is only probability .05 that the null hypothesis is true." Is this an essentially correct explanation of statistical significance? Explain your answer.

8.58. Another student, when asked why statistical significance appears so often in research reports, says, "Because saying that results are significant tells us that they cannot easily be explained by chance variation alone." Do you think that this statement is essentially correct? Explain your answer.

8.59. A new vaccine for a virus that now has no vaccine is to be tested. Since the disease is usually not serious, 1000 volunteers will be exposed to the virus. After some time, the researchers will record whether or not each volunteer has been infected.

 (a) Explain how you would use these 1000 volunteers in a designed experiment to test the vaccine. Include all important details of designing the experiment (but don't actually do any random allocation).

 (b) We hope to show that the vaccine is more effective than a placebo. State H_0 and H_a.

 (c) The experiment gave a P-value of 0.25. Explain carefully what this means.

 (d) The researchers did not consider this evidence strong enough to recommend regular use of the vaccine. Do you agree?

8.60. Statisticians prefer large samples. Describe briefly the effect of increasing the size of a sample (or the number of subjects in an experiment) on each of the following:

 (a) The width of a confidence interval with confidence level held fixed.

 (b) The P-value of a test, when H_0 is false and all facts about the population remain unchanged as n increases.

Table A. Random digits

Line								
101	19223	95034	05756	28713	96409	12531	42544	82853
102	73676	47150	99400	01927	27754	42648	82425	36290
103	45467	71709	77558	00095	32863	29485	82226	90056
104	52711	38889	93074	60227	40011	85848	48767	52573
105	95592	94007	69971	91481	60779	53791	17297	59335
106	68417	35013	15529	72765	85089	57067	50211	47487
107	82739	57890	20807	47511	81676	55300	94383	14893
108	60940	72024	17868	24943	61790	90656	87964	18883
109	36009	19365	15412	39638	85453	46816	83485	41979
110	38448	48789	18338	24697	39364	42006	76688	08708
111	81486	69487	60513	09297	00412	71238	27649	39950
112	59636	88804	04634	71197	19352	73089	84898	45785
113	62568	70206	40325	03699	71080	22553	11486	11776
114	45149	32992	75730	66280	03819	56202	02938	70915
115	61041	77684	94322	24709	73698	14526	31893	32592
116	14459	26056	31424	80371	65103	62253	50490	61181
117	38167	98532	62183	70632	23417	26185	41448	75532
118	73190	32533	04470	29669	84407	90785	65956	86382
119	95857	07118	87664	92099	58806	66979	98624	84826
120	35476	55972	39421	65850	04266	35435	43742	11937
121	71487	09984	29077	14863	61683	47052	62224	51025
122	13873	81598	95052	90908	73592	75186	87136	95761
123	54580	81507	27102	56027	55892	33063	41842	81868
124	71035	09001	43367	49497	72719	96758	27611	91596
125	96746	12149	37823	71868	18442	35119	62103	39244
126	96927	19931	36089	74192	77567	88741	48409	41903
127	43909	99477	25330	64359	40085	16925	85117	36071
128	15689	14227	06565	14374	13352	49367	81982	87209
129	36759	58984	68288	22913	18638	54303	00795	08727
130	69051	64817	87174	09517	84534	06489	87201	97245
131	05007	16632	81194	14873	04197	85576	45195	96565
132	68732	55259	84292	08796	43165	93739	31685	97150
133	45740	41807	65561	33302	07051	93623	18132	09547
134	27816	78416	18329	21337	35213	37741	04312	68508
135	66925	55658	39100	78458	11206	19876	87151	31260
136	08421	44753	77377	28744	75592	08563	79140	92454
137	53645	66812	61421	47836	12609	15373	98481	14592
138	66831	68908	40772	21558	47781	33586	79177	06928
139	55588	99404	70708	41098	43563	56934	48394	51719
140	12975	13258	13048	45144	72321	81940	00360	02428
141	96767	35964	23822	96012	94591	65194	50842	53372
142	72829	50232	97892	63408	77919	44575	24870	04178
143	88565	42628	17797	49376	61762	16953	88604	12724
144	62964	88145	83083	69453	46109	59505	69680	00900
145	19687	12633	57857	95806	09931	02150	43163	58636
146	37609	59057	66967	83401	60705	02384	90597	93600
147	54973	86278	88737	74351	47500	84552	19909	67181
148	00694	05977	19664	65441	20903	62371	22725	53340
149	71546	05233	53946	68743	72460	27601	45403	88692
150	07511	88915	41267	16853	84569	79367	32337	03316

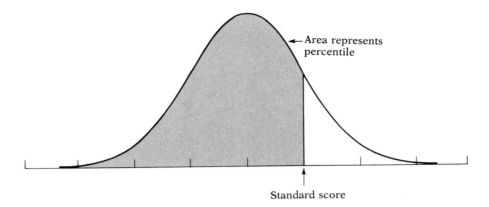

Standard score

Table B. The standard normal distribution

Standard score	Percentile	Standard score	Percentile	Standard score	Percentile
−3.0	0.13	−1.0	15.87	1.0	84.13
−2.9	0.19	−0.9	18.41	1.1	86.43
−2.8	0.26	−0.8	21.19	1.2	88.49
−2.7	0.35	−0.7	24.20	1.3	90.32
−2.6	0.47	−0.6	27.42	1.4	91.92
−2.5	0.62	−0.5	30.85	1.5	93.32
−2.4	0.82	−0.4	34.46	1.6	94.52
−2.3	1.07	−0.3	38.21	1.7	95.54
−2.2	1.39	−0.2	42.07	1.8	96.41
−2.1	1.79	−0.1	46.02	1.9	97.13
−2.0	2.27	0.0	50.00	2.0	97.73
−1.9	2.87	0.1	53.98	2.1	98.21
−1.8	3.59	0.2	57.93	2.2	98.61
−1.7	4.46	0.3	61.79	2.3	98.93
−1.6	5.48	0.4	65.54	2.4	99.18
−1.5	6.68	0.5	69.15	2.5	99.38
−1.4	8.08	0.6	72.58	2.6	99.53
−1.3	9.68	0.7	75.80	2.7	99.65
−1.2	11.51	0.8	78.81	2.8	99.74
−1.1	13.57	0.9	81.59	2.9	99.81
				3.0	99.87

NOTE: The table gives the percentile corresponding to each standard score for the normal distributions. The percentile corresponding to any observation from a normal distribution can be found by converting the observation to a standard score and then looking in the table.

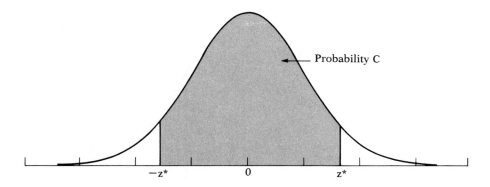

Probability C

Table C-1. Two-sided normal critical values

C	z^*	C	z^*
0.50	0.67	0.80	1.28
0.55	0.76	0.85	1.44
0.60	0.84	0.90	1.64
0.65	0.93	0.95	1.96
0.70	1.04	0.99	2.58
0.75	1.15	0.999	3.29

NOTE: Any normal distribution has probability C within z^* standard deviations on either side of its mean.

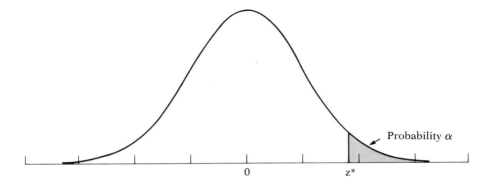

Probability α

0   z^*

Table C-2. One-sided normal critical values

α	z^*	α	z^*
0.20	0.84	0.02	2.05
0.15	1.04	0.01	2.33
0.10	1.28	0.005	2.58
0.05	1.64	0.001	3.09

NOTE: Any normal distribution has probability α more than z^* standard deviations from the mean in one direction.

Index